T0203906

Programming ArcObjects with VBA

A Task-Oriented Approach

Second Edition

Programming ArcObjects with VBA

A Task-Oriented Approach

Second Edition

Kang-Tsung Chang

CRC Press
Taylor & Francis Group
Boca Raton London New York

CRC Press is an imprint of the
Taylor & Francis Group, an informa business

CRC Press
Taylor & Francis Group
6000 Broken Sound Parkway NW, Suite 300
Boca Raton, FL 33487-2742

First issued in paperback 2019

© 2008 by Taylor & Francis Group, LLC
CRC Press is an imprint of Taylor & Francis Group, an Informa business

No claim to original U.S. Government works

ISBN-13: 978-0-8493-9283-2 (hbk)
ISBN-13: 978-0-367-38868-3 (pbk)

Library of Congress Cataloging-in-Publication Data

Krantz, Steven G. (Steven George), 1951-
    Complex variables : a physical approach with applications and Matlab / Steven G. Krantz.
        p. cm. -- (Textbooks in mathematics)
    Includes bibliographical references and index.
    ISBN 978-1-58488-580-1 (alk. paper)
    1. Functions of complex variables. 2. MATLAB. I. Title. II. Series.

QA331.7.K732 2008
515'.9--dc22   2007023147

Visit the Taylor & Francis Web site at
http://www.taylorandfrancis.com

and the CRC Press Web site at
http://www.crcpress.com

Contents

Introduction

This book is designed for ArcGIS users who want to get a quick start on programming ArcObjects. Both ArcGIS and ArcObjects are products developed and distributed by Environmental Systems Research Institute Inc. (ESRI), ArcObjects is the development platform for ArcGIS, a software package for managing geographic information systems (GIS). Ideally, users should learn ArcObjects before using ArcGIS, but that is not the case in reality. Users use ArcGIS first through its toolbars and commands. It is easier to follow the user interface in ArcGIS than to sort out objects, properties, and methods in code. The topic of ArcObjects usually emerges when users realize that programming ArcObjects can actually reduce the amount of repetitive work, streamline the workflow, and even produce functionalities that are not easily available in ArcGIS.

How can users learn programming ArcObjects efficiently and quickly? Perhaps surprising to some, the answer is to apply what users already know about ArcGIS to programming ArcObjects.

THE TASK-ORIENTED APPROACH

GIS activities are task oriented. Users use GIS for data integration, data management, data display, data analysis, and so on. Therefore, an efficient way to learn programming ArcObjects is to take a task-oriented approach, which has at least three main advantages.

First, it connects ArcObjects with what users already know. Take the example of *QueryFilter.* This book first links a *QueryFilter* object to the task of data exploration. After users know that the object can perform the same function as the Select By Attributes command in ArcMap, which users have used many times before, it becomes easy to understand the properties and methods that are associated with the object.

Second, the task-oriented approach introduces objects in a way that is logical to ArcGIS users. With thousands of objects, properties, and methods, it can be difficult, if not impossible, for beginners to navigate the ArcObjects model diagrams. Using the task-oriented approach, users can learn ArcObjects incrementally from one group of tasks to another in an organized fashion.

Third, the task-oriented approach can actually help users gain a better understanding of ArcGIS with their new knowledge of ArcObjects. For example, as a type of *QueryFilter,* the *SpatialFilter* class has its own properties of geometry and spatial relation in addition to the properties that it inherits from the *QueryFilter* class. This class relationship explains why the Select By Location command in ArcMap can accept both attribute and spatial constraints for data query. (Perhaps it is more appropriate to name this command Select By Location and Attributes.)

ABOUT THIS BOOK

This book has fourteen chapters. The first three chapters introduce ArcObjects, programming basics, and customization. This book adopts Visual Basic for Applications (VBA) for programming ArcObjects. Because VBA is already embedded within ArcMap and ArcCatalog, it is convenient for ArcGIS users to program ArcObjects in VBA. The following summarizes the major topics covered in the first three chapters:

- Chapter 1: ArcObjects — Geodatabase, ArcObjects, organization of ArcObjects, the help sources on ArcObjects, and the Geoprocessing object.
- Chapter 2: Programming Basics — Basic elements, writing code, calling subs and functions, Visual Basic Editor, and debugging code.
- Chapter 3: Customization of the User Interface — Creating a toolbar with existing commands, adding a new button and tool, adding a form, and making basic templates.

Chapters 4 through 14 discuss programming ArcObjects for solving common GIS tasks. Organized around a central theme, each chapter has three parts. The first part is a quick review of ArcGIS commands on the topic; the second part discusses objects that are related to the theme; and the third part presents sample macros and Geoprocessing macros for solving common tasks under the theme. This combination of ArcGIS commands, ArcObjects, and sample macros can effectively relate the user's experience of working with ArcGIS to programming ArcObjects.

The CD that accompanies this book contains 95 sample macros stored in the VBA_programs folder by chapter. Each sample macro starts with a short description of its usage and a list of key interfaces and members (properties and methods). These are followed by the listing and explanation of code. Many macros are divided into two or more parts to better connect the code lines and their explanation. Stored as text files, these sample macros can be easily imported to Visual Basic Editor in either ArcMap or ArcCatalog to view and run.

The companion CD also includes 33 Geoprocessing macros that are new in this second edition. These macros are stored in the GP_programs folder by chapter. The *Geoprocessing* object is a new ArcObjects component that supports the execution of hundreds of Geoprocessing tools in a scripting language such as VBA or Python. These tools are the same as in the ArcToolbox application of ArcGIS Desktop. The *Geoprocessing* object is a "coarse-grained" object, which is simpler to use than a "fine-grained" object. Therefore it allows users who do not understand all the details of ArcObjects to run macros. To separate them from "regular" VBA macros, Geoprocessing macros are included in "boxes" in Chapters 4 through 7 and in Chapters 9 through 14.

All sample macros in the text have been run successfully in ArcGIS 9.2. The companion CD contains datasets for the test runs, which are stored by chapter in the Data folder. Two notes must be made about use of the sample macros. First, ArcGIS 9.1 or 9.2 is needed to run the macros. Second, the Data folder is coded in the sample macros as residing on the C drive (for example, c:\data\chap4). If the folder is stored on a different drive (for example, the G drive), then the path should be changed (for example, g:\data\chap4) before running the macros.

The following summarizes the major tasks covered in each chapter:

- Chapter 4: Dataset and Layer Management — Add datasets as layers, manage layers and datasets, and report geographic dataset information.
- Chapter 5: Attribute Data Management — List fields, add or delete fields, calculate field values, and join and relate tables.
- Chapter 6: Data Conversion — Convert shapefile to geodatabase, convert coverage to geodatabase and shapefile, perform rasterization and vectorization, and add XY data.
- Chapter 7: Coordinate Systems — Manipulate on-the-fly projection, define the coordinate system, perform geographic transformation, and project datasets.
- Chapter 8: Data Display — Display vector data, display raster data, and create a layout page.
- Chapter 9: Data Exploration — Perform attribute query, perform spatial query, combine attribute and spatial queries, and derive descriptive statistics.
- Chapter 10: Vector Data Operations — Run buffer, perform overlay, join data by location, and manipulate features.
- Chapter 11: Raster Data Operations — Manage raster data and perform local, neighborhood, zonal, and distance measure operations.
- Chapter 12: Terrain Mapping and Analysis — Derive contour, slope, aspect, and hillshade; perform viewshed analysis; perform watershed analysis; and create and edit triangulated irregular networks (TIN).
- Chapter 13: Spatial Interpolation — Perform spatial interpolation and compare interpolation methods.
- Chapter 14: Binary and Index Models — Build binary and index models, both vector and raster based.

TYPOGRAPHICAL CONVENTIONS

The following lists the typographical conventions used in this book:

- Sample VBA macros are set off from the text and appear in a different typeface.
- Sample Geoprocessing macros are included in boxes and appear in a different typeface.
- Names of sample macros are capitalized and italicized.
- ArcObjects, interfaces, properties, and methods appear in italics.
- Names of datasets and variables appear in italics in the text.

CHAPTER 1

ArcObjects

ArcGIS from Environmental Systems Research Institute (ESRI), Inc. uses a single, scalable architecture. The three versions of ArcGIS (ArcView, ArcEditor, and ArcInfo) share the same applications of ArcCatalog and ArcMap. The geodatabase data model and ArcObjects provide the foundation for these two desktop applications. They also provide the basis for readers of this book to write programs in Visual Basic for Applications (VBA) for customized applications in ArcGIS.

The geodatabase data model replaces the georelational data model that has been used for coverages and shapefiles, two older data formats from ESRI, Inc. These two data models differ in how geographic and attribute data are stored. The georelational data model stores geographic and attribute data separately in a split system: geographic data ("geo") in graphic files and attribute data ("relational") in a relational database. Typically, a georelational data model uses the feature label or ID to link the two components. The two components must be synchronized so that they can be queried, analyzed, and displayed in unison. By contrast, the geodatabase data model stores geographic and attribute data together in a single system and geographic data in a geometry field.

Another important difference that characterizes the geodatabase data model is the use of object-oriented technology. Object-oriented technology treats a spatial feature as an object and groups spatial features of the same type into a class. A class, and by extension an object in the class, can have properties and methods. A property describes a characteristic or attribute of an object. A method carries out an action by an object. Developers of ArcGIS have already implemented properties and methods on thousands of classes in ArcGIS. Therefore, when we work in ArcCatalog and ArcMap, we actually interact with these classes and their properties and methods.

This chapter focuses on the geodatabase data model and ArcObjects. To use ArcObjects programmatically, we must understand how spatial data are structured and stored in a geodatabase and how classes in ArcObjects are designed and organized. Section 1.1 describes the basics of the geodatabase data model, including the types of data that the model covers. Section 1.2 explains the basics of ArcObjects, including classes, relationships between classes, interfaces, properties, and methods.

Section 1.3 outlines the organization of ArcObjects. Section 1.4 covers the help sources on ArcObjects.

1.1 GEODATABASE

A geographic information system (GIS) manages geospatial data. Geospatial data are data that describe both the location and characteristics of spatial features such as roads, land parcels, and vegetation stands on the Earth's surface. The locations of spatial features are measured in geographic coordinates (i.e., longitude and latitude values) or projected coordinates (for example, Universal Transverse Mercator or UTM coordinates). The characteristics of spatial features are expressed as numeric and string attributes. This book uses the term *geographic data* to describe data that include the locations of spatial features, and the term *nongeographic data* to describe data that include only the attributes of spatial features.

A geodatabase uses tables to store geographic data as well as nongeographic data. It is therefore important to distinguish different types of tables. A table consists of rows and columns. Each row corresponds to a feature, and each column or field represents an attribute. A table that contains geographic data has a geometry field, which distinguishes the table from tables that contain only nongeographic data. The following sections describe different types of data, including both geographic and nongeographic data, which can be stored in a geodatabase.

1.1.1   Vector Data

The geodatabase data model represents vector-based spatial features as points, polylines, and polygons.[1] A point feature may be a simple point feature or a multi-point feature with a set of points. A polyline feature is a set of line segments, which may or may not be connected. A polygon feature may be made of one or many rings. A ring is a set of connected, closed, nonintersecting line segments.

A geodatabase organizes spatial features into feature classes and feature datasets. A *feature class* is a collection of spatial features with the same type of geometry. A feature class may therefore contain simple point, line, or polygon features. A *feature dataset* is a collection of feature classes that have the same coordinate system and area extent. A feature dataset can therefore be used for managing different feature classes from the same study area or reserved for feature classes that participate in topological relationships with each other such as in a geometric network or a planar (two-dimensional) topology. A topology is a set of relationships that defines how the features in one or more feature classes share geometry.

A feature class is like a shapefile in that it has simple features. A feature dataset is similar to a coverage in having multiple datasets based on a common coordinate system. However, this kind of analogy does not address other differences between the traditional and geodatabase data models that are driven by advances in computer technology.

In a geodatabase, a feature class can be a standalone feature class or part of a feature dataset. In either case, a feature class is stored as a table. A feature class has two default fields. One is the object or feature ID and the other is the geometry or shape field. A feature class can have other attribute fields, but the geometry field sets a feature class apart from other tables.

ArcGIS users recognize a feature class as a feature attribute table. When we open the attribute table of a feature layer in ArcMap, we see the two default fields in the table and through them, we can locate and highlight spatial features on a map only through a feature attribute table.

Features within a feature class can be further segregated by subtype. For example, a road feature class can have subtypes based on average daily traffic volume. The geodatabase data model provides four general validation rules for the grouping of objects: attribute domains, default values, connectivity rules, and relationship rules.[1] An attribute domain limits an attribute's values to a valid range of values or a valid set of values. A default value sets an expected attribute value. Connectivity rules control how features in a geometric network are connected to one another. Relationship rules determine, for example, how many features can be associated with another.

1.1.2   Raster Data

The geodatabase data model represents raster data as a two-dimensional array of equally spaced cells.[1] The use of arrays and cells for raster data is the same as the ESRI grid model.

A large variety of raster data are available in GIS. They include satellite imagery, digital elevation models (DEMs), digital orthophotos, scanned files, graphic files, and software-specific raster data such as ESRI grids.[2] The geodatabase model treats them equally as *raster datasets,* but a raster dataset may have a single band or multiple bands. An ESRI grid typically contains a single band, whereas a multispectral satellite image typically contains multiple bands.

A multiband raster dataset may also appear as the output from a raster data operation. For example, a cost distance measure operation can produce results showing the least accumulative cost distance, the back link, and the allocation (Chapter 11). These different outputs can be initially saved into a multiband raster dataset, one band per output, and later extracted to create the proper raster datasets.

1.1.3   Triangulated Irregular Networks (TINs)

The geodatabase data model uses a *TIN dataset* to store a set of nonoverlapping triangles that approximate a surface. Elevation values along with x-, y-coordinates are stored at nodes that make up the triangles. In many instances, a TIN dataset is an alternative to a raster dataset for surface mapping and analysis. The choice between the two depends on data flexibility and computational efficiency.[2]

Inputs to a TIN include DEMs, contour lines, GPS (global positioning system) data, LIDAR (light detection and ranging) data, and survey data. We can also modify and improve a TIN by using linear features, such as streams and roads, and area features, such as lakes and reservoirs. Data flexibility is therefore a major advantage of using a TIN. In addition, the triangular facets of a TIN tend to create a sharper image of the terrain than an elevation raster does.

Computational efficiency is the main advantage of using raster datasets. The simple data structure of arrays and cells makes it relatively easy to perform computations that are necessary for deriving slope, aspect, surface curvature, viewshed, and watershed.

1.1.4 Location Data

The term *location data* refers to data that can be converted to point features. Common examples of location data are tables that contain *x*-, *y*-coordinates or street addresses. We can convert a table with *x*-, *y*-coordinates directly into a point feature class, with each feature corresponding to a pair of *x*- and *y*-coordinates. Using a street network as a reference, we can geocode a list of street addresses into a set of point features.

1.1.5 Nongeographic Data

A table that stores nongeographic data does not have a geometry field. The geodatabase data model defines such a table as an *object class*. Examples of object classes include comma-delimited text files and dBASE files. These files or tables contain attributes of spatial features and have keys (i.e., relate fields) to link to geographic data in a relational database environment.

1.2 ARCOBJECTS

ArcObjects is the development platform for ArcGIS Desktop, ArcGIS Engine, and ArcGIS Server. (This book covers only ArcGIS Desktop.) A collection of objects, ArcObjects is behind the menus and icons that we use to perform tasks in ArcGIS. These same objects also allow software developers to access data and to perform tasks programmatically.

1.2.1 Objects and Classes

ArcObjects consists of objects and classes.[3] An *object* represents a spatial feature such as a road or a vegetation stand. In a geodatabase, an object corresponds to a row in a table and the object's attributes appear in columns. A *class* is a set of objects with similar attributes. An ArcObjects class can have built-in interfaces, properties, and methods.
ArcObjects includes three types of classes:

The most common type is the coclass. A *coclass* can be used to create new objects. For example, *FeatureClass* is a coclass that allows new feature class objects to be created as instances of the coclass.
The second type is the abstract class. An *abstract* class cannot be used to create new objects, but it exists so that other classes (i.e., subclasses) can use or share the properties and methods that the class supports. For example, *GeoDataset* is an abstract class. The class exists so that geographic datasets such as feature classes and raster datasets can all share the properties of extent and spatial reference that the *GeoDataset* class supports.

The third type is the class. A *class* cannot be used directly to create new objects; instead, objects of a class can only be created from another class. For example, *EnumInvalidObject* is a noncreatable class because an *EnumInvalidObject* can only be obtained from another object such as a data conversion object. When converting a shapefile from one coordinate system to another, for example, a data conversion object automatically creates an *EnumInvalidObject* to keep track of those objects that have failed to be converted.

1.2.2 Relationships between Classes

Object-oriented technology has introduced different types of relationships that can be established between classes. Developers of ArcObjects have generally followed these relationships. A good reference on relationships between classes in ArcObjects is Zeiler.[3] There are also books such as Larman's[4] that deal with this topic in the general context of object-oriented analysis and design. A basic understanding of class relationships is important for navigating the object model diagrams and for programming ArcObjects as well.

Association describes the relationship between two classes. An association uses multiplicity expressions to define how many instances of one class can be associated with the other class. Common multiplicity expressions are one (1), one or more (1..*). For example, Figure 1.1 shows an association between *Fields* and *Field* and between *Field* and *GeometryDef*. The multiplicity expressions in Figure 1.1 suggest that:

One fields object, which represents a collection of fields in a table, can be associated with one or more field objects.

One field object can be associated with zero or one *GeometryDef* object, which represents a geometry definition.

A field associated with a geometry definition is the geometry field, and a table can have one geometry field at most.

Type inheritance defines the relationship between a superclass and a subclass. A subclass is a member of a superclass and inherits the properties and methods of the superclass. But a subclass can have additional properties and methods to separate itself from other members of the superclass. For example, Figure 1.2 shows *GeographicCoordinateSystem* is a type of *SpatialReference* (an abstract class). *GeographicCoordinateSystem, ProjectedCoordinateSystem,* and *UnknownCoordinateSystem*

Figure 1.1   The association between *Fields* and *Field* is one or more, and between *Field* and *GeometryDef* is zero or one.

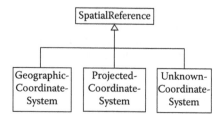

Figure 1.2   *SpatialReference* and its three subclasses.

share the same properties and methods that the *SpatialReference* class supports, but the *GeographicCoordinateSystem* class has additional properties and methods that are unique to the geographic coordinate system.

Composition describes the whole–part relationship between classes. Composition is a kind of association except that the multiplicity at the composite end is typically one and the multiplicity at the other end can be zero or any positive integer. For example, a composition describes the relationship between the *Map* class and the *FeatureLayer* class (Figure 1.3). A map object represents a map or a data frame in ArcMap and a feature layer object represents a feature-based layer in a map. A map can be associated with a number of feature layers. Or, to put it the other way, a feature layer is part of a map.

Aggregation, also called shared aggregation, describes the whole–part relationship between classes. Unlike composition, however, the multiplicity at the composite end of an aggregation relationship is typically more than one. For example, Figure 1.4 shows that a *SelectionSet* object can be created from a *QueryFilter* object and a *Table* object. A table and a query filter together at the composite end can create a selection set (a data subset) at the other end.

Instantiation means that an object of a class can be created from an object of another class. Figure 1.4 shows that, for example, a selection set can be created from a query filter and a table. Another example is an *EnumInvalidObject,* which, as explained earlier, can be created from a *FeatureDataConverter* object (Figure 1.5).

1.2.3   Interfaces

When programming with objects in ArcObjects, one would never work with the object directly but, instead, would access the object via one of its interfaces. An *interface* represents a set of externally visible operations. For example, a *RasterReclassOp* object implements *IRasterAnalysisEnvironment* and *IReclassOp*

Figure 1.3   A *Map* object composes zero, one, or more *FeatureLayer* objects.

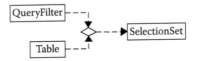

Figure 1.4   A *QueryFilter* object and a *Table* object together can create a *SelectionSet* object.

(Figure 1.6). We can access a *RasterReclassOp* object via either the *IRaster-AnalysisEnvironment* interface or the *IReclassOp* interface, but not the object itself.

An object may support two or more interfaces and, additionally, the same object may inherit interfaces from its superclass. Given multiple interfaces, it is possible to access an interface via another interface, or to jump from an interface to another. This technique is called *QueryInterface* or QI for short. QI simplifies the process of coding. Suppose we want to use a *RasterReclassOp* object to perform raster data classification. First, we use *IRasterAnalysisEnvironment* to set up the analysis environment. Then, we switch, via QI, to *IReclassOp* to perform data reclassification. Chapter 2 on the basics of programming has a more detailed discussion on the QI technique.

Some objects in ArcObjects have two or more similar interfaces. For example, a *FeatureDataConverter* object implements *IFeatureDataConverter* and *IFeatureDataConverter2*. Both interfaces have methods to convert a feature class to a geodatabase feature class. But *IFeatureDataConverter2* has the additional option of working with data subsets. Object-oriented technology allows developers of ArcObjects to add new interfaces to a class without having to remove or update the existing interfaces.

1.2.4   Properties and Methods

An interface represents a set of externally visible operations. More specifically, an interface allows programmers to use the properties and methods that are on the interface. A *property* describes an attribute or characteristic of an object. A *method*, also called behavior, performs a specific action. Figure 1.7, for example, shows the properties and methods on *IRasterAnalysisEnvironment*. These properties and methods are collectively called *members* on the interface.

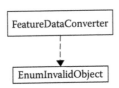

Figure 1.5   An *EnumInvalidObject* can only be created by a *FeatureDataConverter* object.

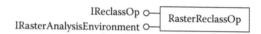

IReclassOp ○─┐
IRasterAnalysisEnvironment ○─┤ RasterReclassOp

Figure 1.6 A *RasterReclassOp* object supports both *IReclassOp* and *IRasterAnalysisEnvironment.*

A property can be for read only, write only, or both read and write. The read property is also called the get property, and the write property the put property. In Figure 1.7, the barbell symbols accompany the properties of *Mask* and *Out-Workspace* on *IRasterAnalysisEnvironment.* The square on the left is for the get property, and the square on the right is for the put property. If the square on the right is open, such as in Figure 1.7, the property is defined as put by reference. If it is solid, the property is defined as put by value. The two put properties differ depending on if a value or an object is assigned to the property. Additionally, the put by reference property requires the keyword *Set,* whereas the put by value property does not. For example, to specify an analysis mask through the *Mask* property on *IRasterAnalysisEnvironment,* we need to use a statement such as *Set pEnv.Mask = pMaskDataset,* where *pEnv* represents an analysis environment and *pMaskDataset* represents an analysis mask object.

To carry out an action, a method on an interface may require some arguments and may return a value or values. In Figure 1.7, the arrow symbols show the methods of *SetCellSize* and *SetExtent* on *IRasterAnalysisEnvironment.* The syntax of the *SetCellSize* method is object.SetCellSize (envType [,cellSizeProvider]). The method has two arguments of which the first is required and the second is optional.

An interface may not have both properties and methods. Some interfaces have properties only, while some have methods only. *IReclassOp,* for example, only has methods. Figure 1.8 shows two of the five methods on *IReclassOp.* These methods all perform reclassification of raster data but use different mechanisms. The *Reclass-ByASCIIFile* method uses an ASCII file, whereas the *ReclassByRemap* method uses a remap that can be built programmatically.

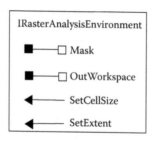

Figure 1.7 Properties and methods on *IRasterAnalysisEnvironment.* Properties are shown with the barbell symbols, and methods are shown with the arrow symbols. Occasionally in this book, properties and methods are shown with the double colon symbols, such as *IRasterAnalysisEnvironment::Mask.*

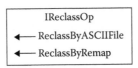

Figure 1.8 Methods on *IReclassOp*.

1.3 ORGANIZATION OF ARCOBJECTS

ArcGIS 9.2 has thousands of coclasses and interfaces. ESRI, Inc. groups ArcObjects into more than 65 libraries. Examples of core libraries are ArcCatalog, ArcCatalogUI, ArcMap, ArcMapUI, Carto, Display, Geodatabase, and Geoprocessing. Examples of extension libraries are 3D Analyst, Spatial Analyst, and Network Analyst. Each library consists of objects that can be diagrammed by their class relationships. For example, the Carto library has objects such as a map document, map, and page layout.

The organization of ArcObjects by library and subsystem resembles that of ArcGIS and its applications and functionalities. This organization provides a good starting point for those who are already familiar with operations in ArcGIS. For example, the Spatial Analyst extension library organizes objects by type of raster data operation. Therefore, the library's object model lists *RasterConditionalOp, RasterExtractionOp, RasterLocalOp, RasterMapAlgebraOp, RasterNeighborhoodOp, RasterZonalOp,* and other objects that closely resemble different functionalities of the Spatial Analyst extension by name. ArcGIS users who are familiar with the Spatial Analyst extension should have no problems using these objects. Objects in other libraries such as Geodatabase, ArcMap, and ArcCatalog are more difficult to relate to because many of them represent new object-oriented concepts and methods.

To use an object in a library, it requires that a reference be made to the library first. This means that the library to be referenced must be available to the user. The availability of ArcObject libraries depends on available licenses. For example, a user will not have access to the ArcScene library and its objects without having a license for the ArcScene extension.

As of ArcGIS 9.2, ArcObjects core libraries and 3D Analyst and Spatial Analyst extension libraries are automatically loaded in VBA. In other words, we do not have to make reference to these libraries before using objects in them. For those libraries that are not automatically loaded, the Tools menu of Visual Basic Editor has a References selection that opens a dialog listing available object libraries and allows the user to make reference to them.

ArcObjects contains objects developed by ESRI, Inc. A recent development is industry-specific objects. Because real-world objects all have different properties and methods, it is impossible to apply, for example, the methods and properties of transportation-related objects to forestry-related objects. ESRI has set up a Web site that supports the development of object models for address, forestry, transportation, hydro, land parcels, environmental regulated facilities, and other fields (http://www.esri.com/software/arcgisdatamodels/). At the same time, increased research activities have developed complex objects for 3D, transportation, and other applications.[5-7] It is worth watching these new objects.

1.4 HELP SOURCES ON ARCOBJECTS

The help sources on ArcObjects include books and documents. ESRI has published two books on ArcObjects: *Exploring ArcObjects*,[3] and *Getting to Know ArcObjects: Programming ArcGIS with VBA*.[8] The former is a two-volume reference on ArcObjects, and the latter is a workbook with 20 hands-on exercises. There are other publications such as *ArcGIS Developer's Guide for VBA*, which covers the basics of developing ArcGIS applications,[9] and *Avenue Wraps*, which is a guide for converting Avenue scripts into VBA code.[10] This section covers electronic and online help documents, which one must regularly consult while programming ArcObjects.

1.4.1 ArcObjects Developer Help

Developer Help on the start-up menu of ArcGIS offers VBA Developer Help, which has links to ArcObjects library reference, Geoprocessing tool reference (Section 1.5), query the samples, and the ESRI Developer Network (EDN) Web site (http://edn.esri.com). EDN maintains the up-to-date ArcGIS development information, including object libraries, sample code, technical documents, and object model diagrams. The first page of EDN Documentation Library lists the following libraries for ArcGIS: Current library, 9.1 library, 9.0 library, and 8.x library. Click on Current library. Then click on ArcObjects Library Reference on the side bar. The library reference lists ArcObjects core and extension libraries. We can look at the Geo-Analyst library as an example by clicking on it. Help on the library is organized into GeoAnalyst Library Overview, GeoAnalyst Library Contents, GeoAnalyst Library Object Model Diagram, Interfaces, CoClasses and Classes, and Constants (i.e., enumerations). The overview page introduces the library and coclasses and classes in it. The contents page lists interfaces, coclasses and classes, and enumerations. The object model diagram is in PDF and shows relationships between classes as well as the interfaces, properties, and methods of the coclasses and classes (Figure 1.9). Both the interfaces page and the coclasses and classes page are sorted by alphabetical order. Suppose we want to get help on *IRasterAnalysisEnvironment*. We can click Interfaces, IR, and then *IRasterAnalysisEnvironment*. The interface and its properties and methods are listed in separate entries. If we click *GetCellSize* Method, it shows the method's syntax for Visual Basic as well as other languages.

1.4.2 Environmental Systems Research Institute, Inc. (ESRI) Object Browser

The ESRI Object Browser, or EOBrowser, is a utility for browsing object libraries. The utility is available through ArcGIS/Develop Tools in the start-up menu. The Object Library References dialog, which can be accessed through the browser's File

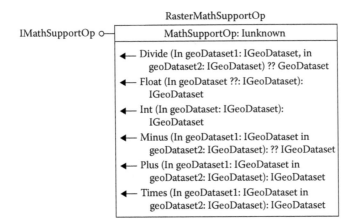

Figure 1.9   A portion of the Spatial Analyst Object Model diagram.

menu, allows the user to add and remove object libraries (Figure 1.10). The EOBrowser window has controls so that the user can select all coclasses and all interfaces in an object library for display and browsing (Figure 1.11). Suppose we want to browse *IRasterAnalysisEnvironment*. First select Object Library References from the browser's File menu. If the window does not list ESRI GeoAnalyst Object Library as an active library, click on the Add button and select ESRI GeoAnalyst Object Library from the Select From Registry dropdown menu. Close the Object Library References dialog. Type irasteranalysisenviron in the Search For box, click the Contains button, uncheck All boxes for Coclasses, Interfaces, Enumerations, and Structures, but check the Interface Name box. Then click the Search button. *IRaster-AnalysisEnvironment* should now appear at the top of the EOBrowser window. Click *IRasterAnalysisEnvironment* and then Show Selected Objects. This displays the properties and methods (sub) of *IRasterAnalysisEnvironment.*

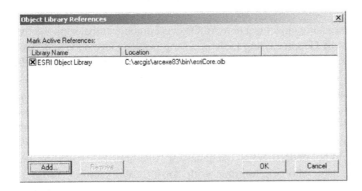

Figure 1.10   The Object Library References dialog box lets the user add and remove object libraries.

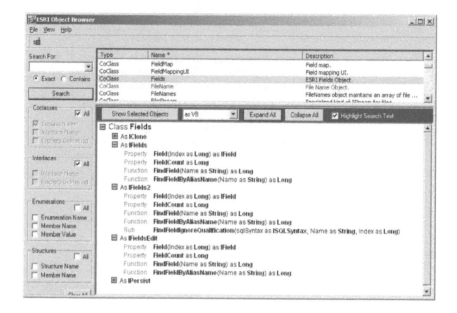

Figure 1.11   The top part of the EOBrowser shows all coclasses in the ESRI Object Library, and the bottom part shows all interfaces that the *Fields* coclass supports.

1.4.3   ESRI Library Locator

The ESRI Library Locator is a tool, available through ArcGIS/Develop Tools in the start-up menu, for finding the object library that contains a specified interface or coclass. The tool opens a dialog for the user to type in an interface, coclass, enumeration, or structure. Then it reports the library that contains the search item.

1.5 GEOPROCESSING OBJECT

The *Geoprocessing* object is a new ArcObjects component that supports the execution of hundreds of Geoprocessing tools in a scripting language such as VBA or Python. These tools correspond to tools in the ArcToolbox application of ArcGIS Desktop. The *Geoprocessing* object differs from other objects because it implements *GpDispatch,* which can pass strings or objects from a script to the *Geoprocessing* object as commands and values. The *Geoprocessing* object is often called a "coarse-grained" object. Unlike other objects, which typically involve lots of little pieces when used in a macro, a coarse-grained object is simpler to use and can do more work, thus allowing users who do not understand all the details of "fine-grained" objects to run programs.

For readers who are familiar with ArcInfo Workstation, programming the Geoprocessing tools is similar to programming ArcInfo commands in AML (Arc Macro Language). As long as the syntax is followed correctly, the object or command will work. How the object or command is pieced together is not a matter of concern to the programmer.

This book covers macros using the *Geoprocessing* (GP) object in Chapters 4 to 7 and 9 to 14. These GP macros are presented in boxes so that they are separated from regular ArcObjects macros. As the name suggests, the *Geoprocessing* object has little to offer in the areas of data display, data query, and layer management. This shortcoming, however, can be remedied by combining GP macros with regular macros. Chapter 12 has an example that combines a GP macro for deriving an aspect layer from a digital elevation model (DEM) and a regular macro for displaying the aspect layer with color symbols.

As a coarse-grained object, the *Geoprocessing* object is most useful to GIS users who must perform repetitive data processing tasks. Software developers, who must work with properties and methods of ArcObjects and combine them in various ways in macros, will find the *Geoprocessing* object less useful.

The Geoprocessing tool reference of ArcGIS Desktop Help Online offers up-to-date information on tools that can be used with the *Geoprocessing* object. The tools are organized in the same way as in ArcToolbox. To get the syntax for the Clip tool, for example, one would select Analysis toolbox, Extract toolset, Tools, and then Clip (Analysis). On the Clip (Analysis) page, scroll down to the command line syntax. The help document lists the syntax as follows:

Clip_analysis <in_features> <clip_features> <out_feature_class> {cluster_tolerance}

The first three parameters are the required parameters, representing the input layer, the clip layer, and the output layer. The last parameter of cluster tolerance is optional. This command line syntax is to be used in a VBA macro. Python script users, on the other hand, must follow the scripting syntax: Clip_analysis (in_features, clip_features, out-feature-class, cluster_tolerance). Python is a text-based, platform-independent language that can be downloaded from http://www.python.org. This book does not cover Python scripting.

ESRI, Inc. has published two documents on the Geoprocessing tools: *Writing Geoprocessing Scripts with ArcGIS* and *Geoprocessing Commands: Quick Reference Guide*. Both documents can be downloaded from their Web site (http://www. esri.com).

REFERENCES CITED

1. Zeiler, M., *Modeling Our World: The ESRI Guide to Geodatabase Design,* Environmental Systems Research Institute (ESRI), Redlands, CA, 1999.
2. Chang, K., *Introduction to Geographic Information Systems,* 4th ed., McGraw-Hill, New York, 2006.
3. Zeiler, M., Ed., *Exploring ArcObjects,* ESRI, Redlands, CA, 2001.
4. Larman, C., *Applying UML and Patterns: An Introduction to Object-Oriented Analysis and Design and the Unified Process,* 2nd ed., Prentice Hall, Upper Saddle River, NJ, 2001.
5. Koncz, N.A. and Adams, T.M., A data model for multi-dimensional transportation applications, *International Journal of Geographic Information Science,* 16, 551, 2002.
6. Huang, B., An object model with parametric polymorphism for dynamic segmentation, *International Journal of Geographic Information Science,* 17, 343, 2003.

7. Shi, W., Yang, B., and Li, Q., An object-oriented data model for complex objects in three-dimensional geographic information systems, *International Journal of Geographic Information Science,* 17, 411, 2003.
8. Burke, R., *Getting to Know ArcObjects: Programming ArcGIS with VBA,* ESRI, Redlands, CA, 2003.
9. Razavi, A.H., *ArcGIS Developer's Guide for VBA,* OnWord Press/Delmar Learning, Clifton Park, NY, 2002.
10. Tonias, C.N. and Tonias, E.C., *Avenue Wraps,* CEDRA Press, Rochester, NY, 2002.

CHAPTER 2

Programming Basics

ArcObjects is the development platform for ArcGIS. Because ArcObjects is built using Microsoft's COM (Component Object Model) technology, it is possible to use any COM-compliant development language with ArcObjects to customize applications in ArcGIS. This book adopts Visual Basic for Applications (VBA), which is already embedded in ArcMap and ArcCatalog of ArcGIS. Other COM-compliant programming languages include Visual Basic and C++.

Writing application programs for ArcGIS requires knowledge of both VBA and ArcObjects: VBA provides the programming language and ArcObjects provides objects and their built-in properties and methods. It may be of interest to some readers to compare ArcObjects with Avenue and AML (Arc Macro Language), two programming languages previously developed by Environmental Systems Research Institute, Inc. (ESRI). Programming ArcObjects is similar to Avenue programming in that both use objects and their built-in properties and methods (called requests in Avenue), but they differ in two important aspects. First, we program ArcObjects using VBA, a common programming language available in Microsoft's products. Second, ArcObjects has many more objects, properties, and methods than Avenue does. Programming ArcObjects is conceptually different from AML programming because AML is a procedural, rather than an object-oriented, language. The exception is the *Geoprocessing* object, which, as explained in Chapter 1, is a coarse-grained object. Programming the Geoprocessing tools is in many ways similar to programming ArcInfo commands in AML.

This chapter deals with the programming language and code writing, although many examples in the chapter do involve ArcObjects. Section 2.1 discusses the basic elements in VBA programming such as procedures, variables, interfaces, and arrays. Section 2.2 offers common techniques for writing code. Section 2.3 explains how to put together a program as a collection of code blocks. Section 2.4 covers Visual Basic Editor, a medium for preparing, compiling, and running macros. Section 2.5 covers the debugging tools that can help identify mistakes in macros.

2.1 BASIC ELEMENTS

This section covers the basic programming elements. Many elements are directly related to VBA. Therefore, additional information on these elements can be found in the Microsoft Visual Basic Help, which is accessible through Visual Basic Editor in either ArcMap or ArcCatalog.

2.1.1 Projects, Modules, Procedures, and Macros

Procedures are the basic units in VBA programming. A *procedure* is a block of code that can perform a specific task such as defining the coordinate system of a geographic dataset. Applications developed using VBA are called *macros* in Microsoft's products such as Word, Excel, and Access. A macro is functionally similar to a procedure. A *module* is a collection of procedures, and a *project* is a collection of modules (Figure 2.1). Most sample macros in this book are procedures, but some are modules. For example, modules, each with several procedures, are used to build binary and index models in Chapter 14.

A procedure can be private or public. A private procedure can only be called or used by another procedure in the same module. By contrast, a public procedure is available to different modules that make up a project.

Three types of procedures exist: events, subs, and functions. *Event* procedures are associated with controls on a form or dialog such as command buttons. *Subs* and *functions,* on the other hand, are not directly associated with controls. A function

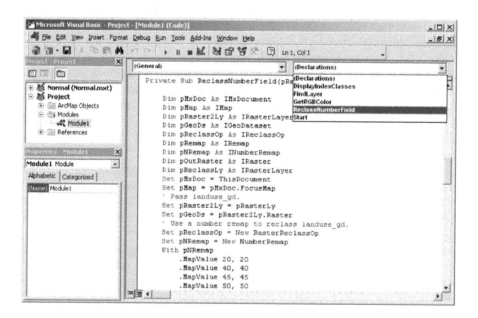

Figure 2.1   On the left of Visual Basic Editor, the Project Explorer shows that Module1 is a module in the Project. On the right, the procedure list shows that ReClassNumberField is the procedure in the Code window.

returns a value, whereas a sub does not. This book uses mainly subs and functions. Chapters 3 and 8 have examples that use event procedures to customize the user interface.

A procedure starts with the keyword of *Sub* or *Function* and ends with the *End Sub* or *End Function* statement. VBA automatically creates the first and last lines of a new procedure, which are called the wrapper lines.

2.1.2 Variables

A *variable* stores a value that can be accessed or changed by a macro. VBA requires that a variable be declared before it can be used. Declaration statements can be all placed at the top of a macro or placed wherever they are needed. This book adopts the style of declaring variables at the top of a macro. If a macro is divided into parts, then variables are declared at the top of each part. To ensure that variables are declared explicitly, the addition of *Option Explicit* at the beginning of a module is recommended. When a variable is not declared in a macro, the Option Explicit statement highlights the line in which an undeclared variable resides and produces an error message stating "compile error: variable not defined."

How to declare a variable in a macro depends on whether the variable refers to an ArcObjects class or not. The following two lines declare a counter variable *n*, which does not refer to an ArcObjects class, and assign 5 to be its value:

```
Dim n As Integer
n = 5
```

Dim is the most often used keyword for declaring a variable. A variable declared with the *Dim* keyword within a procedure is only available in that procedure. But a variable declared with the *Dim* keyword at the head (i.e., the Declarations section) of a module is available to all procedures within the module. Other keywords for declaring variables include *Public* and *Private*. A public variable is available to all modules in a project. A private variable, on the other hand, is available only to the module in which it is declared. A declaration statement usually includes a data type. "Integer" in "Dim n As Integer" represents the data type. Other data types include Boolean, Single, Double, String, and Variant.

If a variable refers to an existing class in ArcObjects, it must be declared by pointing to an interface that the class supports. The properties and methods of an object are hidden according to the encapsulation principle in object-oriented technology. Therefore, the object can only be accessed through the predefined interfaces. Encapsulation also means that the terms "interface" and "object" can be interchangeable.

The following two lines show how to declare a variable by referencing an existing class in ArcObjects:

```
Dim pField As IFieldEdit
Set pField = New Field
```

The first line declares *pField* by pointing the variable to the *IFieldEdit* interface that the *Field* coclass supports (Figure 2.2). The second line creates a new field object by making *pField* an instance of the *Field* class. *p* in *pField* stands for pointer,

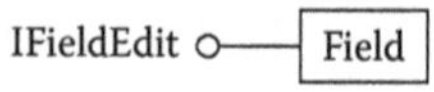

IFieldEdit o——| Field |

Figure 2.2   A field variable can be declared by pointing to *IFieldEdit* that a *Field* object supports.

and *I* in *IFieldEdit* stands for interface. *IFieldEdit* has the uppercase and lowercase letters for better reading. These are the naming conventions in object-oriented programming. The keyword *Set* assigns a value to a variable.

The next example defines the top layer in an active data frame of a running ArcMap.

```
Dim pMxDoc As IMxDocument
Dim pMap As IMap
Dim pFeatureLayer As ILayer
Set pMxDoc = ThisDocument
Set pMap = pMxDoc.FocusMap
Set pFeatureLayer = pMap.Layer(0)
```

The *Dim* statements point *pMxDoc* to *IMxDocument, pMap* to *IMap,* and *pFeatureLayer* to *IFeatureLayer.* The first *Set* statement assigns ThisDocument to *pMxDoc.* ThisDocument is the predefined name of an *MxDocument* object. When we launch ArcMap, *MxDocument* and *Application* are already in use. The alternative to This-Document is Application.Document, which refers to the document of the ArcMap application. The second *Set* statement assigns *FocusMap* or the focus map of the map document to *pMap.* The third statement assigns *Layer(0)* or the top layer in the focus map to *pFeatureLayer.* (0) is called an index, and the index begins with 0 in VBA. *FocusMap* and *Layer()* are both properties, which are covered in the next section.

2.1.3   Use of Properties and Methods

Properties are attributes of an object. As examples, *FocusMap* is a property on *IMxDocument* and *Layer(0)* is a property on *IMap.* The syntax for using a property is *object.property,* such as *pMxDoc.FocusMap.* Both *FocusMap* and *Layer()* happen to be get-only, or read-only, properties. The following example shows the put, or write, properties:

```
Dim pFeatureClass As IFeatureClass
Set pFeatureLayer.FeatureClass = pFeatureClass
PFeatureLayer.Name = "breakstrm"
```

The example shows two methods for putting properties: by reference and by value. The second line statement sets *pFeatureClass* to be the feature class of *pFeatureLayer* by reference, and the third line statement assigns the string "breakstrm" to be the name of *pFeatureLayer* by value.

The difference between put by reference and put by value is the use of the *Set* keyword. How can we tell which method to use? One approach is to consult the ArcObjects Developer Help. The put by reference property has an open square symbol, whereas the put by value property has a solid square symbol. Another approach is to let the VBA compiler catch the error. The error messages are

"Method or data member not found" if the *Set* keyword is missing and "Invalid use of property" if the *Set* keyword is unnecessary.

Methods perform specific actions. A method may or may not return a value. The syntax for calling a method is *object.method*. Many methods require object qualifiers and arguments. The following line, for example, adds a feature layer to a map:

```
PMap.AddLayer pFeatureLayer
```

The *AddLayer* method on *IMap* adds *pFeatureLayer* to *pMap*. The method requires an object qualifier (i.e., *pFeatureLayer*) and does not return a value or an interface.

The next example gets a workspace on disk and then gets a shapefile from the workspace.

```
Dim pFeatureWorkspace As IFeatureWorkspace
Dim pFeatureClass As IFeatureClass
Set pFeatureWorkspace = pWorkspaceFactory.OpenFromFile("c:\data\chap2", 0)
Set pFeatureClass = pFeatureWorkspace.OpenFeatureClass("emidastrm")
```

The *OpenFromFile* method on *IWorkspaceFactory* returns an interface on the specified workspace (i.e., "c:\data\chap2\"). The code then switches to the *IFeature-Workspace* interface and uses the *OpenFeatureClass* method on the interface to open *emidastrm* in the workspace. Both *OpenFromFile* and *OpenFeatureClass* require arguments in their syntax. The first argument for *OpenFromFile* is a workspace, and the second argument of 0 tells VBA to get the ArcMap window handle. The only argument for *OpenFeatureClass* is a string that shows the name of the feature class.

VBA has the automatic code completion feature to work with properties and methods. After an object variable is entered with a dot, VBA displays available properties and methods for the object variable in a dropdown list. We can either scroll through the list to select a property or method, or type the first few letters to come to the property or method to use.

2.1.4 QueryInterface

A class object may support two or more interfaces, and each interface may have a number of properties and methods. When we declare a variable, we point the variable to a specific interface. To switch to a different interface, we can use QueryInterface or QI for short. QI lets the programmer jump from one interface to another.

We can revisit the code fragment from Section 2.1.3 to get a better understanding of QI.

```
Dim pFeatureWorkspace As IFeatureWorkspace
Dim pFeatureClass As IFeatureClass
Set pFeatureWorkspace = pWorkspaceFactory.OpenFromFile("c:\data\chap2", 0)
Set pFeatureClass = pFeatureWorkspace.OpenFeatureClass("emidastrm") ' QI
```

The syntax of the *OpenFromFile* method suggests that the method returns the *IWorkspace* interface that a workspace object supports. But to use the *OpenFeature-Class* method, which is on *IFeatureWorkspace,* the code must perform a QI for *IFeatureWorkspace* that a workspace object also supports (Figure 2.3).

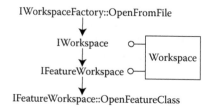

IWorkspaceFactory::OpenFromFile

IWorkspace

IFeatureWorkspace

Workspace

IFeatureWorkspace::OpenFeatureClass

Figure 2.3   The diagram shows how to switch from *IWorkspace* to *IFeatureWorkspace* by using QI.

The next example shows a code fragment for converting feature data to raster data. A *RasterConversionOp* object supports both *IConversionOp* and *IRaster-AnalysisEnvironment* (Figure 2.4). The *IConversionOp* interface has methods for converting feature data to raster data, and the *IRasterAnalysisEnvironment* interface has properties and methods to set the analysis environment. The following example uses QI to define the output cell size as 5000 for a vector to raster data conversion:

```
Dim pConversionOp As IConversionOp
Dim pEnv As IRasterAnalysisEnvironment
Set pConversionOp = New RasterConversionOp
Set pEnv = pConversionOp ' QI
PEnv.SetCellSize esriRasterEnvValue, 5000
```

2.1.5   Comment Lines and Line Continuation

A comment line is a line of text that is added to explain how a code statement or a block of code works. A comment line starts with an apostrophe ('). Except for short comment lines such as QI, which can be placed at the end of a statement, this book typically places a comment line before a statement or a group of statements. By default, comments are displayed as green text in Visual Basic Editor.

A code statement usually fits on one line. A long statement can be divided into two or more lines. An underscore (_) at the end of a line statement means that the statement continues onto the next line.

2.1.6   Arrays

An *array* is a special type of variable that holds a set of values of the same data type, rather than a single value as in the case of a regular variable. Arrays are declared the

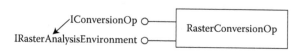

IConversionOp

IRasterAnalysisEnvironment

RasterConversionOp

Figure 2.4   Use QI to jump from *IConversionOp* to *IRasterAnalysisEnvironment*.

same way as other variables using the *Dim, Private,* and *Public* statements, but an array variable must have the additional specification for the size of the array. For example, the following line declares the *AnArray* variable as an array of 11 (0 to 10) integers:

```
Dim AnArray(10) As Integer
```

AnArray is a static array, meaning that it has a predefined size of 11. The other type of array is a dynamic array. A dynamic array has no fixed size but uses VBA keywords (e.g., *ReDim*) to find out information about the array and to change its size. A dynamic array must be declared at the module level.

2.1.7   Collections

A *collection* consists of a set of ordered objects that do not have to be of the same data type. Collections are therefore special arrays. A collection can be created as follows:

```
Dim theList As New Collection
```

The following code fragment uses a loop and the method *Add* on a *Collection* object to add the field names on *theList*:

```
Dim theList As New Collection
For ii = 0 To pFields.FieldCount - 1
    Set aField = pFields.Field(ii)
    fieldName = aField.Name
    theList.Add (fieldName)
Next
```

2.2 WRITING CODE

This section covers programming techniques for handling decision making, branching, repetitive operations, and dialogs.

2.2.1   *If...Then...Else* Statement

A simple way for decision making in a macro is to use the *If...Then...Else* statement. The statement has the following syntax:

```
If condition Then
    [statements]
Else
    [else_statements]
End If
```

If the condition is true, the program executes statements that follow the *Then* keyword. If the condition is false, the program executes statements that follow the *Else* keyword. The *If...Then...Else* statement can therefore handle two possible outcomes. To handle more than two outcomes, one can add the *ElseIf* clause to the statement:

```
If condition Then
    [statements]
ElseIf condition-n Then
    [elseif_statements]
Else
    [else_statements]
End If
```

The following example assigns 5 to the variable *n* if the name of the top layer is idcounty and 3 to *n* if the name is not idcounty:

```
If (pFeatureLayer.Name = "idcounty") Then
    n = 5
Else
    n = 3
End If
```

When the *If...Then...Else* statement is used jointly with the *TypeOf* keyword, the statement can check whether an object supports a specific interface before using the interface. The following code fragment verifies that *pConversionOp* does support *IRasterAnalysisEnvironment* before specifying the output cell size of 5000:

```
Dim pConversionOp As IConversionOp
Dim pEnv As IRasterAnalysisEnvironment
Set pConversionOp = New RasterConversionOp
If TypeOf pConversionOp Is IRasterAnalysisEnvironment Then
    Set pEnv = pConversionOp ' QI
    PEnv.SetCellSize esriRasterEnvValue, 5000
End If
```

Another common use of the *If...Then...Else* statement is to check for a condition that can cause a program error such as division by zero. If such a condition is determined to exist, the *Exit Sub* statement placed after *Else* can terminate the execution of a macro immediately.

2.2.2 *Select Case* Statement

The *If...Then...Else* statement can become confusing and untidy if more than three or four possible outcomes exist. An alternative is to use the *Select Case* statement, which has the following syntax:

```
Select Case test_expression
    [Case expression_list-n
    [statements-n]]...
    [Case Else
    [else_statements]]
End Select
```

ArcObjects codes the data type of a field in numeric values from 0 to 8. A *Select Case* statement can translate these numeric values into text strings. The following example uses a *Select Case* statement to prepare the data type description of a field:

```
Dim fieldType As Integer
Dim typeDes As String
Select Case fieldType
Case 0
    typeDes = "SmallInteger"
Case 1
    typeDes = "Integer"
Case 2
    typeDes = "Single"
Case 3
    typeDes = "Double"
Case 4
    typeDes = "String"
Case 5
    typeDes = "Date"
Case 6
    typeDes = "OID"
Case 7
    typeDes = "Geometry"
Case 8
    typeDes = "Blob"
End Select
```

2.2.3 *Do...Loop* Statement

A *Do...Loop* statement repeats a block of statements in a macro. VBA offers two types of loops. A *Do While* loop continues while the condition is true:

```
Do While condition
    [statements]
Loop
```

The following example uses a *Do While* loop to repeat a block of statement as long as the user provides the name of a shapefile and stops the loop when *pInput* is empty:

```
Dim pInput As String
pInput = InputBox("Enter the name of the input shapefile")
Do While pInput <> ""
    [statements]
Loop
```

A *Do Until* loop continues until the condition becomes true:

```
Do Until condition
    [statements]
Loop
```

The following example uses a *Do Until* loop to count how many cities are in a cursor (that is, a selection set):

```
Dim pCity As IFeature
Dim intCount As Integer
Dim pCityCursor As IFeatureCursor
Set pCity = pCityCursor.NextFeature
Do Until pCity is Nothing
    IntCount = intCount + 1
    Set pCity = pCityCursor.NextFeature
Loop
```

The *FeatureCursor* object holds a set of selected features. The *IFeatureCursor* interface has the *NextFeature* method that advances the position of the feature cursor by one and returns the feature at that position. By using the cursor and the *NextFeature* method, the code increases the *intCount* value by 1 each time *Next-Feature* advances a feature. The loop continues until no feature (i.e., Nothing) is advanced.

2.2.4   *For...Next* Statement

Like the *Do...Loop* statement, the *For...Next* statement also repeats a block of statements. But instead of using a conditional statement for the loops, the *For...Next* statement runs a given number of times as determined by the start, end, and step (with the default of one) values:

```
For counter = start To end [Step step]
    [statements]
Next
```

The following example uses a *For...Next* statement to add the field names of a feature class to an array:

```
Dim pFields As IFields
Dim ii As Long
Dim aField As IField
Dim fieldName As Variant
Dim theList As New Collection
For ii = 0 To pFields.FieldCount - 1
    Set aField = pFields.Field(ii)
    fieldName = aField.Name
    theList.Add (fieldName)
Next
```

The *FieldCount* property on *IFields* returns the number of fields in *pFields*. The code then sets the *For...Next* statement to begin with zero and to end with the number of fields minus 1 so that the *ii* counter corresponds to the index of a field.

The *Exit For* statement provides a way to exit a *For...Next* loop and transfers control to the statement following the *Next* statement. The following example uses an *Exit For* statement to exit the loop if a layer named idcities is located before reaching a fixed number of loops:

```
Dim pMxDoc As IMxDocument
Dim pMap As IMap
Dim pLayer As ILayer
Dim i As Integer
Set pMxDoc = ThisDocument
Set pMap = pMxDoc.FocusMap
For ii = 0 To pMap.LayerCount - 1
    Set pLayer = pMap.Layer(i)
    If pLayer.Name = "idcities" Then
        i = ii
        Exit For
    End If
Next ii
MsgBox "idcounty is at index " & i
```

2.2.5 *For Each...Next* **Statement**

The *For Each...Next* statement repeats a group of statements for each element in an array or collection.

```
For Each element In group
    [statements]
Next
```

The following code fragment uses a *For Each...Next* statement to print each field name in a collection of field names referenced by *theList*:

```
' Display the list of field names in a message box
For Each fieldName In theList
    MsgBox "The field name is " & fieldName
Next fieldName
```

2.2.6 *With* **Statement**

The *With* statement lets the programmer perform a series of statements on a single object. The *With* statement has the following syntax:

```
With object
    [statements]
End With
```

The following code fragment uses a *With* block to edit the name, type, and length properties of a new field:

```
Dim pField As IFieldEdit
Set pField = New Field
With pField
    .Name = "pop2000"
    .Type = esriFieldTypeInteger
    .Length = 8
End With
```

The alternative to the *With* block is to use the following line statements:

```
pField.Name = "pop2000"
pField.Type = esriFieldTypeInteger
pField.Length = 8
```

2.2.7 Dialog Boxes

Dialogs in a macro serve the purpose of getting information from and to the user. This section covers message boxes and input boxes, two simple dialog boxes that are frequently used in VBA macros. Other types of dialogs are covered elsewhere in the book. Chapter 3 covers custom dialogs using Visual Basic forms, Chapters 4 and 14 use browser dialogs for selecting datasets, and Chapter 10 uses a progress dialog for reporting the progress of a spatial join operation. Browser and progress dialogs are examples of dialogs that allow an ArcObjects macro to interact with the user.

A message box can be used as a statement or a function. As a statement, a message box shows text. For example, the following statement displays the quoted text and the value of the *fieldName* variable:

```
MsgBox "The field name is " & fieldName
```

After viewing the field name, the user must acknowledge by clicking the OK button, which is also displayed in the message box. VBA has the following chr$() functions for handling multiline messages: chr$(13) for a carriage return character, and chr$(10) for a linefeed character. Additionally, the constant vbCrLf also functions as chr$(10) in creating a new line. For example, the following statement displays the minimum and maximum values in two separate lines:

```
MsgBox "The minimum is: " & Min & Chr$(10) & "The maximum is: " & Max
```

As a function, a message box returns the ID of the button that the user presses. The message box includes a prompt (for example, Do you want to continue?), the Yes and No buttons, a question mark icon, and a title of Continue. The returned value is 6 for Yes and 7 for No. The following code fragment creates a message box and returns a value to *iAnswer* based on the user's decision:

```
Dim iAnswer As Integer
iAnswer = MsgBox("Do you want to continue?", vbYesNo + vbQuestion, "Continue")
MsgBox "The answer is : " & iAnswer
```

An input box displays a prompt in a dialog box and returns a string containing the user's input. The following example displays the prompt of "Enter the name of the input shapefile" in a dialog box and returns the user's input as a string to the *pInput* variable:

```
Dim pInput As String
pInput = InputBox("Enter the name of the input shapefile")
```

2.3 CALLING SUBS AND FUNCTIONS

A procedure, either a sub or a function, can be called by another procedure. VBA actually provides many simple functions that we use regularly in macros. Both message boxes and input boxes are VBA functions. Other examples include CStr and CInt. The CStr function converts a number to a string, and the CInt function returns an integer number.

This section goes beyond simple VBA functions and deals with the topic in a broader context. A procedure, depending on whether it is private or public, can be called by another procedure in the same module or throughout a project. Therefore, we can think of a sub or a function as a tool and build a module as a collection of tools. The major advantage of organizing code into separate subs and functions is that they can be reused in different modules. Other advantages include ease of debugging in smaller blocks of code and a better organization of code.

In the following example, the *Start* sub uses an input box to get a number from the user and then calls the *Inverse* sub to compute and report the inverse of the number:

```
Private Sub Start ()
    Dim n As Integer
    n = InputBox("Type a number")
    ' Call the Inverse sub.
    Inverse n
End Sub

Private Sub Inverse (m)
    Dim d As Double
    d = 1 / m
    MsgBox "The inverse of the number is: " & d
End Sub
```

The *Start* sub passes *n* entered by the user as an argument to the *Inverse* sub. *Inverse* uses the passed value *m* (same as *n*) to compute its inverse. Notice that the example does not use the *Call* keyword. If the programmer prefers to use the keyword, the *inverse n* statement can be changed to:

```
Call Inverse (n)
```

The next example lets the *Start* sub call a function instead of a sub to accomplish the same task:

```
Private Sub Start ()
    Dim n As Integer
    Dim dd As Double
    n = InputBox("Type a number")
    ' Call the Inverse function and assign the return value to dd.
    dd = Inverse (n)
    MsgBox "The inverse of the number is: " & dd
End Sub
```

```
Private Function Inverse (m) As Double
    Dim d As Double
    d = 1 / m
    Inverse = d ' Return the d value.
End Function
```

A couple of changes are noted when the code calls a function instead of a sub. First, the **Start** sub uses the following line to assign the returned value from the **Inverse** function to *dd,* which has been previously declared as a Double variable:

```
dd = Inverse (n)
```

Second, the code adds the *As Double* clause to the first line of **Inverse***:*

```
Private Function Inverse (m) As Double
```

The clause declares **Inverse** to be a Double procedure. Thus the value returned by the function is also of the Double data type.

Third, the following line assigns *d,* which is the inverse of the passed value *m,* to **Inverse***:*

```
Inverse = d
```

The *d* value is eventually returned to **Start** and assigned to the *dd* variable.

2.4 VISUAL BASIC EDITOR

Visual Basic Editor is a tool for compiling and running programs. To open Visual Basic Editor in either ArcCatalog or ArcMap, one can click the Tools menu, point to Macros, and select Visual Basic Editor.

Figure 2.5 shows Visual Basic Editor in ArcMap. A menu bar, a toolbar, and windows make up the user interface. Several commands ought to be mentioned at this point. Import File and Export File on the File menu allow the user to import and export macros in text file format. The Debug menu has commands for compiling and debugging macros, and the Run menu has commands for running and resetting macros. The same commands of Run Sub/UserForm, Break, and Reset are also available on the toolbar.

Figure 2.5 shows four types of windows: Code, Project, Properties, and Immediate. The Code window is the area for preparing and editing a macro. We can either type a new macro or import a macro. At the top of the Code window are two dropdown lists. On the left is the object list, and on the right is the procedure list. The Project window, also called the Project Explorer, displays a hierarchical list of projects and the contents and references of each project. Normal.mxt is a template for all map documents and is present whenever Visual Basic Editor is launched. Project, on the other hand, is specific to a map document. Macros for specific tasks are typically developed and stored at the current Project level. The Properties window shows the properties of controls, such as command buttons and text boxes on a user form. Chapter 3 on customization of the user interface covers the use of the Properties window. The Immediate window is designed for debugging. When used with a *Debug.Print* statement, the window can show the value of a variable for debugging.

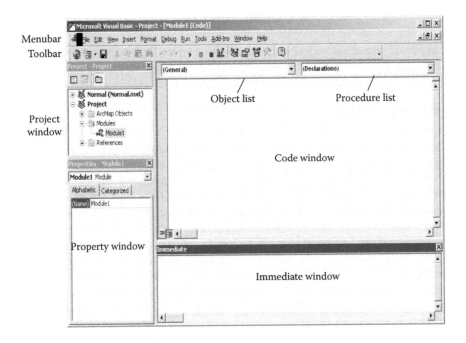

Figure 2.5 Visual Basic Editor consists of a menu bar, a toolbar, the Project window, the Property window, the Code window, the object list, the procedure list, and the Immediate window.

Visual Basic Editor in ArcCatalog is set up the same way as in ArcMap; the only difference is that the Project Explorer in ArcCatalog contains only Normal.gxt. ArcCatalog does not have documents, and all customizations apply to the application.

The following shows how to use Visual Basic Editor to import and use a sample module on the companion CD of this book:

1. Right-click Project in the Project Explorer in ArcMap and select Import File. (In ArcCatalog, right-click Normal in the Project Explorer and select Import File.)
2. Select All Files from the file type dropdown list in the Import File dialog. Navigate to the sample module in text file format. Click Open to import the sample module to Visual Basic Editor.
3. Click the plus sign next to the Modules folder in Project to open its contents.
4. Right-click Module1 and select View Code. The code now appears in the Code window, and the procedure list shows the name of the module.
5. Select Compile Project from the Debug menu to make sure that the module compiles successfully. To run the module, simply click on the Run Sub/UserForm button.

Most macros on the companion CD are designed for ArcMap so that the datasets can be displayed and analyzed immediately. Some macros, such as those for data conversion, can be run in either ArcCatalog or ArcMap.

2.5 DEBUGGING CODE

Every programmer has to deal with programming errors. Some errors are easy to fix, while others may take hours or days to correct. VBA has various debugging tools that can assist programmers in fixing errors. This section covers some of these tools.

2.5.1 Type of Error

There are three possible types of errors in VBA macros: compile, run-time, and logic. VBA stops compiling when it finds a compile error. Compile errors are caused by mistakes with VBA programming syntax. A compile error can occur when a macro misses the *End With* line in a *With* block or the *Loop* keyword in a *Do Until* statement. A compile error can also occur if a macro uses a property or method that is not available on an interface. For example, the *IFeatureWorkspace* interface has the *OpenFeatureClass* method but not *OpenFromFile*. When a macro tries to use *OpenFromFile* to open a feature class, VBA displays a compile error with the message of "Method or data member not found." To make sure that a property or method is available on an interface, one can first highlight the interface in the code window and then press F1. This will open the Help page on the interface from the ArcObjects Developer Help.

A run-time error occurs when a macro, which has been compiled successfully, is running. Run-time errors are more difficult to fix than compile errors. The following code is supposed to report the name of each layer in the active map:

```
Private Sub LayerName()
    Dim pMxDoc As IMxDocument
    Dim pMap As IMap
    Dim pFeatureLayer As IFeatureLayer
    Dim ii As Integer
    Set pMxDoc = ThisDocument
    Set pMap = pMxDoc.FocusMap
    ' Loop through each layer, and report its name.
    For ii = 1 To pMap.LayerCount
        Set pFeatureLayer = pMap.Layer(ii)
        MsgBox "The name of layer is: " & pFeatureLayer.name
    Next
End Sub
```

The macro has no compile errors, but it has a run-time error stating "Run-time error '5': Invalid procedure call or argument." VBA expects to have one more layer than what is available in the active map. To make the macro run successfully, the *For...Next* statement must be changed to

```
For ii = 0 To pMap.LayerCount - 1
```

The next example is similar to the module used previously to derive the inverse of a typed number except that it does not pass the typed number *n* as an argument from the calling sub to the function. Therefore, *m* in the **Inverse** function is treated as 0. The error message in this case is "Run-time error '11': Division by zero."

```
Private Sub Start()
    Dim n As Integer
    Dim dd As Double
    n = InputBox("Type a number")
    ' Call the Inverse function and assign the return value to dd.
    dd = Inverse()
    MsgBox "The inverse of the number is: " & dd
End Sub

Private Function Inverse() As Double
    Dim d As Double
    d = 1 / m
    Inverse = d ' Return the value d.
End Function
```

Logic errors are even more difficult to correct than run-time errors. A logic error does not stop a macro from compiling and running but produces an incorrect result. One type of logical error that every programmer dreads is endless loops. Endless loops can be caused by failing to set the condition in a *Do...Loop* statement correctly.

2.5.2 *On Error* Statement

VBA has a built-in object called *Err.* The *Err* object has properties that identify the number, description, and source of a run-time error. We can use the *On Error* statement to display the properties of the *Err* object when an error occurs:

On Error GoTo line

The code below includes the *On Error* statement to trap the run-time error of division by zero:

```
Private Sub Start()
    On Error GoTo ErrorHandler
    Dim n As Integer
    Dim dd As Double
    n = InputBox("Type a number")
    ' Call the Inverse function and assign the return value to dd.
    dd = Inverse()
    MsgBox "The inverse of the number is: " & dd
    Exit Sub ' Exit to avoid error handler.
ErrorHandler: ' Error-handling routine.
    MsgBox Str(Err.Number) & ": " & Err.Description, , "Error"
End Sub

Private Function Inverse() As Double
    Dim d As Double
    d = 1 / m
    Inverse = d ' Return the value d.
End Function
```

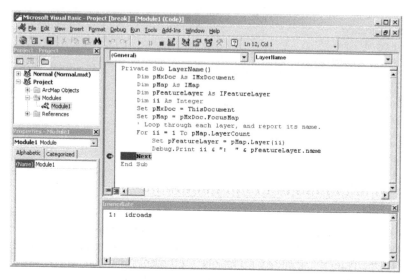

Figure 2.6   The breakpoint at the *Next* line allows the programmer to see that the layer index is 1 and idroads is the name of the layer.

When the error occurs, the *ErrorHandler:* routine displays "11:Division by zero" in a message box with the title of Error. Notice that the *On Error* statement is placed at the top of the code. When a run-time error occurs, the code goes to the *ErrorHandler:* routine and displays the error message. Also notice that the *Exit Sub* line is used right before the *ErrorHandler:* routine to avoid the error message if no errors occurred.

2.5.3   Use of Breakpoint and Immediate Window

A breakpoint suspends execution at a specific statement in a procedure. A breakpoint therefore allows the programmer to examine variables and to make sure that the code is working properly. The following code places a breakpoint at the *Next* line of the *For...Next* statement and uses *Debug.Print* (the *Print* method of the *Debug* object) to print the counter value and the layer's name in the Immediate window (Figure 2.6):

```
Private Sub LayerName()
    Dim pMxDoc As IMxDocument
    Dim pMap As IMap
    Dim pFeatureLayer As IFeatureLayer
    Dim ii As Integer
    Set pMxDoc = ThisDocument
    Set pMap = pMxDoc.FocusMap
    ' Loop through each layer, and report its name.
    For ii = 0 To pMap.LayerCount - 1
        Set pFeatureLayer = pMap.Layer(ii)
        Debug.Print ii & ": " & pFeatureLayer.name
    Next
End Sub
```

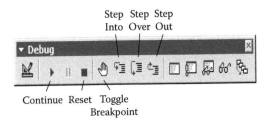

Figure 2.7 The Debug toolbar has the tools of Continue, Reset, Toggle Breakpoint, Step Into, Step Over, and Step Out.

The first time through the loop, the Immediate window shows zero and the name of the top layer in the active map. Click on the Continue button (the same button for Run Macro), and the window shows the next set of values.

Visual Basic Editor has Toggle Breakpoint on the Debug menu, as well as on the Debug toolbar, to add or remove a breakpoint at the current line (Figure 2.7). Other commands on the Debug menu include Step Into for executing code one statement at a time, Step Over for executing a procedure as a unit, Step Out for executing all remaining code in a procedure as if it were a single statement, and Run To Cursor for selecting a statement to stop execution of code.

CHAPTER 3

Customization of the User Interface

As a commercial product, ArcGIS is designed to serve as many users as possible and to meet as many needs as possible. It is no surprise that the software package has a large number of extensions, toolbars, and commands. But most users only use a portion of the available tools at a time. Therefore, a common customization is to simplify the way we interact with ArcGIS. ArcGIS Desktop provides options to view or hide a toolbar. When working in ArcMap, we typically bring those toolbars necessary for a specific task to view and hide the others. Selecting toolbars to view and use is perhaps the easiest form of customization.

Customization can take other forms:

- Streamline the workflow. For example, instead of defining a new field and then calculating the field values in separate steps, we may want to combine them into one step.
- Reduce the amount of repetitive work. For example, rather than repeating for each dataset the same task of defining a common coordinate system, we may write a macro to complete the entire job with a single button click.
- Prevent the user from making unnecessary mistakes. For example, if a project requires distance measures to be in feet, we may choose feet as measurement units in code to prevent use of other units.

The above examples show that customization is most useful if a project has a set of well-defined tasks.

This chapter introduces common methods for customizing the user interface. Section 3.1 describes how to create a new toolbar with existing ArcMap commands. Sections 3.2 and 3.3 discuss how to add a new button and a new tool respectively. Section 3.4 demonstrates the procedure for storing a new toolbar in a template. Section 3.5 explains the design and use of a Visual Basic form. Section 3.6 covers the procedure for storing a password-protected form in a template.

3.1 CREATING A TOOLBAR WITH EXISTING ARCMAP COMMANDS

No code writing is required for creating a new toolbar with existing buttons and tools in ArcMap. It is a simple copy-and-paste process. Suppose an application requires the following commands (buttons and tools) on a new toolbar: Zoom In, Full Extent, Select By Attributes, and Select By Location. The following shows the procedure for completing the task:

1. Select Customize from the Tools menu in ArcMap, or double-click on an empty area of a toolbar, to open the Customize dialog (Figure 3.1). The Customize dialog has three tabs: Toolbars, Commands, and Options. The Toolbars tab shows all toolbars available in ArcMap. The Commands tab shows all commands available in ArcMap by category. The Options tab has options to lock a customization with a password. The Customize dialog in ArcCatalog is set up the same as in ArcMap.
2. Click New in the Customize dialog. In the New Toolbar dialog, enter Selection for the toolbar name and save the toolbar in Untitled. Click OK to dismiss the New Toolbar dialog. A new toolbar now appears in ArcMap.
3. Click the Commands tab in the Customize dialog. Click the category of Pan/Zoom to view its commands (Figure 3.2). After locating the Full Extent command, drag and drop it onto the new toolbar. Next add the Zoom In command.
4. Click the category of Selection. Drag and drop the commands of Select By Attributes and Select By Location onto the new toolbar. As shown in Figure 3.3, the new toolbar now has four commands. These commands can be rearranged in the same way as any graphic elements.
5. Right-click a command on the toolbar to open its context menu. The menu has the options to delete, to change the image icon, to use text only, or to use image and text. The option for View Source is not available.

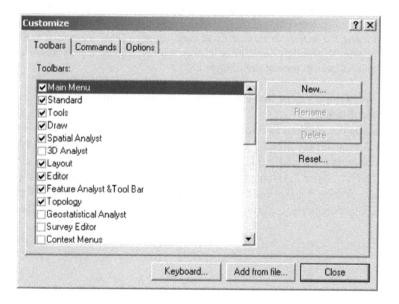

Figure 3.1   The Customize dialog box has the three tabs of Toolbars, Commands, and Options.

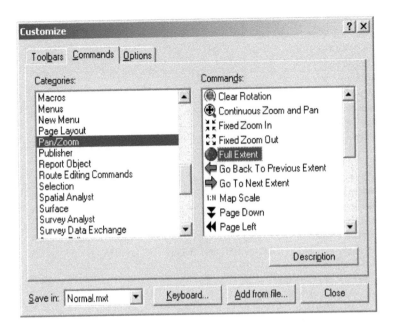

Figure 3.2   On the Commands tab, highlight the category of Pan/Zoom and view existing ArcMap commands in that category.

This new toolbar with four commands can now be used by itself or along with other toolbars in ArcMap.

3.2 ADDING A NEW BUTTON

A new button must be associated with a macro so that the macro can be executed to accomplish a specific task when the button is clicked on. Suppose the task is to report the fields of a geographic dataset in a message box. (Chapter 5 covers macros on managing and reporting fields.) The first step is to prepare an event (Click) procedure as follows:

```
Private Sub UIButtonFields_Click ()
    ' Part 1: Get the feature class and its fields.
    Dim pMxDoc As IMxDocument
    Dim pMap As IMap
```

Figure 3.3   The new Selection toolbar has four commands.

```
Dim pFeatureLayer As IFeatureLayer
Dim pFeatureClass As IFeatureClass
Dim pFields As IFields
Dim count As Long
Set pMxDoc = ThisDocument
Set pMap = pMxDoc.FocusMap
Set pFeatureLayer = pMap.Layer(0)
Set pFeatureClass = pFeatureLayer.FeatureClass
Set pFields = pFeatureClass.Fields

' Part 2: Prepare a list of fields and display the list.
Dim ii As Long
Dim aField As IField
Dim fieldName As Variant
Dim theList As New Collection
Dim NameList As Variant
' Loop through each field, and add the field name to a list.
For ii = 0 To pFields.FieldCount - 1
    Set aField = pFields.Field(ii)
    fieldName = aField.name
    theList.Add (fieldName)
Next
' Display the list of field names in a message box.
For Each fieldName In theList
    NameList = NameList & fieldName & Chr(13)
Next fieldName
MsgBox NameList, , "Field Names"
End Sub
```

After the macro has been compiled and run successfully, the next step is to link the macro to a button by using the following instructions:

1. Select Customize from the Tools menu in ArcMap.
2. Click New in the Customize dialog. In the New Toolbar dialog, enter Thermal for the toolbar name and save the toolbar in Untitled. Click OK to dismiss the New Toolbar dialog. The Thermal toolbar now appears in ArcMap.
3. On the Commands tab of the Customize dialog, select the category of UIControls and then click the New UIControl button (Figure 3.4). In the next dialog, check the option button for UIButtonControl and click Create (Figure 3.5).
4. A new command called Project.UIButtonControl1 appears in the Customize dialog. UIButtonControl1 is a default name. Click the new command and rename the button Project.UIButtonFields. Drag and drop Project.UIButtonFields onto the new toolbar.
5. While the Customize dialog is still open, right-click the new button control and select View Source from its context menu. View Source opens Visual Basic Editor. The first and last wrapper lines of the UIButtonFields_Click() Sub are already in the Code window (Figure 3.6). Copy and paste *UIButtonFields_Click* on the companion CD to the Code window.

To test how the button works, do the following:

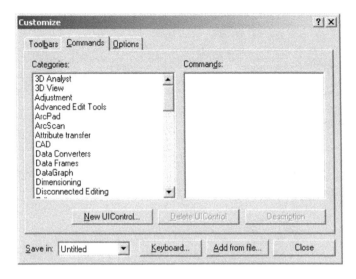

Figure 3.4   The New UIControl button is for creating a new control.

1. Add *thermal.shp* to ArcMap. The shapefile shows thermal springs and wells in Idaho.
2. When the new button is clicked, a message box appears with the fields in *thermal*.

3.3 ADDING A NEW TOOL

A button performs a task as soon as it is clicked on. A tool, on the other hand, requires the user to do something first before the tool can perform a task. The interaction with the user means that a tool has more events to consider and more coding to do than a button. Events associated with a tool include Select, DblClick, MouseDown, MouseUp, and MouseMove.

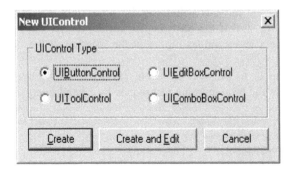

Figure 3.5   The New UIControl dialog shows four types of controls, including Button and Tool.

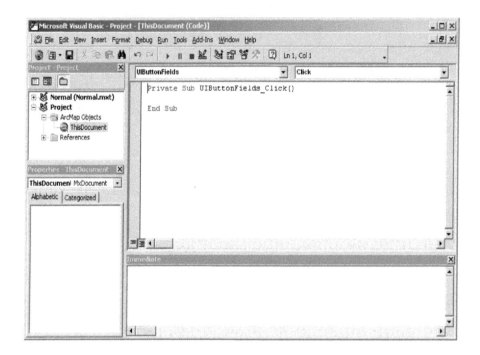

Figure 3.6   Visual Basic Editor automatically adds the wrapper lines of the UIButtonFields_-
Click sub. The object list shows UIButtonFields, and the procedure list shows Click.

This section describes a new tool, which uses a point entered by the user to report the number of features within 16,000 meters of the point. The tool essentially performs a spatial query based on the user's input. The procedure to be associated with the tool is a MouseDown event procedure. A MouseDown event procedure has four variables in its argument list: button, shift, x, and y. The user actually sets the values for these variables by clicking a point on the computer screen. The button value is either 1 or 2: 1 if the user is holding down the left mouse button and 2 if the user is holding down the right mouse button. The shift value is either 0 or 1: 0 if the Shift key is not depressed and 1 if the Shift key is depressed. The x and y values represent the location of the mouse pointer on the map display.

The first step is to prepare the event (MouseDown) procedure as follows (Chapter 9 covers the programming techniques for spatial query):

```
Private Sub UIToolQuery_MouseDown(ByVal button As Long, ByVal shift As Long, ByVal x As Long, ByVal y As Long)
    ' Part 1: Get the point clicked by the user.
    Dim pMxDoc As IMxDocument
    Dim pActiveView As IActiveView
    Dim m_blnMouseDown As Boolean
    Dim pPoint As IPoint
    Set pMxDoc = ThisDocument
```

```
Set pActiveView = pMxDoc.FocusMap
' Convert the entered point from display coordinates to map coordinates.
Set pPoint = pActiveView.ScreenDisplay.DisplayTransformation.ToMapPoint(x, y)

' Part 2: Perform a spatial query of features within 16,000 meters of the entered point.
Dim pLayer As IFeatureLayer
Dim pSpatialFilter As ISpatialFilter
Dim pTopoOperator As ITopologicalOperator
Dim pSelection As IFeatureSelection
Dim pElement As IElement
Dim pSelectionSet As ISelectionSet
Set pLayer = pMxDoc.FocusMap.Layer(0)
' Create a 16,000-meter buffer polygon around the clicked point.
Set pTopoOperator = pPoint
Set pElement = New PolygonElement
pElement.Geometry = pTopoOperator.Buffer(16000)
' Create a spatial filter for selecting features within the buffer polygon.
Set pSpatialFilter = New SpatialFilter
pSpatialFilter.SpatialRel = esriSpatialRelContains
Set pSpatialFilter.Geometry = pElement.Geometry
' Refresh the active view.
pActiveView.PartialRefresh esriViewGeoSelection, Nothing, Nothing
' Perform spatial query.
Set pSelection = pLayer
pSelection.SelectFeatures pSpatialFilter, esriSelectionResultNew, False
' Refresh the active view to highlight the selected features.
pActiveView.PartialRefresh esriViewGeoSelection, Nothing, Nothing
' Create a selection set and report number of features in the set.
Set pSelectionSet = pSelection.SelectionSet
MsgBox pSelectionSet.Count & " thermals selected"
End Sub
```

After the macro has been compiled successfully, the next step is to link the macro to a tool. In this case, the new tool is added to the Thermal toolbar from Section 3.2 as follows:

1. Open the Customize dialog.
2. Click Commands in the Customize dialog. Select the category of UIControls and click New UIControl. In the next dialog, select UIToolControl and then click Create. Rename the new control Project.UIToolQuery. Drag and drop Project.UIToolQuery onto the Thermal toolbar.
3. Right-click UIToolQuery and select View Source. View Source opens Visual Basic Editor. The top of the Code window has the object dropdown list on the left and the (event) procedure list on the right. The object list shows UIToolQuery and the procedure list shows the default procedure of Select. This application, however, uses the MouseDown event. Click the procedure dropdown arrow and choose MouseDown (Figure 3.7). Visual Basic Editor automatically inserts the wrapper lines for the UIToolQuery_MouseDown Sub in the Code window. Copy and paste **UIToolQuery_MouseDown** on the companion CD to the Code window. (The UIToolQuery_Select Sub remains in the Code window. It can be left alone or deleted.) Close Visual Basic Editor.

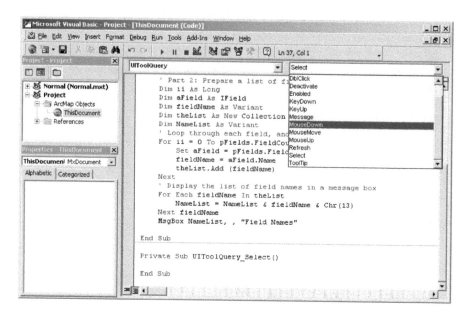

Figure 3.7   The procedure list shows different events, including MouseDown.

The Thermal toolbar now has a button and a tool. To test how the new tool works, do the following:

1. Make sure that *thermal.shp* is still in view. Click the new tool. Then click a point on the map.
2. A message box reports how many thermal wells and springs are within 16,000 meters of the entered point. At the same time, the selected thermal wells and springs are highlighted in the map.
3. Click another point on the map. The tool again reports the number of features selected and refreshes the map to show the newly selected features.

3.4 STORING A NEW TOOLBAR IN A TEMPLATE

A customized toolbar such as the Thermal toolbar can be saved for future use or distributed to other users. ArcMap users can save a customized application at three different levels: the Normal template (Normal.mxt), a base template (.mxt), or the current map document (.mxd). Normal.mxt is used every time ArcMap is launched. An mxd file, on the other hand, is available only in a particular map document. A base template represents an intermediate customization between Normal.mxt and the local mxd file. An mxt file is used whenever the user opts to open the file.

A layout template (e.g., USA.mxt) is one type of template that is familiar to many ArcGIS users; it has a layout design complete with map elements such as a

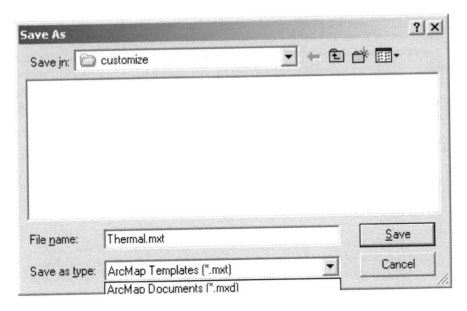

Figure 3.8   Save the new Thermal toolbar as an ArcMap Template.

legend and a scale bar. To make a map based on a layout template, we only have to add data, a title, and any other supporting information. A layout template therefore represents a customized application that is available to any ArcMap users who ask for it.

The following shows how to make a template that contains the Thermal toolbar so that the template can be distributed to other users:

1. Use *thermal* to test that the commands on the Thermal toolbar work correctly. Remove *thermal*. Select Save As from the File menu in ArcMap. In the Save As dialog, select to save as ArcMap Templates (*.mxt) and enter *Thermal.mxt* for the file name (Figure 3.8). Exit ArcMap.
2. Anyone who has access to *Thermal.mxt* can now use the commands on the Thermal toolbar. Launch ArcMap. Click on *Thermal.mxt* in the ArcMap dialog to open the template. (If *Thermal.mxt* does not show up in the dialog, select Open in the File menu to locate and open the template.) The commands on the Thermal toolbar are ready to work with the top layer in the active map.

3.5 ADDING A FORM

A form is a dialog box that uses controls such as text boxes and command buttons for the user interface. A variety of dialog boxes or forms exist. For example, a message box or an input box is actually a form, albeit with only a couple of controls.

Forms are particularly useful for gathering from the user various inputs that are needed for an operation. For example, a form can be used to get a numeric field, the number of classes, and the classification method before making a graduated color map. (Chapter 8 has an example of using a form to gather such inputs.)

As an introduction to forms, this section uses a relatively simple form with four controls: a label, a dropdown list with acres and square miles, a command button to run, and a command button to cancel. The user can use the form to convert area units of a feature class from square meters to either acres or square miles and to save the new area units in a new field.

3.5.1 Designing a Form

Visual Basic Editor provides the environment for designing a user form. The following shows the steps for opening a form and placing controls from the toolbox onto the form in ArcMap:

1. Click the Tools menu in ArcMap, point to Macros, and select Visual Basic Editor.
2. Right-click Project in the Project Explorer, point to Insert, and select UserForm. The Toolbox and UserForm1 now appear in Visual Basic Editor (Figure 3.9). The Properties window shows the default properties of UserForm1. Change the name of the form to *frmAreaUnits*, and change the caption to *New Area Units*. The prefix of frm in *frmAreaUnits* is the recommended naming convention for a form.
3. The Toolbox offers 15 different controls. (If the Toolbox disappears, click the View Object button at the top of the Project Explorer.) The tool tips show that these controls are: Select Object, Label, TextBox, ComboBox, ListBox, Check-Box, OptionButton, ToggleButton, Frame, CommandButton, TabStrip, MultiPage,

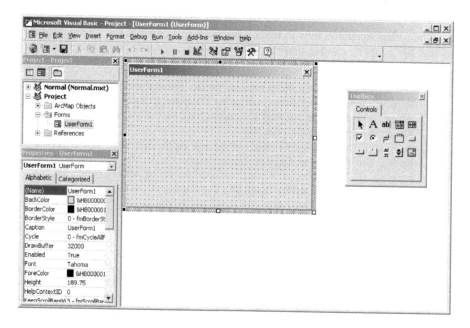

Figure 3.9   Controls in the Toolbox are placed onto UserForm1 to make a form. The Properties window shows the properties of each control, including the form.

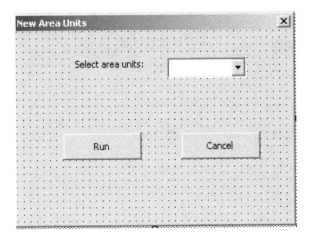

Figure 3.10   The New Area Units form contains four controls.

ScrollBar, SpinButton, and Image. The Microsoft Forms Reference section of Microsoft Visual Basic Help covers each control and its usage. The sample form in this section uses a label, a combo box, and two command buttons.

4. This step is to add controls to the form. Drag the label control from the toolbox and drop it onto the form. Drag and drop a combo box and two command buttons onto the form. At design time, the controls on the form are graphic elements. Therefore they can be added, removed, resized, and repositioned. Arrange the controls so that the form looks like Figure 3.10.

5. The Properties window allows the user to set the properties of a control at design time. The alternative is to set control properties at run time. This step is to set the properties of each control on *frmAreaUnits* at design time. Click Label1 on the form. Rename the label *lblUnits* and change its caption to Select area units. Rename the combo box *cboUnits*, and change the Style property to *2 – fmStyle-DropDownList* on the dropdown list. Rename the first command button *cmdRun* and change its caption to Run. Rename the second command button *cmdCancel* and change its caption to Cancel. Again, the prefixes of lbl, cbo, and cmd are the recommended naming conventions for labels, combo boxes, and command buttons respectively.

3.5.2   Associating Controls with Procedures

After the design of the *frmAreaUnits* form is complete, the next task is to associate the event procedures with the form and its controls:

1. Double-click the form to open the Code window. (An alternative is to click the View Code button in the Project Explorer.) At the top of the Code window are two dropdown lists. On the left is the object (control) list that includes the form and its controls. On the right is the procedure list. VBA automatically adds Private Sub UserForm_Click() to the Code window because Click is the default procedure for a form. For this sample application, choose Initialize from the procedure list

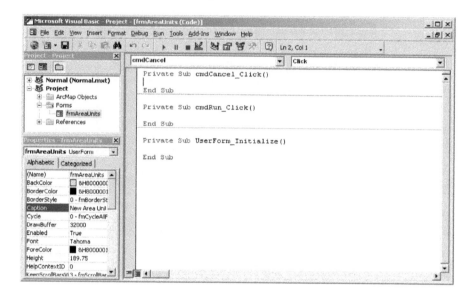

Figure 3.11 The Code window shows the wrapper lines of the three subs to be used for converting area units.

instead. Proceed to select *cmdRun* and *cmdCancel* from the object list. The Code window now has the wrapper lines for UserForm_Initialize, cmdRun_Click, and cmdCancel_Click (Figure 3.11). To complete the task, code must be provided for each of the procedures.

2. Copy and paste the code in ***UserForm_Initialize*** to the UserForm_Initialize procedure to initialize the user form. The code adds two choices of area units to the combo box at run time.

```
Private Sub UserForm_Initialize()
    ' Add items to the dropdown list.
    cboUnits.AddItem "Acres"
    cboUnits.AddItem "SqMiles"
End Sub
```

3. Copy and paste the code in ***cmdRun_Click*** to the cmdRun_Click procedure. At a click of the Run command button, the code adds either Acres or SqMiles as a new field to the feature class of the top layer in the active map, prepares a feature cursor, and calculates the new field values. Notice that cboUnits.ListIndex is used in Parts 2 and 3 of the procedure to determine if the user's choice is Acres or SqMiles. If the value of cboUnits.ListIndex is 0, the user's choice is Acres; if the value is 1, the user's choice is SqMiles.

```
Private Sub cmdRun_Click()
    ' Part 1: Define the feature class.
    Dim pMxDoc As IMxDocument
    Dim pFeatureLayer As IFeatureLayer
    Dim pFeatureClass As IFeatureClass
```

```
Set pMxDoc = ThisDocument
Set pFeatureLayer = pMxDoc.FocusMap.Layer(0)
Set pFeatureClass = pFeatureLayer.FeatureClass

' Part 2: Add Acres or SqMiles as a new field.
Dim pField As IFieldEdit
Set pField = New Field
pField.Type = esriFieldTypeDouble
If cboUnits.ListIndex = 0 Then
    pField.Name = "Acres"
Else
    pField.Name = "SqMiles"
End If
pFeatureClass.AddField pField

' Part 3: Calculate the new field values.
Dim pCursor As ICursor
Dim pCalculator As ICalculator
' Prepare a cursor with all records.
Set pCursor = pFeatureClass.Update(Nothing, True)
' Define a calculator.
Set pCalculator = New Calculator
Set pCalculator.Cursor = pCursor
' Calculate the field values.
If cboUnits.ListIndex = 0 Then
    pCalculator.Expression = "[Area] / 4046.7808"
    pCalculator.Field = "Acres"
    pCalculator.Calculate
Else
    pCalculator.Expression = "([Area] / 1000000) * 0.3861"
    pCalculator.Field = "SqMiles"
    pCalculator.Calculate
End If
End Sub
```

4. Finally, type *End* between the wrapper lines of the cmdCancel_Click procedure. The *End* statement terminates code execution.

```
Private Sub cmdCancel_Click()
    End
End Sub
```

3.5.3 Running a Form

To test how the *frmAreaUnits* form works, do the following:

1. Add *idcounty.shp* to ArcMap. The county shapefile has square meters as area units.
2. Click the Run Sub/UserForm button. The form appears. Select either Acres or SqMiles from the dropdown list. Click Run on the form.
3. Open the attribute table of *idcounty*. A new field has been added to the table and the field values have been calculated.

We can export the *frmAreaUnits* form, after it has been tested successfully, by selecting Export File from the File menu in Visual Basic Editor. The form is saved as a form file with the frm extension. Additionally, an frx file is created to save information about the graphics on the form. (frmAreaUnits_Copy.frm on the companion CD is a copy of the form.)

3.5.4   Linking a Button to a Form

This section shows how to link a customized button to the *frmAreaUnits* form so that when the button is clicked, it will open the form for use.

1. Make sure that the *frmAreaUnits* form is still available in the Project Explorer of Visual Basic Editor. Otherwise, import the form.
2. Select Customize from the Tools menu in ArcMap to open the Customize dialog. Click New in the Customize dialog. Enter Calculate Area Units for the toolbar name and save the toolbar in Untitled.
3. On the Commands tab of the Customize dialog, select the category of UIControls and click the New UIControl button. In the next dialog, check the option button for UIButtonControl and click Create. Change the name of the new command to Project.UIButtonUnits. Drag and drop Project.UIButtonUnits onto the new toolbar.
4. Right-click the new button control and select View Source. Visual Basic Editor opens with the wrapper lines of the UIButtonUnits_Click() Sub in the Code window. Type the following line between the wrapper lines: *frmAreaUnits.Show*. When this line of code runs, the Show method opens the *frmAreaUnits* form.
5. Close Visual Basic Editor. Add *idcounty2.shp* to an active map. Click on the customized button. The New Area Units form appears and is ready for use.

3.6 STORING A FORM IN A TEMPLATE

Similar to a new toolbar with commands, a form and its controls can be stored in the Normal.mxt, a base template, or a map document. The following shows how to store *frmAreaUnits.frm* in a base template:

1. Exit ArcMap so that the template to be created will not have datasets from the previous section. Launch ArcMap, and open Visual Basic Editor. Right-click Project in the Project Explorer and select Import File. Import *frmAreaUnits.frm*.
2. Select Save As from the File menu in ArcMap. In the Save As dialog, select to save as ArcMap Templates (*.mxt) and enter *AreaUnits.mxt* for the file name. Exit ArcMap.
3. ArcMap offers password protection to viewing project properties. This step is to add the password protection to *AreaUnits.mxt*. Launch ArcMap, and open *AreaUnits.mxt*. Open Visual Basic Editor in ArcMap. The Project Explorer lists TemplateProject(AreaUnits.mxt). Select TemplateProject Properties by right-clicking

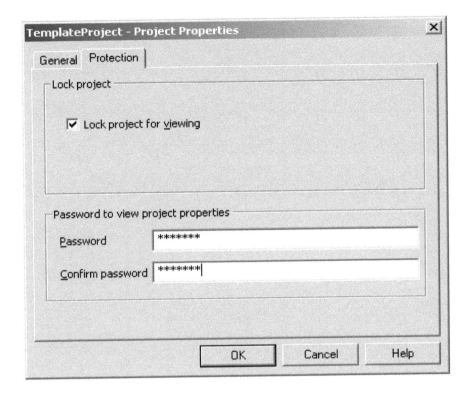

Figure 3.12 The Protection tab of the Template Properties dialog box lets the user enter the protection password.

TemplateProject(AreaUnits.mxt). On the Protection tab of the next dialog, choose to lock the project for viewing and enter a password for protection (Figure 3.12). Select Save AreaUnits.mxt from the File menu of Visual Basic Editor.

4. Next time when *AreaUnits.mxt* is opened in ArcMap, a password is required to view the form and its associated procedures.

VBA users can only save a customization in the Normal.mxt, an mxt file, or an mxd file. To create a dll (dynamic-link library) or an exe (executable) file, we must use standalone Visual Basic, C++, or other programming languages.

CHAPTER 4

Dataset and Layer Management

The Geodatabase data model separates geographic data from nongeographic data. Geographic data have the geometry of spatial features, whereas nongeographic data do not. Geographic data include feature-based and raster-based datasets, and nongeographic data include tables in text, dBASE, and other formats.

The first step in many custom applications is to add geographic datasets as layers in ArcMap. A layer is a reference to a geographic dataset. This definition of layer carries two meanings:

- A layer must be associated with a dataset. A layer can therefore be described as a feature layer if it is associated with a feature-based dataset such as a shapefile, a coverage, or a geodatabase feature class. A raster layer refers to a layer that is associated with a raster dataset.
- A layer is a graphic representation of a geographic dataset. We can therefore use different attributes and different symbols to display a layer without affecting the underlying dataset.

ArcMap organizes layers hierarchically. A map document may consist of one or more data frames, and a data frame may have one or more layers. Within a data frame, a layer can be added, deleted, or changed in the drawing order. A layer can also be saved as a layer file, a cartographic view of a geographic dataset.

Nongeographic data are called tables in ArcMap. Tables are listed in the table of contents on the Source tab, and they can be added and deleted in the same way as layers. To display tabular data in a map, they must be first linked to a feature class (a feature attribute table).

This chapter covers management of datasets and layers. Section 4.1 describes use of datasets in ArcGIS. Section 4.2 reviews objects relevant to datasets and layers in ArcObjects. Section 4.3 includes a series of macros for adding different types of datasets in ArcMap. Section 4.4 offers a macro for managing layers in an active map. Section 4.5 discusses macros and a Geoprocessing (GP) macro for copying and deleting datasets. Section 4.6 includes a macro for reporting the spatial reference and area extent of a geographic dataset. All macros start with the listing of key interfaces and key members (properties and methods) and the usage.

4.1 USING DATASETS IN ARCGIS

ArcCatalog is the ArcGIS Desktop application for managing datasets. The catalog tree groups datasets by the connected folder. Within each folder, different icons represent different types of datasets. The context menu of each dataset, regardless of its type, offers commands to copy, delete, and rename the dataset.

ArcMap is the application for displaying and analyzing datasets. The Add Data button lets the user add geographic datasets as layers and nongeographic datasets as tables to an active data frame. Each data frame has a context menu with commands for removing or activating the data frame. Each data frame also has a table of contents that lists the datasets it contains. The table of contents has two tabs: Display and Source. The Display tab shows the drawing order of the layers. The Source tab organizes the layers and tables by data source. To list tables that have been added as datasets to a data frame, the table of contents must be on the Source tab.

The context menu of a layer has a command to remove the dataset from an active data frame. It also has a command to save the layer as a layer file. The context menu of a table has commands to remove and open the table.

4.2 ARCOBJECTS FOR DATASETS AND LAYERS

Figure 4.1 shows the hierarchical structure of map and layer objects in ArcMap. At the top of the hierarchy is the *Application*, which in this case represents ArcMap. The *Application* is composed of *MxDocument* objects; an *MxDocument* object is composed of *Map* objects, and a *Map* object is composed of *Layer* objects. A data frame in ArcMap represents a map object. Examples of layers include feature layers, raster layers, and TIN (triangulated irregular network) layers.

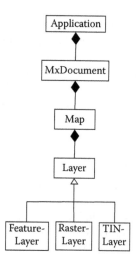

Figure 4.1 The hierarchical structure of the *Application, MxDocument, Map,* and *Layer* classes in ArcMap.

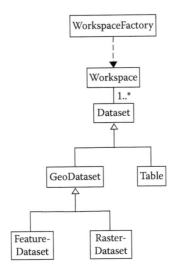

Figure 4.2   The hierarchical structure of the *WorkspaceFactory, Workspace,* and *Dataset* classes in Geodatabase.

Figure 4.2 shows the hierarchical structure of datasets and data source objects in the Geodatabase library. At the top of the hierarchy is the *WorkspaceFactory* abstract class. Many coclasses inherit the properties and methods of the *Workspace-Factory* class. These coclasses include *ShapefileWorkspaceFactory, ArcInfoWorkspaceFactory, RasterWorkspaceFactory, AccessWorkspaceFactory,* and *FileGDB-WorkspaceFactory* for shapefiles, coverages, rasters, personal geodatabases, and file geodatabases respectively.

A workspace factory object can create a new workspace. The *OpenFromFile* method on *IWorkspaceFactory,* for example, returns an interface (i.e., *IWorkspace*) on a workspace by following the pathname of a file or directory. Using the returned *IWorkspace,* we can perform a QueryInterface (QI) for *IFeatureWorkspace* to open feature-based datasets such as shapefiles or feature classes, or for *IRasterWorkspace* to open raster-based datasets. A workspace object is therefore a container of different types of datasets.

The *Dataset* abstract class represents both geographic and nongeographic data (Figure 4.2). Two *Dataset* types are *GeoDataset* and *Table*. A *GeoDataset* object has the two important properties of *Extent* and *SpatialReference,* which describe the area extent and the spatial reference of a geographic dataset respectively. Types of geodataset objects include feature layers, feature classes, raster datasets, and raster layers. A *Table* object is a collection of rows with attributes stored in columns. Examples of table objects include tables in text and dBASE formats as well as feature classes.

Macros dealing with workspaces and datasets often use name objects. A *Name* object identifies and locates a geodatabase object such as a workspace or a dataset. A name object is a lightweight version of an object because it typically has a limited number of properties and methods. Among the methods that a name object has is the *Open* method, which allows the programmer to get an instance of the actual object. This chapter uses the name objects in adding a nongeographic table in ArcMap. Chapter 6 uses a variety of name objects for data conversion.

4.3 ADDING DATASETS AS LAYERS

This section covers adding geographic and nongeographic datasets in ArcMap. The geographic datasets include the shapefile, coverage, geodatabase feature class, and raster. Nongeographic datasets include the layer file and dBASE table. From the programming perspective, different dataset types require the use of different data source objects.

4.3.1 *AddFeatureClass*

AddFeatureClass adds a shapefile to an active map. The macro performs the same function as using the Add Data command in ArcMap. With minor modifications, *AddFeatureClass* can also add a coverage or a geodatabase feature class to an active map.

> **Key Interfaces:** *IMxDocument, IMap, IWorkspaceFactory, IFeatureWorkspace, IFeatureLayer, IFeatureClass*
> **Key Members:** *FocusMap, OpenFromFile, OpenFeatureClass, FeatureClass, Name, AliasName, AddLayer*
> **Usage:** Import *AddFeatureClass* to Visual Basic Editor in ArcMap. Run the macro. The macro adds *emidastrm* to the active map.

```
Private Sub AddFeatureClass()
    Dim pMxDoc As IMxDocument
    Dim pMap As IMap
    Dim pWorkspaceFactory As IWorkspaceFactory
    Dim pFeatureWorkspace As IFeatureWorkspace
    Dim pFeatureLayer As IFeatureLayer
    Dim pFeatureClass As IFeatureClass
    ' Specify the workspace and the feature class.
    Set pWorkspaceFactory = New ShapefileWorkspaceFactory
    Set pFeatureWorkspace = pWorkspaceFactory.OpenFromFile("c:\data\chap4", 0)
    Set pFeatureClass = pFeatureWorkspace.OpenFeatureClass("emidastrm")
    ' Prepare a feature layer.
    Set pFeatureLayer = New FeatureLayer
    Set pFeatureLayer.FeatureClass = pFeatureClass
    pFeatureLayer.Name = pFeatureLayer.FeatureClass.AliasName
    ' Add the feature layer to the active map.
    Set pMxDoc = ThisDocument
    Set pMap = pMxDoc.FocusMap
    pMap.AddLayer pFeatureLayer
    ' Refresh the active view.
    pMxDoc.ActiveView.Refresh
End Sub
```

The macro first creates *pWorkspaceFactory* as an instance of the *ShapefileWork-spaceFactory* class. Next the code uses the *OpenFromFile* method on *IWorkspace-Factory* to return an *IWorkspace*, perform a QI for the *IFeatureWorkspace* interface, and uses the *OpenFeatureClass* method to open a feature class. The feature class is *emidastrm*, which is referenced by *pFeatureClass*. Using *pFeatureClass* and its name, the code creates *pFeatureLayer* as an instance of the *FeatureLayer* class. The last part of the macro adds *pFeatureLayer* to an active map. The code sets *pMxDoc* to be ThisDocument and *pMap* to be the focus map of *pMxDoc*. (This-Document is the predefined name of the *MxDocument* object, which, along with the *Application* object, is already in use when ArcMap is launched.) Then the code uses the *AddLayer* method on *IMap* to add *pFeatureLayer* to the active map, before refreshing the view.

With two minor changes, we can use **AddFeatureClass** to add a coverage to an active map. Suppose we want to add the arcs of the *breakstrm* coverage. The first change relates to the workspace factory and the path to the feature class.

```
' Specify the workspace and the feature class.
Set pWorkspaceFactory = New ArcInfoWorkspaceFactory
Set pFeatureWorkspace = pWorkspaceFactory.OpenFromFile("c:\data\chap4\", 0)
Set pFeatureClass = pFeatureWorkspace.OpenFeatureClass("breakstrm:Arc")
```

ArcInfoWorkspaceFactory is the class that creates workspaces for coverages. Also, the argument for the *OpenFeatureClass* method must be arc (i.e., line) for the feature class.

Arc is one of the feature classes contained in *breakstrm*; the others are node and tic. To avoid having the feature layer named simply as Arc, the second change adds a prefix of breakstrm: to the name property of *pFeatureLayer*.

```
' Add the prefix to the layer name.
pFeatureLayer.Name = "breakstrm: " & pFeatureLayer.FeatureClass.AliasName
```

With one minor change, we can also use **AddFeatureClass** to add a geodatabase feature class to an active map. The feature class can be either standalone or part of a feature dataset. For example, to add the *emidastrum* feature class in *emida.mdb*, we need to make the following change in **AddFeatureClass**:

```
' Specify the workspace and the feature class.
Set pWorkspaceFactory = New AccessWorkspaceFactory
Set pFeatureWorkspace = pWorkspaceFactory.OpenFromFile("c:\data\chap4\emida.mdb", 0)
Set pFeatureClass = pFeatureWorkspace.OpenFeatureClass("emidastrm")
```

AccessWorkspaceFactory is the class that creates workspaces for personal geo-databases. Also, the path to the feature workspace must include the geodatabase (i.e., *emida.mdb*). Each feature class, whether it is standalone or part of a feature dataset, has a unique name so that the *OpenFeatureClass* method can use the name to open the feature class regardless of its type.

4.3.2 *AddFeatureClasses*

AddFeatureClasses lets the user select shapefiles from a dialog box and adds them to an active map. The dialog box is similar to the Add Data tool in ArcMap, except that it only shows shapefiles.

> **Key Interfaces:** *IGxDialog, IGxObjectFilter, IEnumGxObject, IGxDataset*
> **Key Members:** *AllowMultiSelect, ButtonCaption, ObjectFilter, StartingLocation, Title, DoModalOpen, Next, Refresh, UpdateContents*
> **Usage:** Import *AddFeatureClasses* to Visual Basic Editor in ArcMap. Run the macro. A dialog box with the Add Shapefiles caption appears. Choose *idcities.shp* and *idcounty.shp* and click Add. The macro adds the two shapefiles, which are based on the same coordinate system, to the active map.

```
Private Sub AddFeatureClasses()
    ' Part 1: Prepare an Add Shapefiles dialog.
    Dim pGxDialog As IGxDialog
    Dim pGxFilter As IGxObjectFilter
    Set pGxDialog = New GxDialog
    Set pGxFilter = New GxFilterShapefiles
    ' Define the dialog's properties.
    With pGxDialog
        .AllowMultiSelect = True
        .ButtonCaption = "Add"
        Set .ObjectFilter = pGxFilter
        .StartingLocation = "c:\data\chap4"
        .Title = "Add Shapefiles"
    End With
```

Part 1 prepares an Add Shapefiles dialog. The code first creates *pGxDialog* as an instance of the *GxDialog* class and *pGxFilter* as an instance of the *GxFilter-Shapefiles* class. Both *GxDialog* and *GxFilter* are ArcCatalog classes. COM (Component Object Model) technology allows ArcCatalog objects to be used in Arc-Map. A *GxDialog* object is basically a form that has been coded by ArcGIS developers to accept the datasets selected by the user and add them to ArcMap. A *GxFilter* object filters the type of data to be displayed in a *GxDialog* object. *GxFilter* is an abstract class with more than 30 different types. Part 1 uses the *GxFilterShapefiles* class, limiting the data sources to only shapefiles. The rest of Part 1 uses a *With* block to define the properties of *pGxDialog*: the title is Add Shapefiles, the button caption is Add, the object filter is *pGxFilter*, and the starting location is the path to the data sources. The *AllowMultiSelect* property is set to be true, meaning that the user can select multiple datasets. If false, then the user can only select a single dataset at a time.

```
    ' Part 2: Get the datasets from the dialog and add them to the active map.
    Dim pGxObjects As IEnumGxObject
    Dim pMxDoc As IMxDocument
    Dim pMap As IMap
    Dim pGxDataset As IGxDataset
```

```
Dim pLayer As IFeatureLayer
Set pMxDoc = ThisDocument
Set pMap = pMxDoc.FocusMap
' Open the dialog.
pGxDialog.DoModalOpen 0, pGxObjects
Set pGxDataset = pGxObjects.Next
' Exit sub if no dataset has been added.
If pGxDataset Is Nothing Then
    Exit Sub
End If
' Step through the datasets and add them as layers to the active map.
Do Until pGxDataset Is Nothing
    Set pLayer = New FeatureLayer
    Set pLayer.FeatureClass = pGxDataset.Dataset
    pLayer.Name = pLayer.FeatureClass.AliasName
    pMap.AddLayer pLayer
    Set pGxDataset = pGxObjects.Next
Loop
' Refresh the map and update the table of contents.
pMxDoc.ActivatedView.Refresh
pMxDoc.UpdateContents
End Sub
```

Part 2 gets the datasets selected by the user and adds them as layers to the active map. The *DoModalOpen* method on *IGxDialog* opens the dialog box and saves the selected shapefiles into a collection. In the code, the first argument for *DoModalOpen* is set to be zero (i.e., to use the ArcMap window) and the second is *pGxObjects*, a reference to an *EnumGxObject*. An *EnumGxObject* represents a collection of ordered objects. The code uses the *Next* method on *IEnumGxObject* to advance one object at a time and assigns the object to the *pGxDataset* variable, a reference to a *GxDataset* object. A type of *GxObject*, a *GxDataset* object represents a dataset (Figure 4.3). If the *Next* method advances nothing the first time, exit the sub. If the collection contains selected shapefiles, then the code uses a *Do...Loop* to step through each of them. Within each loop, the code creates *pLayer* as an instance of the *FeatureLayer* class, assigns the dataset of *pGxDataset* to be the feature class of *pLayer*, and adds the layer to the active map. Finally, the code refreshes the view and updates the table of contents of the map document.

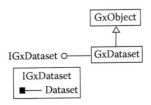

Figure 4.3   A type of *GxObject*, a *GxDataset* object represents a dataset that can be read via the *Dataset* property on *IGxDataset*.

4.3.3 *AddRaster*

AddRaster adds a raster dataset to an active map. The macro performs the same function as using the Add Data command in ArcMap.

> **Key Interfaces:** *IWorkspaceFactory, IRasterWorkspace, IRasterLayer, IRasterDataset*
> **Key Members:** *OpenFromFile, OpenRasterDataset, CreateFromDataset, AddLayer*
> **Usage:** Import *AddRaster* to Visual Basic Editor in ArcMap. Run the macro. The macro adds *emidalat* to the active map.

```
Private Sub AddRaster()
    Dim pMxDoc As IMxDocument
    Dim pMap As IMap
    Dim pWorkspaceFactory As IWorkspaceFactory
    Dim pRasterWorkspace As IRasterWorkspace
    Dim pRasterDS As IRasterDataset
    Dim pRasterLayer As IRasterLayer
    ' Specify the workspace and the raster dataset.
    Set pWorkspaceFactory = New RasterWorkspaceFactory
    Set pRasterWorkspace = pWorkspaceFactory.OpenFromFile("c:\data\chap4\", 0)
    Set pRasterDS = pRasterWorkspace.OpenRasterDataset("emidalat")
    ' Prepare a raster layer.
    Set pRasterLayer = New RasterLayer
    pRasterLayer.CreateFromDataset pRasterDS
    ' Add the raster layer to the active map.
    Set pMxDoc = ThisDocument
    Set pMap = pMxDoc.FocusMap
    pMap.AddLayer pRasterLayer
    pMxDoc.ActiveView.Refresh
End Sub
```

The macro creates *pWorkspaceFactory* as an instance of the *RasterWorkspaceFactory* class and uses the *OpenFromFile* method on *IWorkspaceFactory* to open a raster-based workspace referenced by *pRasterWorkspace*. Next, the code uses the *OpenRasterDataset* method on *IRasterWorkspace* to open a raster dataset named *emidalat* and referenced by *pRasterDS*. Then the code creates *pRasterLayer* as an instance of the *RasterLayer* class and defines its dataset. Finally, the code uses the *AddLayer* method on *IMap* to add *pRasterLayer* to the active map.

The *CreateFromDataset* method is one of the three methods on *IRasterLayer* for creating a raster layer. The other two methods are *CreateFromFilePath* and *CreateFromRaster* (Figure 4.4). The following macro uses the *CreateFromFilePath* method to complete the same task as *AddRaster*.

```
Private Sub AddRaster_2()
    Dim pMxDocument As IMxDocument
    Dim pMap As IMap
    Set pMxDocument = ThisDocument
    Set pMap = pMxDocument.FocusMap
    Dim pRasterLayer As IRasterLayer
```

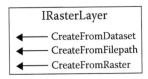

Figure 4.4   Methods on *IRasterLayer*.

```
Set pRasterLayer = New RasterLayer
pRasterLayer.CreateFromFilePath "c:\data\chap4\emidalat"
pMap.AddLayer pRasterLayer
End Sub
```

4.3.4   *AddLayerFile*

AddLayerFile adds a layer file to an active map. The macro performs the same function as using the Add Data command in ArcMap.

> **Key Interfaces:** *IGxFile, IGxLayer*
> **Key Members:** *Path, Layer, AddLayer*
> **Usage:** Import ***AddLayerFile*** to Visual Basic Editor in ArcMap. Run the macro. The macro adds *emidalat.lyr* to the active map. Because *emidalat.lyr* references *emidalat*, the location of *emidalat* must be known before the macro can complete its task.

```
Private Sub AddLayerFile()
    Dim pMxDoc As IMxDocument
    Dim pMap As IMap
    Dim pGxLayer As IGxLayer
    Dim pGxFile As IGxFile
    ' Get the layer file.
    Set pGxLayer = New GxLayer
    Set pGxFile = pGxLayer
    pGxFile.Path = "c:\data\chap4\emidalat.lyr"
    ' Add the layer file to the active map.
    Set pMxDoc = ThisDocument
    Set pMap = pMxDoc.FocusMap
    pMap.AddLayer pGxLayer.Layer
    pMxDoc.ActiveView.Refresh
    ' If red exclamation mark appears, use Set Data Source on the Source tab of
    ' the Properties dialog to set the layer file's data source.
End Sub
```

The macro creates *pGxLayer* as an instance of the *GxLayer* class. Next the code performs a QI for the *IGxFile* interface and uses the *Path* property to define the path to *emidalat.lyr* (Figure 4.5). Then the code uses the *AddLayer* method on *IMap* to add the layer associated with *pGxLayer* to the active map. Both *GxLayer* and *GxFile* are ArcCatalog objects.

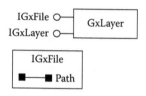

Figure 4.5   *GxLayer* and the interfaces that a *GxLayer* object supports.

4.3.5   *AddTable*

AddTable adds a nongeographic table to an active map and opens the table. The macro performs the same function as using the Add Data command in ArcMap to add a table and using the Open command to open the table. *AddTable* has three parts. Part 1 uses the name objects to define the input table, Part 2 adds the table to the active map, and Part 3 uses a table window to open the table.

> **Key Interfaces:** *IWorkspaceName, IDatasetName, IName, ITable, ITableCollection, ITableWindow*
>
> **Key Members:** *WorkspaceFactoryProgID, PathName, Name, WorkspaceName, AddTable, UpdateContents, Table, Application, Show*
>
> **Usage:** Import *AddTable* to Visual Basic Editor in ArcMap. Run the macro. The macro opens the *comp.dbf* table and adds the table to the active map. Click on the Source tab in the table of contents to see *comp.dbf*.

```
Private Sub AddTable()
    ' Part 1: Define the input table.
    Dim pWSName As IWorkspaceName
    Dim pDatasetName As IDatasetName
    Dim pName As IName
    Dim pTable As ITable
    ' Get the dbf file by specifying its workspace and name.
    Set pDatasetName = New TableName
    Set pWSName = New WorkspaceName
    pWSName.WorkspaceFactoryProgID = "esriCore.ShapefileWorkspaceFactory"
    pWSName.PathName = "c:\data\chap4"
    pDatasetName.Name = "comp.dbf"
    Set pDatasetName.WorkspaceName = pWSName
    Set pName = pDatasetName
    ' Open the dbf table.
    Set pTable = pName.Open
```

Part 1 first creates *pDatasetName* as an instance of the *TableName* class. Next, the code defines the workspace name and name properties of *pDatasetName* by using members on *IWorkspaceName* and *IDatasetName*. Notice that the program ID of the workspace factory is esriCore.ShapefileWorkspaceFactory because the dataset to be added is a dBASE file. (If the dataset to be added is a text file, one would opt for esriCore.TextfileWorkspaceFactory.) The code then switches to the *IName* interface and uses the *Open* method to open *pTable*.

```
' Part 2: Add the table to the active map.
Dim pMxDoc As IMxDocument
Dim pMap As IMap
Dim pTableCollection As ITableCollection
Set pMxDoc = Application.Document
Set pMap = pMxDoc.FocusMap
Set pTableCollection = pMap ' QI
pTableCollection.AddTable pTable
pMxDoc.UpdateContents
```

Part 2 first sets *pMap* to be the active map. The code then accesses the *ITable-Collection* interface and uses the *AddTable* method to add *pTable* to *pMap*. The *UpdateContents* method on *IMxDocument* updates ArcMap's table of contents.

```
' Part 3: Open the table in a table window.
Dim pTableWindow As ITableWindow
' Create a table window and specify its properties and methods.
Set pTableWindow = New TableWindow
With pTableWindow
    Set .Table = pTable
    Set .Application = Application
    .Show True
End With
End Sub
```

Part 3 creates *pTableWindow* as an instance of the *TableWindow* class. Next the code uses a *With* block to define the table window for view. The *Application* property is set to be Application, which represents ArcMap in this case. If the *Application* property is not set, the macro will crash!

4.4 MANAGING LAYERS

Layers in a map are indexed from the top with the base of zero. A common application of layer management is to locate a particular layer by its index so that the layer can be accessed in code.

4.4.1 *FindLayer*

FindLayer finds a layer in the active map and reports its index value. Sample modules in this book frequently use *FindLayer* as a function and call the function to access a layer.

Key Interfaces: *IMap*
Key Members: *LayerCount, Layer( ), Name*
Usage: Add *emidalat* and *emidastrm.shp* to an active map. Import *FindLayer* to Visual Basic Editor in ArcMap. Run the macro. The macro first reports the number of datasets in the active map. After getting a layer name (for example, *emidalat*) from the user, the macro reports the index of the layer.

```
Private Sub FindLayer()
    Dim pMxDoc As IMxDocument
    Dim pMap As IMap
    Dim FindDoc As Variant
    Dim aLName As String
    Dim name As String
    Dim i As Long
    Set pMxDoc = ThisDocument
    Set pMap = pMxDoc.FocusMap
    MsgBox "The active map has " & pMap.LayerCount & " datasets."
    ' Use an input box to get a layer name.
    name = InputBox("Enter a layer name:", "")
    ' Loop through layers in the active map and match the entered name
    ' with the layer name in uppercase.
    For i = 0 To pMap.LayerCount - 1
        aLName = UCase(pMap.Layer(i).name)
        ' Given a match, assign the counter to FindDoc.
        If (aLName = (UCase(name))) Then
            FindDoc = i
            Exit for
        End If
    Next
    MsgBox name & " is at index " & FindDoc
End Sub
```

The macro first uses the *LayerCount* property on *IMap* to report the number of layers in the active map. Next, the code gets a layer name from the user and assigns it to the *Name* variable. Using a *For...Next* loop, the code steps through each layer, converts the name of the layer to its uppercase, and matches the name with the uppercase of the input name. When a match is found, its index is assigned to the *FindDoc* variable. A message box then reports the name of the layer and its index. UCase is a VBA function that converts a string to the uppercase.

4.5 MANAGING DATASETS

This section covers dataset management, such as copying and deleting datasets programmatically. Before deleting a dataset, the layer that uses the dataset should be first removed from ArcMap.

4.5.1 *CopyDataset*

CopyDataset copies the dataset of a layer in the active map and saves the copied dataset in a specified workspace. The macro performs the same function as using the Copy command in ArcCatalog.

> **Key Interfaces:** *IWorkspaceFactory, IFeatureWorkspace, IDataset, IFeatureClass*
> **Key Members:** *OpenFromFile, Copy*

Usage: Add *emidastrm.shp* to an active map. Import **CopyDataset** to Visual Basic Editor in ArcMap. Run the macro. The macro copies *emidastrm.shp* and saves the copy as *emidastrmCopy.shp*.

```
Private Sub CopyDataset()
    ' Part 1: Define the top layer as the dataset to be copied.
    Dim pMxDocument As IMxDocument
    Dim pMap As IMap
    Dim pFeatureLayer As IFeatureLayer
    Dim pFeatureClass As IFeatureClass
    Set pMxDocument = ThisDocument
    Set pMap = pMxDocument.FocusMap
    Set pFeatureLayer = pMap.Layer(0)
    Set pFeatureClass = pFeatureLayer.FeatureClass
```

Part 1 defines *pFeatureClass* as the feature class of the top layer in the active map. This feature class is the geographic dataset to be copied.

```
    ' Part 2: Copy the dataset and add it to the active map.
    Dim pWorkspaceFactory As IWorkspaceFactory
    Dim pFeatureWorkspace As IFeatureWorkspace
    Dim pDataset As IDataset
    Dim pCopyFC As IFeatureClass
    Dim CopyDSName As String
    ' Define the workspace for the copied dataset.
    Set pWorkspaceFactory = New ShapefileWorkspaceFactory
    Set pFeatureWorkspace = pWorkspaceFactory.OpenFromFile("c:\data\chap4\", 0)
    ' Copy the dataset.
    Set pDataset = pFeatureClass
    CopyDSName = pFeatureLayer.name & "Copy"
    Set pCopyFC = pDataset.Copy(CopyDSName, pFeatureWorkspace)
End Sub
```

Part 2 creates *pWorkspaceFactory* as an instance of the *ShapefileWorkspaceFactory* class and uses the *OpenFromFile* method to open a feature-based workspace. Then the code performs a QI for the *IDataset* interface, and uses the *Copy* method to make a copy of *pFeatureClass*.

Box 4.1   CopyDataset_GP

CopyDataset_GP uses the CopyFeatures tool in the Data Management toolbox to make a copy of *idcounty.shp* and save the copy as *idcountycopy2.shp*. A macro that uses a Geoprocessing tool must first create the *Geoprocessing* object. The name of a tool (CopyFeatures) is usually followed by the name of the toolbox (Management for Data Management) in which the tool resides. This naming convention becomes necessary if two or more tools with the same name exist. Run the macro in ArcEditor, and check for the copied dataset in the Catalog tree. *CopyDataset_GP* performs the same task as *CopyDataset*.

```
Private Sub CopyDataset_GP()
    ' Create the Geoprocessing object.
    Dim GP As Object
    Set GP = CreateObject("esriGeoprocessing.GpDispatch.1")
    ' CopyFeatures <in_features> <out_feature_class> {configuration_keyword}
    ' {spatial_grid_1} {spatial_grid_2} {spatial_grid_3}
    ' Execute the copyfeatures tool, which has two required parameters.
    GP.CopyFeatures_management "c:\data\chap4\idcounty.shp", "c:\data\chap4\idcountycopy2.shp"
End Sub
```

4.5.2 *DeleteDataset*

DeleteDataset removes a layer from an active map and deletes the layer's dataset. The macro performs the same function as using the Remove command in ArcMap to remove a layer first and then, using the Delete command in ArcCatalog, to delete the layer's dataset.

> **Key Interfaces:** *IMap, IDataset, IActiveView*
> **Key Members:** *DeleteLayer, Delete, Refresh*
> **Usage:** Add *emidastrmCopy.shp* to an active map. Import *DeleteDataset* to Visual Basic Editor in ArcMap. Run the macro. The macro removes the *emidastrmCopy* layer from the active map and deletes *emidastrmCopy.shp*.

```
Private Sub DeleteDataset()
    Dim pMxDocument As IMxDocument
    Dim pMap As IMap
    Dim pFeatureLayer As IFeatureLayer
    Dim pFeatureClass As IFeatureClass
    Dim pDataset As IDataset
    Dim pActiveView As IActiveView
    Set pMxDocument = ThisDocument
    Set pMap = pMxDocument.FocusMap
    ' Define the dataset to be deleted.
    Set pFeatureLayer = pMap.Layer(0)
    Set pFeatureClass = pFeatureLayer.FeatureClass
    ' Remove the layer from the active map.
    pMap.DeleteLayer pFeatureLayer
    ' Delete the dataset.
    Set pDataset = pFeatureClass
    pDataset.Delete
    ' Refresh the map.
    Set pActiveView = pMap
    pActiveView.Refresh
End Sub
```

The macro sets *pFeatureLayer* to be the top layer in the active map and *pFeatureClass* to be its feature class. Next, the code uses the *DeleteLayer* method on *IMap* to remove *pFeatureLayer* from the active map, and then the code accesses *IDataset* and uses the *Delete* method to delete *pFeatureClass*. Finally, the code refreshes the map.

4.6 REPORTING GEOGRAPHIC DATASET INFORMATION

This section shows how we can report the spatial reference and area extent properties of a *GeoDataset* object by using a macro.

4.6.1 *SpatialRef*

SpatialRef reports the spatial reference and the extent of a geographic dataset. The macro performs the same function as looking up the metadata of a geographic dataset in ArcCatalog or getting the information on the Source tab of the Layer Properties dialog in ArcMap.

> **Key Interfaces:** *IGeoDataset, ISpatialReference, IEnvelope*
> **Key Members:** *SpatialReference, Extent, Xmin, YMin, XMax, YMax*
> **Usage:** Add *emidastrm.shp* to an active map. Import *SpatialRef* to Visual Basic Editor in ArcMap. Run the macro. The macro reports the dataset's coordinate system and area extent.

```
Private Sub SpatialRef()
    Dim pMxDocument As IMxDocument
    Dim pMap As IMap
    Dim pFeatureLayer As IFeatureLayer
    Dim pGeoDataset As IGeoDataset
    Dim pSpatialRef As ISpatialReference
    Dim pEnvelope As IEnvelope
    Dim MinX, MaxX, MinY, MaxY As Double
    Set pMxDocument = ThisDocument
    Set pMap = pMxDocument.FocusMap
    ' Define the input geodataset.
    Set pFeatureLayer = pMap.Layer(0)
    Set pGeoDataset = pFeatureLayer 'QI
    ' Derive the spatial reference and extent of the geodataset.
    Set pSpatialRef = pGeoDataset.SpatialReference
    Set pEnvelope = pGeoDataset.Extent
    ' Get MinX, MinY, MaxX, and MaxY.
    MinX = pEnvelope.XMin
    MinY = pEnvelope.YMin
    MaxX = pEnvelope.XMax
    MaxY = pEnvelope.YMax
    ' Report the geodataset information.
    MsgBox "The layer's spatial reference is: " & pSpatialRef.Name
    MsgBox "Minimum X is: " & MinX & " Minimum Y is: " & MinY & Chr$(10) & "Maximum X is: " & MaxX & _
    "Maximum Y is: " & MaxY
End Sub
```

The macro first defines *pFeatureLayer* as the top layer in the active map. Next, the code accesses the *IGeoDataset* interface and derives the *SpatialReference* and *Extent* properties of *pFeatureLayer*. The extent of a geographic dataset is an envelope or a rectangular object. The code assigns the *XMin, YMin, XMax*, and *YMax* values of the envelope to the variables of *MinX, MinY, MaxX*, and *MaxY* respectively. Finally, the code uses the dialog boxes to report the spatial information of *pFeatureLayer*.

CHAPTER **5**

Attribute Data Management

A geographic information system (GIS) involves both geographic data and attribute data. Geographic data relate to the geometry of spatial features, whereas attribute data describe the characteristics of the features. The geodatabase data model uses tables to store both types of data in a relational database environment. A table with a geometry field is a feature class, a feature attribute table, or simply a geographic dataset. A table with attribute data only is a nongeographic dataset.

A table, either a feature class or a nongeographic table, consists of rows and columns. Each row represents a feature, and each column represents a characteristic. A row is also called a record, and a column a field. The intersection of a column and a row shows the value of a particular characteristic for a particular feature.

Attribute data management takes place at either the field level or the table level. At the field level, common tasks include deriving the field information, adding fields, deleting fields, and calculating the field values. These tasks require working with the properties of a field, such as name, type, and length. At the table level, common tasks typically involve joining and relating tables in a relational database environment. A join brings together two tables. A relate connects two tables but keeps the tables separate. Both operations use keys and relationships to link tables.

This chapter covers attribute data management. Section 5.1 reviews management of attribute data using ArcGIS. Section 5.2 discusses objects relevant to tables, fields, and relationship classes. Section 5.3 includes macros for listing fields and the field properties. Section 5.4 offers a macro and a Geoprocessing (GP) macro for adding and deleting fields. Section 5.5 offers macros and a GP macro for calculating the field values. Section 5.6 discusses four macros for joining and relating tables and a GP macro for relating a table to a layer. All macros start with the listing of key interfaces and key members (properties and methods) and the usage.

5.1 MANAGING ATTRIBUTE DATA IN ARCGIS

An ArcGIS user can add and delete fields in either ArcCatalog or ArcMap. To add a field, we must first define the field properties. Depending on the field type, the definition may include length, precision, and scale. Length is the maximum length, in bytes, reserved for the field. Precision is the number of digits reserved for a numeric field. Scale is the number of decimal digits reserved for a field of the Double data type.

Field Calculator, available through a field's context menu in ArcMap, is a tool for calculating the field values. To use Field Calculator, we must prepare a calculation expression with fields and mathematical functions.

Joins and relates are available through the context menu or the properties of a feature layer or a table in ArcMap. To add a join or relate, we must specify the tables to join or relate and the fields on which the join or relate is based. The fields used in a join or relate are called keys. A primary key represents a field whose values can uniquely identify a record in a table. Its counterpart in another table for the purpose of linkage is a foreign key. Joins and relates can only be built one at a time, but existing joins and relates can be removed individually or as a group.

There are four possible relationships, also called cardinalities, in connecting two tables. The one-to-one relationship means that one and only one record in a table is related to one and only one record in another table. The one-to-many relationship means that one record in a table may be related to many records in another table. The many-to-one relationship means that many records in a table may be related to one record in another table. And the many-to-many relationship means that many records in a table may be related to many records in another table.

Joins are usually recommended for the one-to-one or many-to-one relationship. Given a one-to-one relationship, two tables are joined by record. Given a many-to-one relationship, many records in the base table have the same value from a record in the other table. Relates, on the other hand, are appropriate for all four relationships.

5.2 ARCOBJECTS FOR ATTRIBUTE DATA MANAGEMENT

This section covers objects that are related to table, fields, field, and relationship classes.

5.2.1   Tables

Figure 5.1 shows the hierarchical structure of *Table*, *ObjectClass*, and *FeatureClass*. The *ObjectClass* is a type of the *Table* class whose rows represent entities, and the *FeatureClass* is a type of the *ObjectClass* whose rows represent features. A feature class object has two default fields: the shape field that stores the geometry of features, and the FID field that stores the feature IDs.

Chapter 4 has shown how to access a feature class through its data source. Another way to access a feature class is through a feature layer that is already present

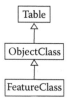

Figure 5.1 *FeatureClass* is a type of *ObjectClass,* and *ObjectClass* is a type of *Table.*

in an active map. For example, we can access a feature layer by using the *Layer()* property of *IMap* and then access the layer's feature class by using the *FeatureClass* property on *IFeatureLayer* (Figure 5.2).

Chapter 4 has also shown how to access a dBASE file or a text file through its data source. An alternative for accessing a file is through a standalone table that is already present in an active map. A *StandaloneTable* object is not associated with a feature class but is based on a nongeographic table. Figure 5.3 shows how to access the table underlying a standalone table. First, use the *IStandaloneTableCollection* interface that a map object supports to access the standalone table. Second, use the *Table* property on *IStandaloneTable* to access the table. Conceptually, a standalone table is like a feature layer and the *Table* property of a standalone table is like the *FeatureClass* property of a feature layer.

5.2.2 Fields and Field

A *Fields* object is a collection of fields such as a feature class or a nongeographic table. In either case, a fields object is associated with a table object (Figure 5.4). We can therefore add, delete, or find a field through a fields or table object. The *AddField* and *DeleteField* methods are available on both *ITable* and *IFieldsEdit,* and the *FindField* method is available on both *ITable* and *IFields.*

A Fields object consists of one or more *Field* objects (Figure 5.4). Each field has an index or a numbered position. A Field object implements *IField* and *IFieldEdit.*

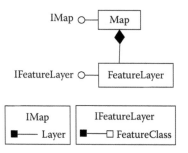

Figure 5.2 A feature class can be accessed through a feature layer in an active map.

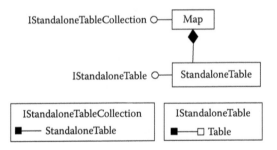

Figure 5.3   A table can be accessed through a standalone table in an active map.

IField has the read-only field properties such as name, length, precision, and type. *IFieldEdit*, on the other hand, has write-only field properties and is therefore useful for defining a new field.

ArcObjects has the *Calculator* coclass for calculating the field values. *ICalculator* has the properties of *Cursor*, *Expression*, and *Field* as well as the *Calculate* method (Figure 5.5). A cursor is a data-access object, which allows a macro to step through a set of records in a table. The *Calculate* method uses an expression defined by the user to calculate the values of a specified field.

5.2.3   Relationship Classes

In ArcObjects, a join or relate is defined as a relationship class linking two tables. Relationship classes can be stored in a geodatabase or, as in this chapter, built in

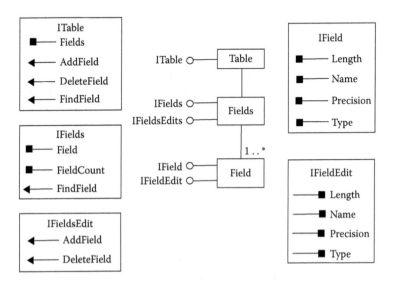

Figure 5.4   The relationship between *Table*, *Fields*, and *Field* objects as well as properties and methods of these objects.

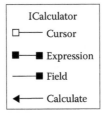

Figure 5.5   Properties and methods on *ICalculator*.

code between tables that are in use. To build a join or relate in code requires the relationship class be prepared as a memory (i.e., virtual) relationship class first (Figure 5.6). The *MemoryRelationshipClassFactory* coclass implements *IMemory-RelationshipClassFactory*, which has the *Open* method to create memory relationship class objects.

A join or relate can be set up after a memory relationship class is opened. The setup is made through a feature layer object, but the method differs between a join and a relate. A feature layer object implements *IDisplayRelationshipClass*, which has the *DisplayRelationshipClass* method to set up a join between two tables and to get the joined table ready for use (Figure 5.7).

To join more than two tables, ArcObjects stipulates that a *RelQueryTable* object be created from the first join and the object be used as the source to create another *RelQueryTable* object for the second join (Figure 5.8). Representing a joined pair of tables, a *RelQueryTable* object is obtained through the *RelQueryTableFactory* coclass. *IRelQueryTableFactory* provides the *Open* method that can create a new *RelQueryTable* object.

A feature layer object also implements *IRelationshipClassCollection*, and *IRelationshipClassCollectionEdit* is used for managing relates (Figure 5.9). *IRelationshipClassCollection* has members for deriving and finding relates, and *IRelationshipClassCollectionEdit* has methods for adding or removing a relate.

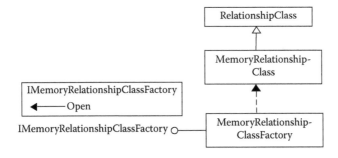

Figure 5.6   A *MemoryRelationshipClassFactory* can create a *MemoryRelationshipClass* object, which is a type of *RelationshipClass*.

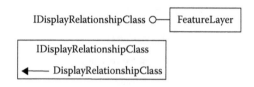

Figure 5.7   *IDisplayRelationshipClass* can set up a join for a feature layer.

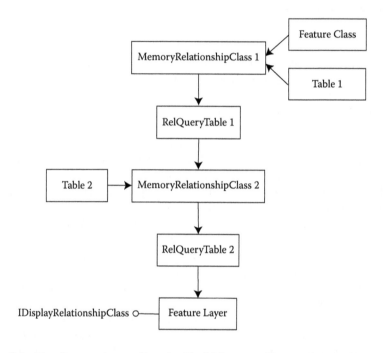

Figure 5.8   The diagram shows a flow chart for joining two tables to a feature class.

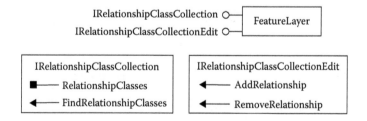

Figure 5.9   A *FeatureLayer* object supports interfaces that work with relates.

5.3 LISTING FIELDS AND FIELD PROPERTIES

This section introduces macros for reporting the number of fields and field properties in a dataset.

5.3.1 *ListOfFields*

ListOfFields reports the number of fields in a feature class and the field names. The macro performs the same function as using the Fields tab in a feature layer's Properties dialog. *ListOfFields* has two parts. Part 1 gets the feature class and reports the number of fields in the feature class. Part 2 steps through each field, gets the field name, adds the field name to a list, and reports the list.

> **Key Interfaces:** *IFields, IField*
> **Key Members:** *Fields, FieldCount, Field( ), Name, Add*
> **Usage:** Add *idcounty.shp* to an active map. *Idcounty* shows 44 counties in Idaho and has some demographic attributes. Import *ListOfFields* to Visual Basic Editor. Run the macro. The first message box reports the number of fields in *idcounty*, and the second message lists the field names.

```
Private Sub ListOfFields()
    ' Part 1: Get the feature class and its fields.
    Dim pMxDoc As IMxDocument
    Dim pMap As IMap
    Dim pFeatureLayer As IFeatureLayer
    Dim pFeatureClass As IFeatureClass
    Dim pFields As IFields
    Dim count As Long
    Set pMxDoc = ThisDocument
    Set pMap = pMxDoc.FocusMap
    Set pFeatureLayer = pMap.Layer(0)
    Set pFeatureClass = pFeatureLayer.FeatureClass
    Set pFields = pFeatureClass.Fields
    ' Get the number of fields.
    count = pFields.FieldCount
    MsgBox "There are " & count & " fields"
```

Part 1 sets *pFeatureClass* to be the feature class of the top layer in the active map. Next the code sets *pFields* to be the fields of *pFeatureClass*, and the *count* variable to be the Fieldcount property on *IFields*. A message box then reports the number of fields.

```
    ' Part 2: Prepare a list of fields and display the list.
    Dim ii As Long
    Dim aField As IField
    Dim fieldName As Variant
    Dim NameList As Variant
    ' Loop through each field, and add the field name to a list.
    For ii = 0 To pFields.FieldCount - 1
```

```
    Set aField = pFields.Field(ii)
    fieldName = aField.name
    NameList = NameList & fieldName & Chr(13)
  Next
  ' Display the list of field names in a message box.
  MsgBox NameList, , "Field Names"
End Sub
```

Part 2 uses a *For...Next* statement to step through each field in *pFields*. Because the loop starts with the base of zero, the predefined number of iterations is based on the *FieldCount* value minus one. Each time through the loop, the code assigns the name of the field to the *fieldName* variable and adds the name to *Namelist*. Finally, the code displays each field name in a message box. The constant Chr(13) adds a carriage return.

5.3.2   *ListFieldProps*

ListFieldProps reports the field name, type, length, precision, and scale of each field in a dataset. The macro performs the same function as using the Fields tab in a feature layer's Properties dialog. *ListFieldProps* has two parts. Part 1 gets the feature class and its fields. Part 2 gets the field properties of each field and reports them.

> **Key Interfaces:** *IFields, IField*
> **Key Members:** *Fields, FieldCount, Field( ), Name, Type, Length, Precision, Scale, Add*
> **Usage:** Add *idcounty.shp* to an active map. Import *ListFieldProps* to Visual Basic Editor. Run the macro. The message box lists the properties of each field in *idcounty*.

```
Private Sub ListFieldProps()
  ' Part 1: Get the feature class and its fields.
  Dim pMxDoc As IMxDocument
  Dim pMap As IMap
  Dim pFeatureLayer As IFeatureLayer
  Dim pFeatureClass As IFeatureClass
  Dim pFields As IFields
  Dim count As Long
  Set pMxDoc = ThisDocument
  Set pMap = pMxDoc.FocusMap
  Set pFeatureLayer = pMap.Layer(0)
  Set pFeatureClass = pFeatureLayer.FeatureClass
  Set pFields = pFeatureClass.Fields
```

Part 1 is the same as *ListOfFields*. The code derives the number of fields from the feature class of the top layer and saves the fields in *pFields*.

```
  ' Part 2: Get the field properties for each field, and report them.
  Dim ii As Long
  Dim aField As IField
```

```
        Dim fieldName As String
        Dim fieldType As Integer
        Dim fieldLength As Integer
        Dim fieldPrecision As Integer
        Dim fieldScale As Integer
        Dim typeDes As String
        Dim out As Variant
        Dim theList As New Collection
        Dim NameList As Variant
        ' Set up a do loop.
        For ii = 0 To pFields.FieldCount - 1
        Set aField = pFields.Field(ii)
        ' Derive properties of the field.
        fieldName = aField.name
        fieldType = aField.Type
        fieldLength = aField.Length
        fieldPrecision = aField.Precision
        fieldScale = aField.Scale
        ' Determine the field type.
        Select Case fieldType
            Case 0
                typeDes = "SmallInteger"
            Case 1
                typeDes = "Integer"
            Case 2
                typeDes = "Single"
            Case 3
                typeDes = "Double"
            Case 4
                typeDes = "String"
            Case 5
                typeDes = "Date"
            Case 6
                typeDes = "OID"
            Case 7
                typeDes = "Geometry"
            Case 8
                typeDes = "Blob"
            End Select
            ' Save the field properties to the variable out.
            out = fieldName & "" & typeDes & "" & fieldLength & "" & fieldPrecision &"" & fieldScale
            ' Add the variable out to a list.
            theList.Add out
        Next
        ' Display the list in a message box.
        For Each out In theList
            NameList = NameList & out & Chr(13)
        Next out
        MsgBox NameList, , "Field Name, Type, Length, Precision, & Scale"
End Sub
```

Part 2 uses a *For...Next* statement to step through each field in *pFields* and to derive its field property values. Besides the field name, the code works with the additional field properties of type, length, precision, and scale. Because the field type value can range from zero to eight, the code uses a *Select Case* statement to translate the value into a description such as integer or double. The field length is the maximum length in bytes. The field precision is the number of digits reserved for a numeric field. The field scale is the number of decimal digits reserved for a double field. After the field properties are derived, they are strung together and assigned to the variable *out*. A variant-type variable can take any type of data. The code then adds *out* to *theList*. Finally, a message box reports *theList*, one field per line.

5.3.3 *UseFindLayer*

UseFindLayer consists of a sub and a function. The **FindLayer** function uses an input box to get the name of a layer from the user and returns the index of the layer to the **Start** sub. **Start** then reports the number of fields in the layer.

> **Key Interfaces:** *IFields, IField*
> **Key Members:** *Fields, FieldCount, Field, Name, Add*
> **Usage:** Add *idcounty.shp* and *idcities.shp* to an active map. Import *UseFindLayer* to Visual Basic Editor. Run the module. Enter the name of a layer in the input box. The module reports the number of fields in the layer.

```
Private Sub Start()
    Dim pMxDoc As IMxDocument
    Dim pMap As IMap
    Dim i As Long
    Dim pFeatureLayer As IFeatureLayer
    Dim pFeatureClass As IFeatureClass
    Dim pFields As IFields
    Dim count As Long
    Set pMxDoc = ThisDocument
    Set pMap = pMxDoc.FocusMap
    ' Run the FindLayer function.
    i = FindLayer()
    ' Use the returned Id to locate the layer.
    Set pFeatureLayer = pMap.Layer(i)
    Set pFeatureClass = pFeatureLayer.FeatureClass
    Set pFields = pFeatureClass.Fields
    ' Get the number of fields.
    count = pFields.FieldCount
    MsgBox "There are " & count & " fields"
End Sub
```

Start assigns the returned value from **FindLayer** to *i*, and uses *i* as the index to locate *pFeatureLayer*. Next the code sets *pFeatureClass* to be the feature class of *PFeatureLayer* and *pFields* to be the fields of *pFeatureClass*. A message box reports the number of fields in *pFields*.

```
Private Function FindLayer() As Long
    Dim pMxDoc As IMxDocument
    Dim pMap As IMap
    Dim FindDoc As Variant
    Dim aLName As String
    Dim name As String
    Dim i As Long
    Set pMxDoc = ThisDocument
    Set pMap = pMxDoc.FocusMap
    ' Use an input box to get a layer name.
    name = InputBox("Enter a layer name:", "")
    ' Locate the layer in the active map.
    For i = 0 To pMap.LayerCount - 1
        aLName = UCase(pMap.Layer(i).name)
        If (aLName = (UCase(name))) Then
            FindDoc = i
            Exit For
        End If
    Next
    FindLayer = FindDoc
End Function
```

FindLayer receives a layer name from an input box and assigns it to the *Name* variable. Next, the code steps through layers in the active map and matches the upper case of *Name* with the upper case of the layer's name. When a match is found, the code assigns the layer's index value *i* to the *FindDoc* variable. The function then returns *FindDoc* as the value of **FindLayer** to the **Start** sub.

5.4 ADDING OR DELETING FIELDS

This section covers a macro for adding fields to, and deleting fields from, a dataset.

5.4.1 *AddDeleteField*

AddDeleteField adds two new fields to, and deletes a field from, a feature attribute table. The macro performs the same function as using a dataset's Properties dialog in ArcCatalog to add and delete fields. *AddDeleteField* is organized into three parts. Part 1 defines the feature class, Part 2 defines two new fields and adds the fields to the feature class, and Part 3 deletes a field from the feature class.

> **Key Interfaces:** *IFeatureClass, IFieldEdit, IFields, IField*
> **Key Members:** *FeatureClass, Name, Type, Length, AddField, Fields, FindField, Field(), DeleteField*
> **Usage:** Add *idcounty2.shp* to an active map. Open the attribute table of *idcounty2*. The table shows a field named Pop94, which will be deleted by the macro. Two new fields, Pop1990 and Pop2000, will be added to the table. Import *AddDeleteField* to Visual Basic Editor. Run the macro. Check the attribute table of *idcounty2* again to make sure that the task is done correctly.

```
Private Sub AddDeleteField()
    ' Part 1: Get a handle on the feature class.
    Dim pMxDoc As IMxDocument
    Dim pFeatureLayer As IFeatureLayer
    Dim pFeatureClass As IFeatureClass
    Set pMxDoc = ThisDocument
    Set pFeatureLayer = pMxDoc.FocusMap.Layer(0)
    Set pFeatureClass = pFeatureLayer.FeatureClass
```

Part 1 sets *pFeatureClass* to be the feature class of the top layer in the active map.

```
    ' Part 2: Add two new fields.
    Dim pField1 As IFieldEdit
    Dim pField2 As IFieldEdit
    ' Define the first new field.
    Set pField1 = New Field
    pField1.name = "Pop1990"
    pField1.Type = esriFieldTypeInteger
    pField1.Length = 8
    ' Add the first new field.
    pFeatureClass.AddField pField1
    ' Define the second new field.
    Set pField2 = New Field
    pField2.name = "Pop2000"
    pField2.Type = esriFieldTypeInteger
    pField2.Length = 8
    ' Add the second new field.
    pFeatureClass.AddField pField2
```

Part 2 creates *pField1* as an instance of the *Field* class and uses the *IFieldEdit* interface to define its field properties of name, type, and length. Next, the code uses the *AddField* method on *IFeatureClass* to add *pField1* to *pFeatureClass*. The code uses the same procedure to add *pField2* to *pFeatureClass*.

```
    ' Part 3: Delete a field.
    Dim pFields As IFields
    Dim ii As Integer
    Dim pField3 As IField
    Set pFields = pFeatureClass.Fields
    ii = pFields.FindField("Pop94")
    Set pField3 = pFields.Field(ii)
    pFeatureClass.DeleteField pField3
End Sub
```

Part 3 first sets *pFields* to be the fields of *pFeatureClass*. Next, the code uses the *FindField* method on *IFields* to find the index of Pop94 among the fields, assigns the index value to the *ii* variable, and sets *pField3* to be the field at the index *ii*. Then the code uses the *DeleteField* method on *IFeatureClass* to delete *pField3*.

Box 5.1 AddDeleteField_GP

AddDeleteField_GP uses DeleteField and AddField, two tools in the Data Management toolbox, to first delete a field (Pop94) and then add two new fields (Pop1990 and Pop2000) in *idcounty5.shp*. Run the macro in ArcCatalog. The macro performs the same tasks as *AddDeleteField*.

```
Private Sub AddDeleteField_GP()
    ' Make sure that chap5 is not the active folder in the catalog tree.
    ' Create the Geoprocessing object and define its workspace.
    Dim GP As Object
    Set GP = CreateObject("esriGeoprocessing.GpDispatch.1")
    ' Specify GP's workspace.
    Dim filepath As String
    filepath = "c:\data\chap5\"
    GP.Workspace = filepath
    ' DeleteField <in_table> <drop_field;drop_field...>
    ' AddField <in_table> <field_name> <LONG | TEXT | FLOAT | DOUBLE | SHORT | DATE | BLOB>
    ' {field_precision} {field_length} {field_alias} {NULLABLE | NON_NULLABLE}
    ' {NON_REQUIRED | REQUIRED} {field_domain}
    ' Execute the deletefield and addfield tools.
    GP.DeleteField_management "idcounty5.shp", "Pop94"
    GP.AddField_management "idcounty5.shp", "Pop1990", "SHORT"
    GP.AddField_management "idcounty5.shp", "Pop2000", "SHORT"
End Sub
```

5.5 CALCULATING FIELD VALUES

This section focuses on the field value. The first macro shows how to use an expression to calculate the field values programmatically. The second macro shows how to update the field values of a data subset.

5.5.1 *CalculateField*

CalculateField calculates the values of a field by using the values of two existing fields in a feature class. The macro performs the same function as using Field Calculator for calculation. *CalculateField* has two parts. Part 1 finds the field to be calculated, and Part 2 calculates the field value for each record (feature) in a cursor.

   Key Interfaces: *IFeatureClass, IFields, ICursor, ICalculator*
   Key Members: *FeatureClass, Fields, FindField, Update, Cursor, Expression, Field, Calculate*
   Usage: Add *idcounty3.shp* to an active map. The attribute table of *idcounty3* contains three fields: Pop1990, county population in 1990; Pop2000, county population in 2000; and Change. Import *CalculateField* to Visual Basic Editor. Run the macro. The macro calculates the field values of Change from Pop1990 and Pop2000.

```
Private Sub CalculateField()
    ' Par 1: Find the field to be calculated.
    Dim pMxDoc As IMxDocument
    Dim pFeatLayer As IFeatureLayer
    Dim pFeatureClass As IFeatureClass
    Dim pFields As IFields
    Set pMxDoc = ThisDocument
    Set pFeatLayer = pMxDoc.FocusMap.Layer(0)
    Set pFeatureClass = pFeatLayer.FeatureClass
```

Part 1 sets *pFeatureClass* to be the feature class of the top layer in the active map.

```
    ' Part 2: Calculate the field values by using a cursor.
    Dim pCursor As ICursor
    Dim pCalculator As ICalculator
    ' Prepare a cursor with all records.
    Set pCursor = pFeatureClass.Update(Nothing, True)
    ' Define a calculator.
    Set pCalculator = New Calculator
    With pCalculator
        Set .Cursor = pCursor
        .Expression = "(([Pop2000] - [Pop1990]) / [Pop1990]) * 100"
        .Field = "Change"
    End With
    ' Calculate the field values.
    pCalculator.Calculate
End Sub
```

Part 2 first creates a cursor from *pFeatureClass* by using the *Update* method on *IFeatureClass*. Because the *Update* method uses Nothing in place of a query filter object, all records are included in the cursor. (Chapter 9 covers query filter objects.) Next the code creates *pCalculator* as an instance of the *Calculator* class and defines its properties in a *With* statement. The cursor is set to be *pCursor*, the expression is an equation to calculate the percent change of county population between 1990 and 2000, and the field is Change. Finally, the code uses the *Calculate* method on *ICalculator* to complete the task.

Box 5.2   CalculateField_GP

CalculateField_GP uses the CalculateField tool in the Data Management toolbox to calculate the field values of Change in *idcounty6.shp*. Run the macro in ArcCatalog. The macro performs the same task as *CalculateField*.

```
Private Sub CalculateField_GP()
    ' Create the Geoprocessing object.
    Dim GP As Object
    Set GP = CreateObject("esriGeoprocessing.GpDispatch.1")
```

```
' CalculateField <in_table> <field> <expression>
' Execute the calculatefield tool.
GP.CalculateField_management "c:\data\chap5\idcounty6.shp", "Change", "([Pop2000] - [Pop1990])_
/[Pop1990] * 100"
End Sub
```

5.5.2   *UpdateValue*

UpdateValue differs from *CalculateField* in two aspects. First, *UpdateValue* calculates the field values for a data subset instead of every record. Second, the macro
does not use *ICalculator* to calculate the field values.

   UpdateValue performs the same function as using the Select By Attributes
command in the Selection menu to first select a data subset, and then using Field
Calculator to calculate the field values. The macro has three parts. Part 1 defines the
feature class, Part 2 selects a data subset, and Part 3 calculates the field value for
each record in the data subset.

>   **Key Interfaces:** *IFeatureClass, IFields, IQueryFilter, IFeatureCursor, IFeature*
>   **Key Members:** *FeatureClass, WhereClause, Update, FindField, NextFeature, Value,
>   UpdateFeature*
>   **Usage:** Add *idcounty4.shp* to an active map. The attribute table of *idcounty4* contains
>   the field Change, which shows the rate of population change between 1990 and
>   2000 for Idaho counties. Import *UpdateValue* to Visual Basic Editor. Run the
>   macro. The macro populates the field Class with the value of 1 for high-growth
>   counties (i.e., Change > 30%) and 0 for other counties.

```
Private Sub UpdateValue()
' Part 1: Define the feature class.
Dim pMxDoc As IMxDocument
Dim pFeatLayer As IFeatureLayer
Dim pFeatureClass As IFeatureClass
Dim pFields As IFields
Dim ii As Integer
Set pMxDoc = ThisDocument
Set pFeatLayer = pMxDoc.FocusMap.Layer(0)
Set pFeatureClass = pFeatLayer.FeatureClass
```

   Part 1 sets *pFeatureClass* to be the feature class of the top layer in the active map.

```
' Part 2: Prepare a feature cursor.
Dim pQFilter As IQueryFilter
Dim pUpdateFeatures As IFeatureCursor
' Prepare a query filter.
Set pQFilter = New QueryFilter
pQFilter.WhereClause = "Change > 30"
' Create a feature cursor for updating.
Set pUpdateFeatures = pFeatureClass.Update(pQFilter, False)
```

   Part 2 creates *pQFilter* as an instance of the *QueryFilter* class and defines its
WhereClause condition as "Change > 30." A query filter object filters a data subset
that meets its *WhereClause* expression. The code then uses *pQFilter* as an object

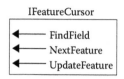

Figure 5.10   *IFeatureCursor* has methods for managing fields and features.

qualifier and the *Update* method on *IFeatureClass* to create a cursor of selected features. A feature cursor is a data-access object designed for the updating, deleting, and inserting of features. A feature cursor object has the *FindField*, *NextFeature*, and *UpdateFeature* methods (Figure 5.10).

```
' Part 3: Calcuate the Class value.
Dim indexClass As Integer
Dim pFeature As IFeature
indexClass = pUpdateFeatures.FindField("Class")
Set pFeature = pUpdateFeatures.NextFeature
' Loop through each feature and update its Class value.
Do Until pFeature Is Nothing
    pFeature.Value(indexClass) = 1
    pUpdateFeatures.UpdateFeature pFeature
    Set pFeature = pUpdateFeatures.NextFeature
Loop
End Sub
```

Part 3 first uses the *FindField* method on *IFeatureCursor* to find the index of the field Class and assigns the index value to the *indexClass* variable. The rest of the code uses a loop to step through each selected feature, assigns 1 as the value to the field at *indexClass*, and uses the *UpdateFeature* method on *IFeatureCursor* to update the feature.

5.6 JOINING AND RELATING TABLES

This section covers joins and relates with four sample macros, two on join and two on relate. By having two macros on each type of relationship class, this section shows how to link a nongeographic table to a feature class as well as how to link two or more nongeographic tables to a feature class.

5.6.1 *JoinTableToLayer*

JoinTableToLayer joins a dBASE file to a feature class. The macro performs the same function as using Join in a feature layer's context menu in ArcMap. *JoinTableToLayer* has three parts. Part 1 defines the feature class, Part 2 defines the attribute table to be joined, and Part 3 joins the attribute table to the feature class.

Key Interfaces: *IFeatureClass, ITable, IStandaloneTableCollection, IStandalone-Table, IMemoryRelationshipClassFactory, IRelationshipClass, IDisplayRelationshipClass*

Key Members: *FeatureClass, StandaloneTable( ), Table, Open, DisplayRelationshipClass*

Usage: Add *wp.shp* and *wpdata.dbf* to an active map. *wp* is a shapefile of vegetation stands, and *wpdata* is a dBASE file containing attributes of the vegetation stands. The common field in *wp*'s attribute table and *wpdata* is ID. Import ***JoinTableTo-Layer*** to Visual Basic Editor. Run the macro. Enter ID for the name of the join field. The macro joins the attributes in *wpdata* to *wp*'s attribute table. To verify that the macro works, open the attribute table of *wp*. The table should have two sets of attributes: one set has the prefix of wp and the other set has the prefix of wpdata.

```
Private Sub JoinTableToLayer()
    ' Part 1: Get a handle on the layer's attribute table.
    Dim pMxDoc As IMxDocument
    Dim pMap As IMap
    Dim pFeatureLayer As IFeatureLayer
    Dim pFeatureClass As IFeatureClass
    Set pMxDoc = ThisDocument
    Set pMap = pMxDoc.FocusMap
    Set pFeatureLayer = pMap.Layer(0)
    Set pFeatureClass = pFeatureLayer.FeatureClass
```

Part 1 sets *pFeatureClass* to be the feature class of the top layer in the active map.

```
    ' Part 2: Define the table to be joined.
    Dim pTabCollection As IStandaloneTableCollection
    Dim pStTable As IStandaloneTable
    Dim pFromTable As ITable
    Set pTabCollection = pMap
    Set pStTable = pTabCollection.StandaloneTable(0)
    Set pFromTable = pStTable.Table
```

Part 2 first performs a QueryInterface (QI) for the *IStandaloneTableCollection* interface and sets *pStTable* to be the first standalone table in *pMap*. The code then sets *pFromTable* to be the table underlying *pStTable*.

```
    ' Part 3: Join the table to the layer's attribute table.
    Dim strJnField As String
    Dim pMemRelFact As IMemoryRelationshipClassFactory
    Dim pRelClass As IRelationshipClass
    Dim pDispRC As IDisplayRelationshipClass
    ' Prompt for the join field.
    strJnField = InputBox("Provide the name of the join field:", "Joining a table to a layer", "")
    ' Create a memory relationship class.
    Set pMemRelFact = New MemoryRelationshipClassFactory
    ' The underscore _ means continuation of a line statement.
    Set pRelClass = pMemRelFact.Open("TableToLayer", pFromTable, strJnField, pToTable, strJnField, "wp", _
    "wpdata", esriRelCardinalityOneToOne)
    ' Perform a join.
```

```
    Set pDispRC = pFeatureLayer
    pDispRC.DisplayRelationshipClass pRelClass, esriLeftOuterJoin
End Sub
```

Part 3 creates a memory relationship class (i.e., virtual join) before joining the two tables. The code first gets the join field from an input box. Next, the code creates *pMemRelFact* as an instance of the *MemoryRelationshipClassFactory* class and uses the *Open* method on *IMemoryRelationshipClassFactory* to create a relationship class referenced by *pRelClass*. The *Open* method uses eight object qualifiers and arguments. The two object qualifiers of *pFromTable* and *pFeatureClass* have been defined. The name of the relationship class is "TableToLayer." The origin primary key and the origin foreign key are both *strJnField*. The forward path label and the backward path label are wp and wpdata respectively, and the cardinality or the type of relationship is specified as one-to-one. Finally, the code accesses the *IDisplay-RelationshipClass* interface and uses the *DisplayRelationshipClass* method to perform the join. Besides *pRelClass*, *DisplayRelationshipClass* uses the argument of *esriLeftOuterJoin*, which stipulates that the join operation includes all rows, rather than the matched rows only.

5.6.2   *JoinMultipleTables*

JoinMultipleTables joins two dBASE tables to a feature class. Conceptually, ***JoinMultipleTables*** is similar to ***JoinTableToLayer,*** but programmatically, ***JoinMultipleTables*** requires use of *IRelQueryTable* objects to carry out two joins. The macro performs the same function as using the Join command in a feature layer's context menu in ArcMap. ***JoinMultipleTables*** has five parts. Part 1 defines the feature class, Part 2 defines the two tables to be joined, Part 3 creates the first virtual join, followed by the second virtual join in Part 4, and Part 5 performs the join operation.

> **Key Interfaces:** *IStandaloneTableCollection, IStandaloneTable, ITable, IMemoryRelationshipClassFactory, IRelationshipClass, IRelQueryTableFactory, IRelQueryTable, IDisplayRelationshipClass*
>
> **Key Members:** *FeatureClass, StandaloneTable( ), Table, Open, DisplayRelationshipClass*
>
> **Usage:** Add *idcounty4.shp, change.dbf,* and *population.dbf* to an active map. *idcounty4* is a shapefile showing 44 Idaho counties. *change* and *population* are two dBASE files that contain demographic data of Idaho counties. (This macro will not work with text files.) The keys for joining *change* to *idcounty* are both co_name. The keys for joining *population* to the joined table are change.co_name and co_name. The keys have been hard-coded in the macro. Import ***JoinMultipleTables*** to Visual Basic Editor. Run the macro. The macro joins the attributes in *change* and *population* to the attribute table of *idcounty4*. To verify that the macro works, open the attribute table of *idcounty4*. The table should include attributes from *change* and *population*. Each field in the table should have a prefix to identify its source.

```
Private Sub JoinMultipleTables()
    ' Part 1: Define the feature class.
    Dim pMxDoc As IMxDocument
    Dim pMap As IMap
```

```
Dim pFeatureLayer As IFeatureLayer
Dim pFeatureClass As IFeatureClass
Set pMxDoc = ThisDocument
Set pMap = pMxDoc.FocusMap
Set pFeatureLayer = pMap.Layer(0)
Set pFeatureClass = pFeatureLayer.FeatureClass
```

Part 1 sets *pFeatureClass* to be the feature class of the top layer in the active map.

```
' Part 2: Define the two tables to be joined.
Dim pTabCollection As IStandaloneTableCollection
Dim pStTable1 As IStandaloneTable
Dim pFromTable1 As ITable
Dim pStTable2 As IStandaloneTable
Dim pFromTable2 As ITable
Set pTabCollection = pMap
' Define the first table.
Set pStTable1 = pTabCollection.StandaloneTable(0)
Set pFromTable1 = pStTable1.Table
' Define the second table.
Set pStTable2 = pTabCollection.StandaloneTable(1)
Set pFromTable2 = pStTable2.Table
```

Part 2 performs a QI for the *IStandaloneTableCollection* interface and sets *pStTable1* and *pStTable2* to be the first and second standalone tables respectively in *pMap*. The code then defines *pFromTable1* and *pFromTable2* to be the underlying tables of *pStTable1* and *pStTable2* respectively.

```
' Part 3: Join the first table to the feature class.
Dim pMemRelFact As IMemoryRelationshipClassFactory
Dim pRelClass1 As IRelationshipClass
Dim pRelQueryTableFact As IRelQueryTableFactory
Dim pRelQueryTab1 As IRelQueryTable
' Create the first virtual join.
Set pMemRelFact = New MemoryRelationshipClassFactory
Set pRelClass1 = pMemRelFact.Open("Table1ToLayer", pFeatureClass, "co_name", pFromTable1, "co_name", _
"forward", "backward", esriRelCardinalityOneToOne)
' Create the first relquerytable.
Set pRelQueryTableFact = New RelQueryTableFactory
Set pRelQueryTab1 = pRelQueryTableFact.Open(pRelClass1, True, Nothing, Nothing, "", True, True)
```

Part 3 creates *pMemRelFact* as an instance of the *MemoryRelationshipClassFactory* class and uses the *Open* method to create a memory relationship class object referenced by *pRelClass1*. The two tables specified for the *Open* method are *pFeatureClass* and *pFromTable1*. Next, the code creates *pRelQueryTableFact* as an instance of the *RelQueryTableFactory* class and uses the *Open* method on *IRelQueryTableFactory* and *pRelClass1* as an argument to create *pRelQueryTab1*, a reference to an *IRelQueryTable* object.

```
' Part 4: Join the second table to the joined table.
Dim pRelClass2 As IRelationshipClass
```

```
Dim pRelQueryTab2 As IRelQueryTable
' Create the second virtual join.
Set pMemRelFact = New MemoryRelationshipClassFactory
Set pRelClass2 = pMemRelFact.Open("Table2ToLayer", pRelQueryTab1, "change.co_name", _
pFromTable2, "co_name", "forward", "backward", esriRelCardinalityOneToOne)
' Create the second relquerytable.
Set pRelQueryTableFact = New RelQueryTableFactory
Set pRelQueryTab2 = pRelQueryTableFact.Open(pRelClass2, True, Nothing, Nothing, "", True, True)
```

Part 4 follows the same procedure as in Part 3: use *pRelQueryTab1* and *pFromTable2* as inputs to create *pRelClass2*, and use *pRelClass2* as an input to create *pRelQueryTab2*.

```
' Part 5: Perform the join operation.
Dim pDispRC2 As IDisplayRelationshipClass
Set pDispRC2 = pFeatureLayer
pDispRC2.DisplayRelationshipClass pRelQueryTab2.RelationshipClass, esriLeftOuterJoin
End Sub
```

Part 5 accesses the *IDisplayRelationshipClass* interface and uses the *DisplayRelationshipClass* method to perform the join operation. Notice that the method uses *RelQueryTab2* as an object qualifier.

5.6.3 *RelateTableToLayer*

RelateTableToLayer creates a relate between a dBASE file and a feature class. The macro performs the same function as using Relate in a feature layer's context menu in ArcMap. ***RelateTableToLayer*** has three parts. Part 1 defines the feature class, Part 2 defines the nongeographic table for a relate, and Part 3 asks for the keys to establish a relate, creates a virtual relate, and performs a relate.

> **Key Interfaces:** *IFeatureClass, ITable, IStandaloneTableCollection, IStandaloneTable, IMemoryRelationshipClassFactory, IRelationshipClass, IRelationshipClassCollectionEdit*
>
> **Key Members:** *FeatureClass, StandaloneTable( ), Table, Open, AddRelationshipClass*
> **Usage:** Add *wp.shp* and *wpdata.dbf* to an active map. ID in both the attribute table of *wp* and *wpdata* can be used as the key. Import ***RelateTableToLayer*** to Visual Basic Editor. Run the macro. Enter ID in both input boxes. The macro creates a relate between *wpdata* and the attribute table. To verify that the macro works, open the attribute table and select some records from the table. Click Options in the attribute table and point to Related Tables. TableToLayer: wpdata should appear on the side bar. Click on the side bar, and *wpdata* appears with highlighted records that correspond to the selected records in *wp*.

```
Private Sub RelateTabletoLayer()
    ' Part 1: Define the feature class.
    Dim pMxDoc As IMxDocument
    Dim pMap As IMap
    Dim pFeatureLayer As IFeatureLayer
    Dim pFeatureClass As IFeatureClass
```

```
Set pMxDoc = ThisDocument
Set pMap = pMxDoc.FocusMap
Set pFeatureLayer = pMap.Layer(0)
Set pFeatureClass = pFeatureLayer.FeatureClass
```

Part 1 sets *pFeatureClass* to be the feature class of the top layer in the active map.

```
' Part 2: Define the table for a relate.
Dim pTabCollection As IStandaloneTableCollection
Dim pStTable As IStandaloneTable
Dim pFromTable As ITable
Set pTabCollection = pMap
Set pStTable = pTabCollection.StandaloneTable(0)
Set pFromTable = pStTable.Table
```

Part 2 defines *pFromTable* to be the underlying table of the first standalone table in the active map.

```
' Part 3: Perform the relate.
Dim strLayerField As String
Dim strTableField As String
Dim pMemRelFact As IMemoryRelationshipClassFactory
Dim pRelClass As IRelationshipClass
Dim pRelClassCollEdit As IRelationshipClassCollectionEdit
' Prompt for the keys.
strLayerField = InputBox("Provide the key from the layer: ", "Relating a table to a layer", "")
strTableField = InputBox("Provide the key from the table: ", "Relating a table to a layer", "")
' Create a virtual relate.
Set pMemRelFact = New MemoryRelationshipClassFactory
Set pRelClass = pMemRelFact.Open("TableToLayer", pFromTable, strTableField, pFeatureClass, _
strLayerField, "wp", "wpdata", esriRelCardinalityOneToOne)
' Add the relate to the collection.
Set pRelClassCollEdit = pFeatureLayer
pRelClassCollEdit.AddRelationshipClass pRelClass
End Sub
```

Part 3 starts by prompting for the keys for establishing the relate. Next, the code creates *pMemRelFact* as an instance of the *MemoryRelationshipClassFactory* class. The code then uses the *Open* method on *IMemoryRelationshipClassFactory* to create a virtual relate referenced by *pRelClass*. Finally, the code switches to the *IRelationshipClassCollectionEdit* interface and uses the *AddRelationshipClass* method to add *pRelClass* to the relation class collection of the feature layer.

Box 5.3   RelateTableToLayer_GP

RelateTableToLayer_GP uses the CreateRelationshipClass tool in the Data Management toolbox to create a relate between a feature class and a table. Both datasets must reside in a geodatabase. For this example, the geodatabase is *relation.mdb*, the feature class is *wp*, and the table is *wpdata*. Run the macro in

ArcCatalog. Enter wp in the first input box, ID in the second, wpdata in the third, and ID in the fourth. ID is the key for the relate. When it is done, a relationship class (*wp_wpdata*) should appear in *relation.mdb*. Except for using a personal geodatabase, **RelateTableToLayer_GP** performs the same task as **RelateTableToLayer**.

```
Private Sub RelateTableToLayer_GP()
    ' This macro can only work with Geodatabase feature classes and tables.
    ' Create the geoprocessing object and define its workspace.
    Dim GP As Object
    Set GP = CreateObject("esriGeoprocessing.GpDispatch.1")
    ' Specify a geodatabase for GP's workspace.
    Dim filepath As String
    filepath = "c:\data\chap5\relation.mdb"
    GP.Workspace = filepath
    ' CreateRelationshipClass <origin_table> <destination_table> <out_relationship_class>
    ' <SIMPLE | COMPOSITE> <forward_label> <backward_label>
    ' <NONE | FORWARD | BACKWARD | BOTH>
    ' <ONE_TO_ONE | ONE_TO_MANY | MANY_TO_MANY>
    ' <NONE | ATTRIBUTED> <origin_primary_key> <origin_foreign_key>
    ' {destination_primary_key}{destination_foreign_key}
    Dim table1 As String
    table1 = InputBox("Enter the name of the origin table")
    Dim key1 As String
    key1 = InputBox("Enter the name of the origin key")
    Dim table2 As String
    table2 = InputBox("Enter the name of the destination table")
    Dim key2 As String
    key2 = InputBox("Enter the name of the destination key")
    ' Execute the createrelationshipclass tool.
    GP.CreateRelationshipClass_management table1, table2, "wp_wpdata", "SIMPLE", "wpdata", "wp", _
    "NONE", "ONE_TO_ONE", "NONE", key1, key2
End Sub
```

5.6.4   *RelationalDatabase*

RelationalDatabase creates relates between a feature class and three dBASE files from a relational database. The macro performs the same function as using Relate in ArcMap three times. **RelationalDatabase** has four parts. Part 1 defines the feature class, Part 2 defines the three dBASE files for relates, Part 3 creates three virtual relates, and Part 4 adds the relates to the relationship class collection of the feature layer.

> **Key Interfaces:** *IFeatureClass, ITable, IStandaloneTableCollection, IStandalone-Table, IMemoryRelationshipClassFactory, IRelationshipClass, IRelationshipClassCollectionEdit*
> **Key Members:** *FeatureClass, StandaloneTable( ), Table, Open, AddRelationshipClass*
> **Usage:** Add *mosoils.shp, comp.dbf, forest.dbf,* and *plantnm.dbf* to an active map. The macro assumes that the order of the three tables is *comp, forest,* and *plantnm* from

top to bottom in the table of contents. *Mosoils* is a soil shapefile, and the three dBASE files contain soil attributes from a relational database. The key relating *mosoils* and *comp* is MUSYM, the key relating *comp* and *forest* is MUID, and the key relating *forest* and *plantnm* is PLANTSYM. Import ***RelationalDatabase*** to Visual Basic Editor. Run the macro. Enter the proper key in each of the three input boxes. To verify that the macro works, select some records from the attribute table of *mosoils* and then open *comp* to see the corresponding records.

```
Private Sub RelationalDatabase()
    ' Part 1: Get a handle on the mosoils attribute table.
    Dim pMxDoc As IMxDocument
    Dim pMap As IMap
    Dim pFeatureLayer As IFeatureLayer
    Dim pFeatureClass As IFeatureClass
    Set pMxDoc = ThisDocument
    Set pMap = pMxDoc.FocusMap
    Set pFeatureLayer = pMap.Layer(0)
    Set pFeatureClass = pFeatureLayer.FeatureClass
```

Part 1 sets *pFeatureClass* to be the feature class of the top layer in the active map.

```
    ' Part 2: Define the tables for relates.
    Dim pTabCollection As IStandaloneTableCollection
    Dim pStTable1 As IStandaloneTable
    Dim pCompTable As ITable
    Dim pStTable2 As IStandaloneTable
    Dim pForestTable As ITable
    Dim pStTable3 As IStandaloneTable
    Dim pPlantnmTable As ITable
    Set pTabCollection = pMap
    ' Define the first table.
    Set pStTable1 = pTabCollection.StandaloneTable(0)
    Set pCompTable = pStTable1.Table
    ' Define the second table.
    Set pStTable2 = pTabCollection.StandaloneTable(1)
    Set pForestTable = pStTable2.Table
    ' Define the third table.
    Set pStTable3 = pTabCollection.StandaloneTable(2)
    Set pPlantnmTable = pStTable3.Table
```

Part 2 sets the tables underlying the three standalone tables as *pCompTable*, *pForestTable*, and *pPlantnmTable* respectively.

```
    ' Part 3: Create three virtual relates.
    Dim strField1 As String
    Dim pMemRelFact As IMemoryRelationshipClassFactory
    Dim pRelClass1 As IRelationshipClass
    Dim strField2 As String
    Dim pRelClass2 As IRelationshipClass
    Dim strField3 As String
    Dim pRelClass3 As IRelationshipClass
```

```
' Create the first virtual relate.
strField1 = InputBox("Provide the common field in Mosoils and Comp: ", "Relate Mosoils to Comp", "")
Set pMemRelFact = New MemoryRelationshipClassFactory
Set pRelClass1 = pMemRelFact.Open("Mosoils-Comp", pCompTable, strField1, pFeatureClass, strField1, _
    "Mosoils", "Comp", esriRelCardinalityManyToMany)
' Create the second virtual relate.
strField2 = InputBox("Provide the common field in Comp and Forest: ", "Relate Comp to Forest", "")
Set pRelClass2 = pMemRelFact.Open("Comp-Forest", pForestTable, strField2, pCompTable, strField2, _
    "Comp", "Forest", esriRelCardinalityManyToMany)
' Create the third virtual relate.
strField3 = InputBox("Provide the common field in Forest and Plantnm: ", "Relate Forest to Plantnm", "")
Set pRelClass3 = pMemRelFact.Open("Forest-Plantnm", pPlantnmTable, strField3, pForestTable, _
    strField3, "Forest", "Plantnm", esriRelCardinalityManyToMany)
```

Part 3 uses the proper keys and tables to create three memory relationship classes, referenced by *pRelClass1*, *pRelClass2*, and *pRelClass3* respectively. Notice that the *Open* method specifies many-to-many for the type of relationship (cardinality) argument.

```
' Part 4: Add the relates to the collection.
Dim pRelClassCollEdit As IRelationshipClassCollectionEdit
Set pRelClassCollEdit = pFeatureLayer
pRelClassCollEdit.AddRelationshipClass pRelClass1
pRelClassCollEdit.AddRelationshipClass pRelClass2
pRelClassCollEdit.AddRelationshipClass pRelClass3
End Sub
```

Part 4 performs a QI for the *IRelationshipClassCollectionEdit* interface and uses the *AddRelationshipClass* method to add the three virtual relates to the relationship class collection of *pFeatureLayer*.

CHAPTER **6**

Data Conversion

A major application of geographic information systems (GIS) is integration of data from different sources and in different formats. To allow for data integration, a GIS must be able to read different data formats and to convert data from one format into another. Data conversion can take place in a variety of ways of which the three most common are:

Between vector data of different types
Between vector and raster data
From x-, y-coordinates to point data

Environmental Systems Research Institute Inc. (ESRI), has introduced a new vector data model with the release of each major product for the past 20 years: coverages with ArcInfo, shapefiles with ArcView, and geodatabases with ArcGIS. Although ArcGIS users can use all three types of vector data, many have converted traditional coverages and shapefiles into geodatabases to take advantage of object-oriented technology and new developments from ESRI, Inc.

Conversion between vector and raster data is important for digitizing (e.g., tracing from scanned files) and data analysis. Vector to raster data conversion, or rasterization, converts points, lines, and polygons to cells and fills the cells with values from an attribute. Raster to vector data conversion, or vectorization, extracts points and lines from cells and usually requires generalization and weeding of the extracted features.

Tables that record the location of weather stations or a hurricane track typically contain x-, y-coordinates. Representing the geographic locations, these x and y coordinates can be converted into point features. It is an alternative to digitizing the point locations on a digitizing tablet.

This chapter deals with the three common types of data conversion. Section 6.1 reviews data conversion operations in ArcGIS. Section 6.2 discusses objects that are related to data conversion. Section 6.3 includes macros and Geoprocessing (GP) macros for converting shapefiles into standalone feature classes or feature classes in a feature dataset. Section 6.4 has macros for converting coverages into shapefiles

91

or geodatabases. Section 6.5 discusses macros for rasterization and vectorization and a GP macro for rasterization. Section 6.6 has a macro and a GP macro for converting *x*-, *y*-coordinates into point data. All macros start with the listing of key interfaces and key members (properties and methods) and the usage.

6.1 CONVERTING DATA IN ARCGIS

ArcGIS has data conversion commands in all three applications of ArcCatalog, ArcMap, and ArcToolbox. ArcToolbox is probably the first choice among many users because it has a large set of data conversion tools in one place. Conversion Tools in ArcToolbox offers the following seven categories:

From Raster
To CAD (computer-aided design)
To Coverage
To dBASE
To Geodatabase
To Raster
To Shapefile

3D Analyst Tools also offers conversion tools between raster and TIN (triangulated irregular network), and Coverage Tools has tools for converting from and to coverage.

ArcCatalog incorporates data conversion commands into the context menus. The context menu of a personal geodatabase has the Import command for importing feature classes, tables, and raster datasets to geodatabase. The context menu of a shapefile or a coverage has the Export command for exporting the dataset to geodatabase and other formats. The context menu of an event table has the Create Features command that can create point features from *x*- and *y*-coordinates and the Export command that can export the table to dBASE and geodatabase.

In ArcMap, conversion of vector data from one format into another is available through the Data/Export Data command in the dataset's context menu. Conversion between raster and vector data, on the other hand, is available through the Convert command in Spatial Analyst and 3D Analyst. The Tools menu in ArcMap has the Add XY Data command for converting a table with *x*-, *y*-coordinates into point features.

6.2 ARCOBJECTS FOR DATA CONVERSION

This section covers objects for feature data conversion, rasterization and vectorization, and XY event.

6.2.1 Objects for Feature Data Conversion

The principal ArcObjects component for converting data between geodatabases, shapefiles, and coverages is the *FeatureDataConverter* class (Figure 6.1). A feature

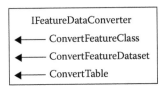

Figure 6.1   *IFeatureDataConverter* has methods for converting feature datasets.

data converter object implements *IFeatureDataConverter* and *IFeatureDataConverter2*. Both interfaces have methods for converting feature classes, feature datasets, and tables. *IFeatureDataConverter2* has the additional functionality of working with data subsets. An alternative to *IFeatureDataConverter* is *IExportOperation*, which, although with fewer options, also has methods for converting vector data of different formats (Figure 6.2).

Three name objects appear frequently in this chapter. A *WorkspaceName* object can identify and locate a workspace. To create a workspace name object, one must define the type of workspace factory and the path name (Figure 6.3). The *WorkspaceFactoryProgID* property on *IWorkspaceName* lets the programmer specify the type of workspace factory. A path name specifies the path of a workspace, which can be coded as a string or based on a generic *PropertySet* object. *IPropertySet* has methods that can set and hold properties for objects such as workspaces.

A *DatasetName* object can represent various types of datasets (Figure 6.4). *IDatasetName* has the properties of *Name* and *WorkspaceName* that let the programmer set the names of the dataset and the workspace of a dataset name object respectively.

A *FeatureClassName* object can identify and locate an object, which may represent a shapefile, a coverage, or a geodatabase feature class. Because *FeatureClassName* is a type of the *DatasetName* class, we can QueryInterface (QI) for the *IDatasetName* interface to specify the name and the workspace of the feature class. Likewise, because *FeatureClassName* is a type of the *Name* class, we can perform a QI for the *IName* interface and use the *Open* method to open the feature class (Figure 6.5).

6.2.2   Objects for Rasterization and Vectorization

The principal component for conversion between vector and raster data is the *RasterConversionOp* class (Figure 6.6). A *RasterConversionOp* object implements

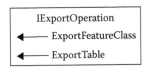

Figure 6.2   *IExportOperation* has methods for exporting feature datasets.

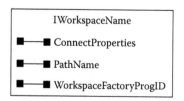

WorkspaceFactoryProgID:
esricore.AccessWorkspaceFactory
esricore.ArcInfoWorkspaceFactory
esricore.RasterWorkspaceFactory
esricore.ShapefileWorkspaceFactory

Figure 6.3 *IWorkspaceName* has properties for identifying workspace factory and path.

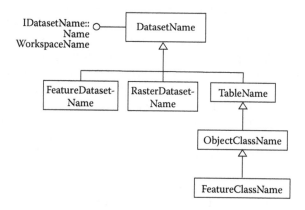

Figure 6.4 Types of *DatasetName* can share the properties of *Name* and *WorkspaceName* on *IDatasetName*.

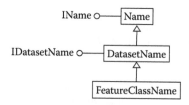

Figure 6.5 A *FeatureClassName* object inherits *IName* and *IDatasetName*.

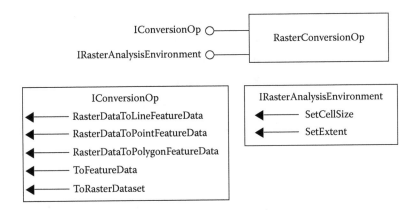

Figure 6.6   A *RasterConversionOp* object supports *IConversionOp* and *IRasterAnalysis-Environment.*

IConversionOp and *IRasterAnalysisEnvironment. IConversionOp* offers methods for converting raster data to point, line, and polygon feature data as well as for converting vector data to raster data. *IRasterAnalysisEnvironment* has members for the conversion environment such as the output cell size.

A *RasterDatasetName* object is the raster equivalent of a *FeatureClassName* object. But macros for rasterization or vectorization do not use raster dataset name objects because methods on *IConversionOp* require use of the actual datasets rather than the name objects.

6.2.3   Objects for XY Event

XYEventSource, XYEventSourceName, and *XYEvent2FieldsProperties* are the primary components for converting *x*-, *y*-coordinates into point features (Figure 6.7). An XY event source object is unique in the following ways:

- An XY event source object is created through an XY event source name object.
- An XY event source object is a type of a point feature class.
- The point feature class represented by an XY event source object is dynamic, meaning that the feature class is generated from a table rather than a physical dataset. Once a dynamic feature class is created, it can be exported to a shapefile or a geodatabase feature class.

An *XYEvent2FieldsProperties* object implements *IXYEvent2FieldsProperties*, which can define the properties of the *x* field, *y* field, and *z* field (optional) of a point feature class. These properties can be passed on to an XY event source name object. *IXYEventSourceName* also has members for specifying the source table and the spatial reference.

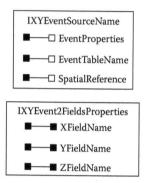

Figure 6.7 *XYEventSourceName* and *XYEvent2FieldProperties* have properties that can define an *XYEventSource* object.

6.3 CONVERTING SHAPEFILE TO GEODATABASE

A shapefile contains one set of spatial data, which may represent point, line, or area features. When converted into the geodatabase format, a shapefile becomes a feature class, which can be a standalone feature class or part of a feature dataset in the geodatabase. This section covers both types of conversion. This section also shows two ways of getting multiple shapefiles for conversion: one uses a dialog box and the other uses the input box.

6.3.1 *ShapefileToAccess*

ShapefileToAccess converts a shapefile into a feature class and saves the output as a standalone feature class in a new geodatabase. The macro performs the same function as creating a new personal geodatabase in ArcCatalog and using the Import/Feature Class (Single) command to import a shapefile to the geodatabase. *ShapefileToAccess* is organized into three parts. Part 1 defines the output, including its workspace and name, Part 2 defines the input, and Part 3 performs the data conversion. *Shapefile-ToAccess* uses the lightweight name objects throughout the code.

> **Key Interfaces:** *IWorkspaceName, IFeatureClassName, IDatasetName, IFeature-DataConverter*
> **Key Members:** *WorkspaceFactoryProgID, PathName, Name, WorkspaceName, ConvertFeatureClass*
> **Usage:** Import *ShapefileToAccess* to Visual Basic Editor in ArcCatalog. Run the macro. The macro reports "Shapefile conversion complete" when the conversion is done. The macro adds *Trial.mdb* to the Catalog tree and *Soils* as a feature class in the geodatabase. *Soils* is converted from *soils.shp*.

```
Private Sub ShapefileToAccess()
    ' Part 1: Define the output.
    Dim pWorkspaceName As IWorkspaceName
    Dim pFeatureClassName As IFeatureClassName
    Dim pDatasetName As IDatasetName
```

```
' Define the workspace.
Set pWorkspaceName = New WorkspaceName
pWorkspaceName.WorkspaceFactoryProgID = "esricore.AccessWorkspaceFactory"
pWorkspaceName.PathName = "c:\data\chap6\Trial.mdb"
' Define the dataset.
Set pFeatureClassName = New FeatureClassName
Set pDatasetName = pFeatureClassName
Set pDatasetName.WorkspaceName = pWorkspaceName
pDatasetName.name = "Soils"
```

Part 1 first creates *pWorkspaceName* as an instance of the *WorkspaceName* class and defines its *WorkspaceFactoryProgID* and *PathName* properties. Next, the code creates *pFeatureClassName* as an instance of the *FeatureClassName* class and uses the *IDatasetName* interface to set its *WorkspaceName* and *Name* properties.

```
' Part 2: Define the input.
Dim pInShpWorkspaceName As IWorkspaceName
Dim pFCName As IFeatureClassName
Dim pShpDatasetName As IDatasetName
' Define the workspace.
Set pInShpWorkspaceName = New WorkspaceName
pInShpWorkspaceName.PathName = "c:\data\chap6"
pInShpWorkspaceName.WorkspaceFactoryProgID = "esriCore.ShapefileWorkspaceFactory"
' Define the dataset.
Set pFCName = New FeatureClassName
Set pShpDatasetName = pFCName
pShpDatasetName.name = "soils.shp"
Set pShpDatasetName.WorkspaceName = pInShpWorkspaceName
```

Part 2 creates *pInShpWorkspaceName* as an instance of the *WorkspaceName* class and specifies its *PathName* and *WorkspaceFactoryProgID* properties. Next, the code creates *pFCName* as an instance of the *FeatureClassName* class and accesses the *IDatasetName* interface to specify its *Name* and *WorkspaceName* properties.

```
' Part 3: Perform data conversion.
Dim pShpToFC As IFeatureDataConverter
Set pShpToFC = New FeatureDataConverter
pShpToFC.ConvertFeatureClass pFCName, Nothing, Nothing, pFeatureClassName, Nothing, Nothing, "", 1000, 0
MsgBox "Shapefile conversion complete!"
End Sub
```

Part 3 creates *pShpToFC* as an instance of the *FeatureDataConverter* class and uses the *ConvertFeatureClass* method on *IFeatureDataConverter* to convert *pFCName* into *pFeatureClassName*. Besides the two object qualifiers, the *ConvertFeatureClass* method specifies 1000 for the flush interval. The flush interval dictates the interval for committing data, which may be important for loading large amounts of data into a geodatabase. An alternative to *IFeatureDataConverter* for data conversion is *IExport-Operation*, which also has methods for exporting feature classes and tables. For converting a shapefile into a geodatabase feature class, *IExportOperation* functions in nearly the same way as *IFeatureDataConverter*. The following bit of code replaces *IFeatureDataConverter* with *IExportOperation* in Part 3 of **ShapefileToAccess**:

```
' Part 3: Use a new ExportOperation to complete the conversion task.
Dim pShpToFC As IExportOperation
Set pShpToFC = New ExportOperation
pShpToFC.ExportFeatureClass pFCName, Nothing, Nothing, Nothing, pFeatureClassName, 0
```

Box 6.1   ShapefileToAccess_GP

ShapefileToAccess_GP uses the CreatePersonalGDB tool in the Management toolbox and the FeatureclassToFeatureclass tool in the Conversion toolbox to first create a personal geodatabase (*lochsa.mdb*) and then convert a shapefile (*deer.shp*) to a standalone feature class (*deer*) in the geodatabase. Run the macro in Arc-Catalog. The output can be examined in the Catalog tree.

```
Private Sub ShapefileToAccess_GP()
    ' Create the Geoprocessing object and define its workspace.
    Dim GP As Object
    Set GP = CreateObject("esriGeoprocessing.GpDispatch.1")
    Dim filepath As String
    filepath = "c:\data\chap6\"
    GP.Workspace = filepath
    ' CreatePersonalGDB <out_folder_path> <out_name>
    ' Execute the createpersonalGDB tool.
    GP.CreatePersonalGDB_management filepath, "lochsa.mdb"
    ' FeatureclassToFeatureclass <input_features> < output_location>
    ' <output_feature_class_name> {expression} {field_info} {SAME_AS_TEMPLATE |
    ' DISABLED | ENABLED} {SAME_AS_TEMPLATE | DISABLED | ENABLED}
    ' {configuration_keyword} {first_spatial_grid}
    ' Execute the featureclasstofeatureclass tool.
    GP.FeatureclassToFeatureclass_conversion "deer.shp", "lochsa.mdb", "deer"
End Sub
```

6.3.2   *MultipleShapefilesToAccess*

MultipleShapefilesToAccess converts two or more shapefiles into standalone feature classes and saves the feature classes in a new geodatabase. The macro performs the same function as creating a new personal geodatabase in ArcCatalog and using the Import/Feature Class (Multiple) command to import two or more shapefiles to the geo-database. *MultipleShapefilesToAccess* has three parts. Part 1 defines the output and input workspace name objects, Part 2 uses a dialog box to get the shapefiles to be converted, and Part 3 uses a *Do...Loop* to convert shapefiles into standalone feature classes.

> **Key Interfaces:** *IWorkspacename, IGxDialog, IGxObjectFilter, IEnumGxObject, IGx-Dataset, IFeatureClassName, IDatasetName, IFeatureDataConverter*
> **Key Members:** *WorkspaceFactoryProgID, PathName, AllowMultiSelect, ButtonCaption, ObjectFilter, StartingLocation, Title, DoModalOpen, Next, Name, Workspace-Name, ConvertFeatureClass*
> **Usage:** Import *MultipleShapefilesToAccess* to Visual Basic Editor in ArcCatalog. Run the macro. Select *landuse.shp*, *sewers.shp*, and *soils.shp* from the dialog box for conversion. The macro adds *SiteAnalysis.mdb* to the Catalog tree and *landuse, sewers,*

and *soils*, converted from the shapefiles, as feature classes in the geodatabase. The macro uses the prefix of the shapefile (e.g., landuse) to name the output feature class.

```
Private Sub MultipleShapefilesToAccess()
    ' Part 1: Define the output and input workspaces.
    Dim pWorkspaceName As IWorkspaceName
    Dim pInShpWorkspaceName As IWorkspaceName
    ' Define the output workspace.
    Set pWorkspaceName = New WorkspaceName
    pWorkspaceName.WorkspaceFactoryProgID = "esricore.AccessWorkspaceFactory"
    pWorkspaceName.PathName = "c:\data\chap6\SiteAnalysis.mdb"
    ' Define the input workspace.
    Set pInShpWorkspaceName = New WorkspaceName
    pInShpWorkspaceName.pathname = "c:\data\chap6"
    pInShpWorkspaceName.WorkspaceFactoryProgID = "esriCore.ShapefileWorkspaceFactory"
```

Part 1 creates *pWorkspaceName* as an instance of the *WorkspaceName* class and defines its *WorkspaceFactoryProgID* and *PathName* properties. The code creates and defines *pInShpWorkspaceName* in the same way.

```
    ' Part 2: Prepare a dialog box for selecting shapefiles to convert.
    Dim pGxDialog As IGxDialog
    Dim pGxFilter As IGxObjectFilter
    Dim pGxObjects As IEnumGxObject
    Dim pGxDataset As IGxDataset
    Set pGxDialog = New GxDialog
    Set pGxFilter = New GxFilterShapefiles
    ' Define the dialog's properties.
    With pGxDialog
        .AllowMultiSelect = True
        .ButtonCaption = "Add"
        Set .ObjectFilter = pGxFilter
        .StartingLocation = pInShpWorkspaceName.PathName
        .Title = "Select Shapefiles to Convert"
    End With
    ' Open the dialog.
    pGxDialog.DoModalOpen 0, pGxObjects
    ' Exit sub if no shapefile has been selected.
    Set pGxDataset = pGxObjects.Next
    If pGxDataset Is Nothing Then
        Exit Sub
    End If
```

Part 2 creates *pGxDialog* as an instance of the *GxDialog* class and *pGxFilter* as an instance of the *GxFilterShapefiles* class. Next, the code defines the properties of *pGxDialog* to allow for multiple selections, to start at the input workspace location, and to show only shapefiles. The code then uses the *DoModalOpen* method on *IGxDialog* to open the dialog box and to save the selected shapefiles into a collection. If no shapefile has been selected, exit the sub.

```
    ' Part 3: Loop through each selected shapefile and convert it to a feature class.
    Dim pFeatureClassName As IFeatureClassName
    Dim pDatasetName As IDatasetName
```

```
Dim pFCName As IFeatureClassName
Dim pShpDatasetName As IDatasetName
Dim pShpToFC As IFeatureDataConverter
Do Until pGxDataset Is Nothing
    ' Define the output dataset.
    Set pFeatureClassName = New FeatureClassName
    Set pDatasetName = pFeatureClassName
    Set pDatasetName.WorkspaceName = pWorkspaceName
    pDatasetName.name = pGxDataset.Dataset.name
    ' Define the input dataset.
    Set pFCName = New FeatureClassName
    Set pShpDatasetName = pFCName
    pShpDatasetName.name = pGxDataset.Dataset.name & ".shp"
    Set pShpDatasetName.WorkspaceName = pInShpWorkspaceName
    ' Perform data conversion.
    Set pShpToFC = New FeatureDataConverter
    pShpToFC.ConvertFeatureClass pFCName, Nothing, Nothing, pFeatureClassName, Nothing, Nothing, "", 1000, 0
    Set pGxDataset = pGxObjects.Next
Loop
MsgBox "Shapefile conversions complete!"
End Sub
```

Part 3 steps through each shapefile selected for conversion. The code performs a QI for the *IDatasetName* interface to specify the *Name* and *WorkspaceName* properties of the output dataset and the input dataset. Then the code uses the *ConvertFeatureClass* method to convert the input dataset into the output dataset. The loop stops when nothing is advanced from *pGxObjects*.

6.3.3   *ShapefilesToFeatureDataset*

ShapefilesToFeatureDataset converts one or more shapefiles into feature classes and saves the feature classes in a specified feature dataset of a geodatabase. The output from the macro therefore follows a hierarchical structure of geodatabase, feature dataset, and feature class. Because all feature classes in a feature dataset must share the same spatial reference (i.e., the same coordinate system and extent), a feature dataset is typically reserved for feature classes that are from the same study area or participate in topological relationships with each other.

ShapefilesToFeatureDataset performs the same function as creating a new personal geodatabase, creating a new feature dataset, and using the Import/Feature Class (Multiple) command to import one or more shapefiles to the geodatabase. By default, a feature dataset uses the extent of the first shapefile as its area extent. If the first shapefile has a smaller area extent, then portions of other shapefiles will disappear after conversion.

ShapefilesToFeatureDataset is organized into two parts. Part 1 defines the output workspace, the output feature dataset, and the input workspace; and Part 2 uses the input box and a *Do...Loop* statement to get shapefiles for conversion.

Key Interfaces: *IWorkspacename, IFeatureDatasetName, IDatasetName, IFeature-ClassName, IFeatureDataConverter*

Key Members: *WorkspaceFactoryProgID, PathName, Name, WorkspaceName, ConvertFeatureClass*

Usage: Import ***ShapefilesToFeatureDataset*** to Visual Basic Editor in ArcCatalog. Run the macro. Enter *landuse, sewers,* and *soils* sequentially in the input box. Click Cancel in the input box to dismiss the dialog. The macro adds to the Catalog tree a hierarchy of *SiteAnalysis2.mdb, StudyArea1,* and *landuse, sewers,* and *soils.* The feature classes in the geodatabase are converted from the shapefiles.

```
Private Sub ShapefilesToFeatureDataset()
    ' Part 1: Define the output and input workspaces.
    Dim pWorkspaceName As IWorkspaceName
    Dim pFeatDSName As IFeatureDatasetName
    Dim pDSName As IDatasetName
    Dim pInShpWorkspaceName As IWorkspaceName
    ' Define the output workspace.
    Set pWorkspaceName = New WorkspaceName
    pWorkspaceName.WorkspaceFactoryProgID = "esricore.AccessWorkspaceFactory"
    pWorkspaceName.PathName = "c:\data\chap6\SiteAnalysis2.mdb"
    ' Specify the output feature dataset.
    Set pFeatDSName = New FeatureDatasetName
    Set pDSName = pFeatDSName
    Set pDSName.WorkspaceName = pWorkspaceName
    pDSName.Name = "StudyArea1"
    ' Specify the input workspace.
    Set pInShpWorkspaceName = New WorkspaceName
    pInShpWorkspaceName.pathname = "c:\data\chap6"
    pInShpWorkspaceName.WorkspaceFactoryProgID = "esriCore.ShapefileWorkspaceFactory"
```

Part 1 defines the output and input workspaces in the same way as ***Multiple-ShapefilesToAccess***. The change is with the feature dataset for the output. The code creates *pFeatDSName* as an instance of the *FeatureDatasetName* class and switches to the *IDatasetName* interface to define its *WorkspaceName* and *Name* properties.

```
    ' Part 2: Loop through every shapefile to be converted.
    Dim pFeatureClassName As IFeatureClassName
    Dim pDatasetName As IDatasetName
    Dim pFCName As IFeatureClassName
    Dim pShpDatasetName As IDatasetName
    Dim pShpToFC As IFeatureDataConverter
    Dim pInput As String
    ' Use the input box to get an input shapefile.
    pInput = InputBox("Enter the name of the input shapefile")
    ' Loop through each input shapefile while the input box is not empty.
    Do While pInput <> ""
        ' Define the output dataset.
        Set pFeatureClassName = New FeatureClassName
```

```
        Set pDatasetName = pFeatureClassName
        Set pDatasetName.WorkspaceName = pWorkspaceName
        pDatasetName.Name = pInput
        ' Define the input dataset.
        Set pFCName = New FeatureClassName
        Set pShpDatasetName = pFCName
        pShpDatasetName.Name = pInput & ".shp"
        Set pShpDatasetName.WorkspaceName = pInShpWorkspaceName
        ' Perform data conversion.
        Set pShpToFC = New FeatureDataConverter
        pShpToFC.ConvertFeatureClass pFCName, Nothing, pFeatDSName, pFeatureClassName, Nothing, _
        Nothing, "", 1000, 0
        pInput = InputBox("Enter the name of the input shapefile")
    Loop
    MsgBox "Shapefile conversions complete!"
End Sub
```

Part 2 uses the *InputBox* function and a *Do...Loop* statement to loop through every shapefile to be converted. The *ConvertFeatureClass* method uses an additional object qualifier of *pFeatDSName* for the feature dataset.

Box 6.2   ShapefilesToFeatureDataset_GP

ShapefilesToFeatureDataset_GP uses the CreatePersonalGeodatabase tool to create *trial.mdb*, the CreateFeatureDataset tool to create *analysis1*, and the FeatureclassToGeodatabase tool to convert three shapefiles (*landuse*, *sewers*, and *soils*) into feature classes and store them in *analysis1*. Parameter 1 for the FeatureclassToGeodatabase tool allows multiple input features; therefore, the macro does not need a Do Loop. Run the macro in ArcCatalog, and examine the output in the Catalog tree.

```
Private Sub ShapefilesToFeatureDataset_GP()
    ' Create the Geoprocessing object and define its workspace.
    Dim GP As Object
    Set GP = CreateObject("esriGeoprocessing.GpDispatch.1")
    ' CreatePersonalGDB <out_folder_path> <out_name>
    ' Execute the createpersonalGDB tool.
    GP.CreatePersonal GDB_management "c:\data\chap6", "trial.mdb"
    ' CreateFeatureDataset <out_dataset_path> <out_name> {spatial_reference}
    ' Execute the createfeaturedataset tool.
    GP.CreateFeatureDataset_management "trial.mdb", "analysis1"
    ' Define two parameters: one for input features and the other for geodatabase.
    Dim parameter1 As String
    parameter1 = "landuse.shp;sewers.shp;soils.shp"
    Dim parameter2 As String
    parameter2 = "trial.mdb\analysis1"
    ' FeatureclassToGeodatabase <input_features; input_features...> < output_geodatabase>
    ' Execute the featureclasstogeodatabase tool.
    GP.FeatureclassToGeodatabase_conversion parameter1, parameter2
End Sub
```

6.4 CONVERTING COVERAGE TO GEODATABASE
AND SHAPEFILE

This section covers conversion of traditional coverages into geodatabase feature classes and shapefiles. A coverage consists of different datasets such as arc, tic, and labels. A coverage based on the regions or dynamic segmentation data model has even more datasets. Therefore, to convert a coverage, we must start by identifying the dataset to be converted.

6.4.1 *CoverageToAccess*

CoverageToAccess converts a coverage to a feature class and saves the feature class in a specified geodatabase. The macro performs the same function as creating a new personal geodatabase in ArcCatalog and using the Import/Feature Class (Single) command to import a coverage into the geodatabase. *CoverageToAccess* has three parts. Part 1 defines the output including its workspace and name, Part 2 defines the input including the workspace and the specific dataset of the coverage to be converted, and Part 3 performs the data conversion.

> **Key Interfaces:** *IPropertySet, IWorkspaceFactory, IWorkspaceName, IFeatureClass-Name, IDatasetName, IFeatureDataConverter*
> **Key Members:** *SetProperty, Create, WorkspaceName, Name, PathName, Workspace-FactoryProgID, ConvertFeatureClass*
> **Usage:** Import *CoverageToAccess* to Visual Basic Editor in ArcCatalog. Run the macro. The macro adds *emida.mdb* to the Catalog tree and *breakstrm* as a feature class in the geodatabase. The feature class in the geodatabase is converted from the arcs of the *breakstrm* coverage.

```
Private Sub CoverageToAccess()
    ' Part 1: Define the output.
    Dim pPropset As IPropertySet
    Dim pOutAcFact As IWorkspaceFactory
    Dim pOutAcWorkspaceName As IWorkspaceName
    Dim pOutAcFCName As IFeatureClassName
    Dim pOutAcDSName As IDatasetName
    ' Set the connection property.
    Set pPropset = New PropertySet
    pPropset.SetProperty "Database", "c:\data\chap6"
    ' Define the workspace.
    Set pOutAcFact = New AccessWorkspaceFactory
    Set pOutAcWorkspaceName = pOutAcFact.Create("c:\data\chap6", "emida", pPropset, 0)
    ' Define the dataset.
    Set pOutAcFCName = New FeatureClassName
    Set pOutAcDSName = pOutAcFCName
    Set pOutAcDSName.WorkspaceName = pOutAcWorkspaceName
    pOutAcDSName.name = "breakstrm"
```

Part 1 defines the input. The code first creates *pPropset* as an instance of the *PropertySet* class. The *Create* method on *IWorkspaceFactory* uses *pPropset* and other arguments to create *pOutAcWorkspaceName*. Next, the code creates *pOutAcFCName* as an instance of the *FeatureClassName* class and performs a QI for the *IDatasetName* interface to set its *WorkspaceName* and *Name* properties.

```
' Part 2: Specify the Input.
Dim pInCovWorkspaceName As IWorkspaceName
Dim pFCName As IFeatureClassName
Dim pCovDatasetName As IDatasetName
' Define the workspace.
Set pInCovWorkspaceName = New WorkspaceName
pInCovWorkspaceName.PathName = "c:\data\chap6"
pInCovWorkspaceName.WorkspaceFactoryProgID = "esriCore.ArcInfoWorkspaceFactory.1"
' Define the dataset of the coverage to be converted.
Set pFCName = New FeatureClassName
Set pCovDatasetName = pFCName
pCovDatasetName.name = "breakstrm:arc"
Set pCovDatasetName.WorkspaceName = pInCovWorkspaceName
```

Part 2 specifies the input. The code creates *pInCovWorkspaceName* as an instance of the *WorkspaceName* class and defines its *PathName* and *WorkspaceFactoryProgID* properties. Then the code creates *pFCName* as an instance of the *FeatureClassName* class and uses the *IDatasetName* interface to set the *Name* and *WorkspaceName* properties. Notice that the name is "*breakstrm:arc*," which refers to the arcs of *breakstrm*.

```
' Part 3: Perform data conversion.
Dim pCovtoFC As IFeatureDataConverter
Set pCovtoFC = New FeatureDataConverter
pCovtoFC.ConvertFeatureClass pFCName, Nothing, Nothing, pOutAcFCName, Nothing, Nothing, "", 1000, 0
MsgBox "Coverage conversion complete!"
End Sub
```

Part 3 uses the *ConvertFeatureClass* method on *IFeatureDataConverter* to convert *pFCName* to *pOutAcFCName*.

6.4.2 *CoverageToShapefile*

CoverageToShapefile converts a coverage into a shapefile. The macro performs the same function as using the Coverage to Shapefile tool in ArcToolbox or the Data/Export Data command in the context menu of a coverage in ArcMap. *Coverage-ToShapefile* has three parts. Part 1 defines the input coverage's workspace, dataset, and geometry; Part 2 specifies the output workspace and dataset; and Part 3 converts the coverage to a shapefile and adds the shapefile as a feature layer in the active map.

> **Key Interfaces:** *IWorkspaceName, IFeatureClassName, IDatasetName, IPropertySet, IName, IFeatureDataConverter*
>
> **Key Members:** *PathName, WorkspaceFactoryProgID, Name, WorkspaceName, SetProperty, ConnectionProperties, ConvertFeatureClass*
>
> **Usage:** Add *breakstrm.arc*, the arc layer of the coverage *breakstrm*, to an active map. Import *CoverageToShapefile* to Visual Basic Editor. Run the macro. The macro creates *breakstrm.shp* and adds the shapefile as a feature layer to the active map.

```
Private Sub CoverageToShapefile()
    ' Part 1: Define the input coverage.
    Dim pInCovWorkspaceName As IWorkspaceName
    Dim pInFCName As IFeatureClassName
    Dim pCovDatasetName As IDatasetName
    ' Define the workspace.
    Set pInCovWorkspaceName = New WorkspaceName
    pInCovWorkspaceName.pathname = "c:\data\chap6"
    pInCovWorkspaceName.WorkspaceFactoryProgID = "esriCore.ArcInfoWorkspaceFactory.1"
    ' Define the dataset.
    Set pInFCName = New FeatureClassName
    Set pCovDatasetName = pInFCName
    pCovDatasetName.Name = "breakstrm:arc"
    Set pCovDatasetName.WorkspaceName = pInCovWorkspaceName
```

Part 1 defines the input coverage. The code first creates *pInCovWorkspaceName* as an instance of the *WorkspaceName* class and defines its *PathName* and *WorkspaceFactoryProgID* properties. Then the code creates *pInFCName* as an instance of the *FeatureClassName* class and accesses the *IDatasetName* interface to define its *Name* and *WorkspaceName* properties.

```
    ' Part 2: Define the output.
    Dim pPropset As IPropertySet
    Dim pOutShWSName As IWorkspaceName
    Dim pOutShFCName As IFeatureClassName
    Dim pOutshDSName As IDatasetName
    ' Set the connection property.
    Set pPropset = New PropertySet
    pPropset.SetProperty "DATABASE", "c:\data\chap6"
    ' Define the workspace.
    Set pOutShWSName = New WorkspaceName
    pOutShWSName.ConnectionProperties = pPropset
    pOutShWSName.WorkspaceFactoryProgID = "esriCore.shapefileWorkspaceFactory.1"
    ' Define the dataset.
    Set pOutShFCName = New FeatureClassName
    Set pOutshDSName = pOutShFCName
    Set pOutshDSName.WorkspaceName = pOutShWSName
    pOutshDSName.Name = "breakstrm.shp"
```

Part 2 defines the output. The code first creates *pPropset* as an instance of the *PropertySet* class and uses *pPropset*, along with the shapefile workspace factory, to define an instance of the *WorkspaceName* class referenced by *pOutShWSName*. Next, the code creates *pOutShFCName* as an instance of the *FeatureClassName* class and accesses the *IDatasetName* interface to define its *WorkspaceName* and *Name* properties.

```
    ' Part 3: Convert the coverage to a shapefile and display the shapefile.
    Dim pName As IName
    Dim pCovtoshape As IFeatureDataConverter
    Dim pOutShFC As IFeatureClass
    Dim pOutSh As IFeatureLayer
```

```
Dim pMxDoc As IMxDocument
Dim pMap As IMap
' Convert the coverage to a shapefile.
Set pCovtoshape = New FeatureDataConverter
pCovtoshape.ConvertFeatureClass pInFCName, Nothing, Nothing, pOutShFCName, Nothing, Nothing, "", 1000, 0
' Create a feature layer from the output feature class name object.
Set pName = pOutShFCName
Set pOutShFC = pName.Open
Set pOutSh = New FeatureLayer
Set pOutSh.FeatureClass = pOutShFC
pOutSh.Name = "breakstrm.shp"
' Add the feature layer to the active map.
Set pMxDoc = ThisDocument
Set pMap = pMxDoc.FocusMap
pMap.AddLayer pOutSh
MsgBox "Coverage conversion complete!"
End Sub
```

Part 3 converts *pInFCName* to *pOutShFCName* by using the *ConvertFeatureClass* method on *IFeatureDataConverter*. Next, the code opens a feature class referenced by *pOutShFC* from its name object, and creates a feature layer from *pOutShFC*. Finally, the code adds the feature layer referenced by *pOutSh* to the active map.

6.5 PERFORMING RASTERIZATION AND VECTORIZATION

This section covers rasterization and vectorization. Rasterization creates a raster and populates the cells of the raster with values from an attribute of a feature class. The raster is an ESRI grid, a software-specific raster. The attribute can be a specified field or feature ID by default. Vectorization creates a feature class from a raster. The cell values of the input raster are stored in a field of the feature class called GRIDCODE. Other fields may be added through code.

6.5.1 *FeatureToRaster*

FeatureToRaster converts a feature layer into a permanent raster and adds the raster to the active map. The input feature layer may represent a shapefile, a coverage, or a geodatabase feature class. The output raster contains cell values that correspond to the feature IDs of the input dataset.

The macro performs the same function as using the Convert/Features to Raster command in Spatial Analyst. *FeatureToRaster* has three parts. Part 1 defines the input feature layer, Part 2 performs the conversion and saves the output raster in a specified workspace, and Part 3 creates a raster layer from the output and adds the layer to the active map.

Key Interfaces: *IWorkspace, IWorkspaceFactory, IConversionOp, IRasterAnalysisEnvironment, IRasterDataset*

Key Members: *OpenFromFile, SetCellSize, ToRasterDataset, CreateFromDataset, Name*

Usage: Add *nwroads.shp*, an interstate highway shapefile of the Pacific Northwest, to an active map. Import *FeatureToRaster* to Visual Basic Editor in ArcMap. Run the macro. *FeatureToRaster* adds *roadsgd* as a raster layer to the active map.

```
Private Sub FeatureToRaster()
    ' Part 1: Define the input dataset.
    Dim pMxDoc As IMxDocument
    Dim pMap As IMap
    Dim pInputFL As IFeatureLayer
    Dim pInputFC As IFeatureClass
    Set pMxDoc = ThisDocument
    Set pMap = pMxDoc.FocusMap
    Set pInputFL = pMap.Layer(0)
    Set pInputFC = pInputFL.FeatureClass
```

Part 1 sets *pInputFC* to be the feature class of the top layer in the active map.

```
    ' Part 2: Convert feature class to raster.
    Dim pWS As IWorkspace
    Dim pWSF As IWorkspaceFactory
    Dim pConversionOp As IConversionOp
    Dim pEnv As IRasterAnalysisEnvironment
    Dim pOutRaster As IRasterDataset
    ' Set the output workspace.
    Set pWSF = New RasterWorkspaceFactory
    Set pWS = pWSF.OpenFromFile("c:\data\chap6", 0)
    ' Prepare a raster conversion operator, and perform the conversion.
    Set pConversionOp = New RasterConversionOp
    Set pEnv = pConversionOp
    pEnv.SetCellSize esriRasterEnvValue, 5000
    Set pOutRaster = pConversionOp.ToRasterDataset(pInputFC, "GRID", pWS, "roadsgd")
```

Part 2 performs the conversion. The code first defines the output workspace referenced by *pWS*. Next, the code creates *pConversionOp* as an instance of the *RasterConversionOp* class and uses the *IRasterAnalysisEnvironment* interface to set the output cell size of 5000 (meters). The code then uses the *ToRasterDataset* method on *IConversionOp* to convert *pInputFC* into *pOutRaster* and to save the output in *pWS*.

```
    ' Part 3: Create a new raster layer and add the layer to the active map.
    Dim pOutRasterLayer As IRasterLayer
    Set pOutRasterLayer = New RasterLayer
    pOutRasterLayer.CreateFromDataset pOutRaster
    pOutRasterLayer.Name = "roadsgd"
    pMap.AddLayer pOutRasterLayer
End Sub
```

Part 3 uses the *CreateFromDataset* method on *IRasterLayer* to create a new raster layer from *pOutRaster*, and adds the layer to the active map.

Box 6.3 FeatureToRaster_GP

FeatureToRaster_GP uses the FeatureToRaster tool in the Conversion toolbox to convert *nwroads.shp* into a raster (*roadsgd2*). Enter RTE_NUM1 in the input box for the field to be used and 5000 (meters) for the output cell size. Run the macro in ArcCatalog and examine the output in the Catalog tree.

```
Private Sub FeatureToRaster_GP()
    ' Create the Geoprocessing object.
    Dim GP As Object
    Set GP = CreateObject("esriGeoprocessing.GpDispatch.1")
    ' FeatureToRaster <input_features> <field> <output_raster> {cell_size}
    Dim value As String
    value = InputBox("Enter the field name for the cell value")
    Dim size As Integer
    size = InputBox("Enter the output cell size in meters")
    ' Execute the featuretoraster tool.
    GP.FeatureToRaster_conversion "c:\data\chap6\nwroads.shp", value, "c:\data\chap6\roadsgd2", size
End Sub
```

6.5.2 *FCDescriptorToRaster*

FeatureToRaster produces an output raster that has feature IDs as cell values. In some cases, it is more useful to have cell values that correspond to county names or highway numbers instead of feature IDs. ***FCDescriptorToRaster***, which uses a feature class descriptor as the input dataset, is useful for those cases. A *Feature-ClassDescriptor* object implements *IFeatureClassDescriptor*, which has methods for creating a feature class descriptor based on a specific field of a dataset (Figure 6.8).

FCDescriptorToRaster performs the same function as using the Convert/Features to Raster command in Spatial Analyst. The Convert/Features to Raster dialog allows the user to specify a field that corresponds to the cell value. ***FCDescriptor-ToRaster*** has three parts. Part 1 creates a feature class descriptor from a feature layer, Part 2 converts the feature class descriptor into a permanent raster, and Part 3 creates a new raster layer from the output raster and adds the layer to the active map.

> **Key Interfaces:** *IFeatureClassDescriptor, IWorkspace, IWorkspaceFactory, IConversionOp, IRasterAnalysisEnvironment, IRasterDataset*
>
> **Key Members:** *Create, OpenFromFile, SetCellSize, ToRasterDataset, CreateFromDataset, Name*
>
> **Usage:** Add *nwroads.shp* to an active map. Import ***FCDescriptorToRaster*** to Visual Basic Editor in ArcMap. Run the macro. ***FCDescriptorToRaster*** adds *rtenumgd*

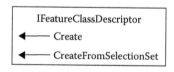

Figure 6.8 Methods on *IFeatureClassDescriptor*.

as a raster layer to the active map. The attribute table of *rtenumgd* has the Rte_num1 field that lists the highway numbers.

```
Private Sub FCDescriptorToRaster()
    ' Part 1: Get a handle on the descriptor of a shapefile to be converted.
    Dim pMxDoc As IMxDocument
    Dim pMap As IMap
    Dim pInputFL As IFeatureLayer
    Dim pInputFC As IFeatureClass
    Dim pFCDescriptor As IFeatureClassDescriptor
    Dim sFieldName As String
    Set pMxDoc = ThisDocument
    Set pMap = pMxDoc.FocusMap
    Set pInputFL = pMap.Layer(0)
    Set pInputFC = pInputFL.FeatureClass
    ' Create a feature class descriptor based on the field RTE_NUM1.
    Set pFCDescriptor = New FeatureClassDescriptor
    sFieldName = "RTE_NUM1"
    pFCDescriptor.Create pInputFC, Nothing, sFieldName
```

Part 1 sets *pInputFC* to be the feature class of the top layer in the active map. Next, the code creates *pFCDescriptor* as an instance of the *FeatureClassDescriptor* class and assigns RTE_NUM1 to the string variable *sFieldName*. The code then uses the *Create* method on *IFeatureClassDescriptor* and *sFieldName* as an argument to create *pFCDescriptor*.

```
    ' Part 2: Convert the feature class descriptor to a raster.
    Dim pWS As IWorkspace
    Dim pWSF As IWorkspaceFactory
    Dim pConversionOp As IConversionOp
    Dim pEnv As IRasterAnalysisEnvironment
    Dim pOutRaster As IRasterDataset
    Set pWSF = New RasterWorkspaceFactory
    Set pWS = pWSF.OpenFromFile("c:\data\chap6", 0)
    Set pConversionOp = New RasterConversionOp
    Set pEnv = pConversionOp
    pEnv.SetCellSize esriRasterEnvValue, 5000
    ' Use the feature class descriptor as the input for conversion.
    Set pOutRaster = pConversionOp.ToRasterDataset(pFCDescriptor, "GRID", pWS, "rtenumgd")
```

Part 2 sets *pWS* as a reference to the output workspace and *pConversionOp* as an instance of the *RasterConversionOp* class. The code then uses the *ToRaster-Dataset* method and *pFCDescriptor* as an object qualifier to create *pOutRaster*.

```
    ' Part 3: Create a new raster layer and add the layer to the active map.
    Dim pOutRasterLayer As IRasterLayer
    Set pOutRasterLayer = New RasterLayer
    pOutRasterLayer.CreateFromDataset pOutRaster
    pOutRasterLayer.Name = "rtenumgd"
    pMap.AddLayer pOutRasterLayer
End Sub
```

Part 3 creates a new raster layer from *pOutRaster* and adds the layer to the active map.

6.5.3 *RasterToShapefile*

RasterToShapefile converts a raster into a shapefile. The macro performs the same function as using the Convert/Raster to Features command in Spatial Analyst. The output shapefile contains a default field called GRIDCODE that stores the cell values of the input raster.

RasterToShapefile is organized into three parts. Part 1 defines the input raster, Part 2 performs the conversion and saves the output shapefile in a specified workspace, and Part 3 creates a feature layer from the output and adds the layer to the active map.

> **Key Interfaces:** *IWorkspace, IWorkspaceFactory, IConversionOp, IFeatureClass*
> **Key Members:** *OpenFromFile, RasterDataToLineFeatureData, FeatureClass, Name*
> **Usage:** Add *nwroads_gd*, a raster showing interstate highways in the Pacific Northwest, to an active map. Import *RasterToShapefile* to Visual Basic Editor in ArcMap. Run the macro. *RasterToShapefile* adds *roads.shp* as a feature layer to the active map.

```
Private Sub RasterToShapefile()
    ' Part 1: Get a handle on the input raster.
    Dim pMxDoc As IMxDocument
    Dim pMap As IMap
    Dim pInputRL As IRasterLayer
    Dim pInputRaster As IRaster
    Set pMxDoc = ThisDocument
    Set pMap = pMxDoc.FocusMap
    Set pInputRL = pMap.Layer(0)
    Set pInputRaster = pInputRL.Raster
```

Part 1 sets *pInputRaster* to be the raster of the top layer in the active map.

```
    ' Part 2: Convert the raster to a shapefile.
    Dim pWS As IWorkspace
    Dim pWSF As IWorkspaceFactory
    Dim pConversionOp As IConversionOp
    Dim pOutFClass As IFeatureClass
    ' Set the output workspace.
    Set pWSF = New ShapefileWorkspaceFactory
    Set pWS = pWSF.OpenFromFile("c:\data\chap6", 0)
    ' Prepare a raster conversion operation, and perform the conversion.
    Set pConversionOp = New RasterConversionOp
    Set pOutFClass = pConversionOp.RasterDataToLineFeatureData(pInputRaster, pWS, "roads.shp", True, True, 5000)
```

Part 2 sets *pWS* to be the workspace of the output shapefile. The code then creates *pConversionOp* as an instance of the *RasterConversionOp* class and uses the *RasterDataToLineFeatureData* method on *IConversionOp* to create *pOutFClass*. The method requires *pInputRaster* and *pWS* as the object qualifiers. Additionally, it has four arguments. The first is the name of the output dataset. The second, if true, means that cells with values of <= 0 or no data belong to the background. The third, if true, means applying line generalization or weeding to the output. The last is the minimum length of dangling arcs on the output.

The *IConversionOp* interface also has a method called *ToFeatureData*, which can convert a raster to a polyline shapefile. But, because *ToFeatureData* does not include the options of line generalization and dangling arcs, the output is typically less desirable than that from *RasterDataToLineFeatureData*.

```
' Part 3: Create a new raster layer and add the layer to the active map.
Dim pOutputFeatLayer As IFeatureLayer
Set pOutputFeatLayer = New FeatureLayer
Set pOutputFeatLayer.FeatureClass = pOutFClass
pOutputFeatLayer.Name = pOutFClass.AliasName
pMap.AddLayer pOutputFeatLayer
End Sub
```

Part 3 creates a new feature layer from *pOutFClass* and adds the layer to the active map.

6.5.4   *RasterDescriptorToShapefile*

RasterDescriptorToShapefile uses a raster descriptor as an input and adds a field in addition to GRIDCODE to the output shapefile. Like a feature class descriptor, a raster descriptor is based on a specific field in a raster. A *RasterDescriptor* object implements *IRasterDescriptor*, which has methods for creating raster descriptors (Figure 6.9).

RasterDescriptorToShapefile has three parts. Part 1 creates a raster descriptor from the input raster, Part 2 performs the conversion and saves the output shapefile in a specified workspace, and Part 3 creates a feature layer from the output and adds the layer to the active map.

Key Interfaces: *IRasterDescriptor, IWorkspace, IWorkspaceFactory, IConversionOp, IFeatureClass*

Key Members: Create, *OpenFromFile, RasterDataToPolygonFeatureData, Feature-Class, Name*

Usage: Add *nwcounties_gd*, a raster showing counties in the Pacific Northwest, to an active map. Import *RasterDescriptorToShapefile* to Visual Basic Editor in ArcMap. Run the macro. *RasterDescriptorToShapefile* adds *Fips.shp* as a feature layer to the active map. The attribute table of *Fips* has FIPS (federal information processing codes) as a field in addition to the default field of GRIDCODE.

Figure 6.9   Methods on *IRasterDescriptor*.

```
Private Sub RasterDescriptorToShapefile()
    ' Part 1: Get a handle on the descriptor of the raster to be converted.
    Dim pMxDoc As IMxDocument
    Dim pMap As IMap
    Dim pInputRL As IRasterLayer
    Dim pInputRaster As IRaster
    Dim pRasterDescriptor As IRasterDescriptor
    Dim sFieldName As String
    Set pMxDoc = ThisDocument
    Set pMap = pMxDoc.FocusMap
    Set pInputRL = pMap.Layer(0)
    Set pInputRaster = pInputRL.Raster
    ' Create a raster descriptor based on the field Fips.
    Set pRasterDescriptor = New RasterDescriptor
    sFieldName = "Fips"
    pRasterDescriptor.Create pInputRaster, Nothing, sFieldName
```

Part 1 sets *pInputRaster* to be the raster of the top layer in the active map. Next, the code creates *pRasterDescriptor* as an instance of the *RasterDescriptor* class and assigns Fips to the string variable *sFieldName*. The code then uses the *Create* method on *IRasterDescriptor* to create *pRasterDescriptor*.

```
    ' Part 2: Convert the raster descriptor to a shapefile.
    Dim pWS As IWorkspace
    Dim pWSF As IWorkspaceFactory
    Dim pConversionOp As IConversionOp
    Dim pOutFClass As IFeatureClass
    Set pWSF = New ShapefileWorkspaceFactory
    Set pWS = pWSF.OpenFromFile("c:\data\chap6", 0)
    Set pConversionOp = New RasterConversionOp
    ' Use the raster descriptor as the input for conversion.
    Set pOutFClass = pConversionOp.RasterDataToPolygonFeatureData(pRasterDescriptor, pWS, "Fips.shp", True)
```

Part 2 first sets *pWS* as a reference to the workspace for the output shapefile. The code then creates *pConversionOp* as an instance of the *RasterConversionOp* class and uses the *RasterDataToPolygonFeatureData* method on *IConversionOp* to create *pOutFClass*. Notice that the method uses *pRasterDescriptor* as an object qualifier.

```
    ' Part 3: Create a new raster layer and add the layer to the active map.
    Dim pOutputFeatLayer As IFeatureLayer
    Set pOutputFeatLayer = New FeatureLayer
    Set pOutputFeatLayer.FeatureClass = pOutFClass
    pOutputFeatLayer.Name = pOutFClass.AliasName
    pMap.AddLayer pOutputFeatLayer
End Sub
```

Part 3 creates a feature layer from *pOutFClass* and adds the layer to the active map.

6.6 ADDING XY EVENTS

This section covers conversion of a table with *x*-, *y*-coordinates into point features. The coordinates can be either geographic or projected. After conversion, each pair of *x*-, *y*-coordinates becomes a point.

6.6.1 *XYEvents*

XYEvents adds a point layer from a table with *x*-, *y*-coordinates to an active map. The source of the table is a text file. The layer can later be exported to a shapefile based on the coordinate system that has already been defined for the *x*-, *y*-coordinates. The macro performs the same function as using the Add XY Data command in ArcMap's Tools menu.

XYEvents is organized into three parts. Part 1 defines the text file in terms of its workspace and name, Part 2 defines the event source including its event-to-fields properties and spatial reference, and Part 3 creates a new feature layer from the event source and adds the layer to the active map.

> **Key Interfaces:** *IWorkspaceName, ITableName, IDatasetName, IXYEvent-2FieldsProperties, ISpatialReferenceFactory, IProjectedCoordinateSystem, IXYEventSourceName, IXYEventSource*
>
> **Key Members:** *PathName, WorkspaceFactoryProgID, Name, WorkspaceName, XFieldName, YFieldName, ZFieldName, CreateProjectedCoordinateSystem, EventProperties, SpatialReference, EventTableName, Open, FeatureClass*
>
> **Usage:** Import *XYEvents* to ArcMap's Visual Basic Editor. Run the macro. The macro adds a layer named *XYEvents* to the active map. The source for *XYEvents* is *events.txt*, a table with *x*-, *y*-coordinates.

```
Private Sub XYEvents()
    ' Part 1: Define the text file.
    Dim pWorkspaceName As IWorkspaceName
    Dim pTableName As ITableName
    Dim pDatasetName As IDatasetName
    ' Define the input table's workspace.
    Set pWorkspaceName = New WorkspaceName
    pWorkspaceName.PathName = "c:\data\chap6"
    pWorkspaceName.WorkspaceFactoryProgID = "esriCore.TextFileWorkspaceFactory.1"
    ' Define the dataset.
    Set pTableName = New TableName
    Set pDatasetName = pTableName
    pDatasetName.Name = "events.txt"
    Set pDatasetName.WorkspaceName = pWorkspaceName
```

Part 1 defines the input text file. The code creates *pWorkspaceName* as an instance of the *WorkspaceName* class and defines its *PathName* and *WorkspaceFactoryProgID* properties. Notice that the ProgID is esriCore.TextFileWorkspaceFactory. Next, the code creates *pTableName* as an instance of the *TableName* class and accesses the *IDatasetName* interface to specify its properties of name and workspace.

```
' Part 2: Define the XY events source name object.
Dim pXYEvent2FieldsProperties As IXYEvent2FieldsProperties
Dim pSpatialReferenceFactory As ISpatialReferenceFactory
Dim pProjectedCoordinateSystem As IProjectedCoordinateSystem
Dim pXYEventSourceName As IXYEventSourceName
' Set the event to fields properties.
Set pXYEvent2FieldsProperties = New XYEvent2FieldsProperties
With pXYEvent2FieldsProperties
    .XFieldName = "easting"
    .YFieldName = "northing"
    .ZFieldName = ""
End With
' Set the spatial reference.
Set pSpatialReferenceFactory = New SpatialReferenceEnvironment
Set pProjectedCoordinateSystem = _
pSpatialReferenceFactory.CreateProjectedCoordinateSystem(esriSRProjCS_NAD1927UTM_11N)
' Specify the event source and its properties.
Set pXYEventSourceName = New XYEventSourceName
With pXYEventSourceName
    Set .EventProperties = pXYEvent2FieldsProperties
    Set .SpatialReference = pProjectedCoordinateSystem
    Set .EventTableName = pTableName
End With
```

Part 2 defines the event source. First, the code creates *pXYEvent2FieldsProperties* as an instance of the *XYEvent2FieldsProperties* class and specifies the *XFiledName*, *YFieldName*, and *ZFieldName* properties on *IXYEvent2FieldsProperties* in a *With* statement. Next, the code creates a projected coordinate system referenced by *pProjectedCoordinateSystem*. The code then creates *pXYEventSourceName* as an instance of the *XYEventSourceName* class and defines the properties of *EventProperties*, *SpatialReference*, and *EventTableName* in a *With* block. The referenced objects for these three properties have all been defined previously.

```
' Part 3: Display XY events.
Dim pMxDoc As IMxDocument
Dim pMap As IMap
Dim pName As IName
Dim pXYEventSource As IXYEventSource
Dim pFLayer As IFeatureLayer
Set pMxDoc = ThisDocument
Set pMap = pMxDoc.FocusMap
' Open the events source name object.
Set pName = pXYEventSourceName
Set pXYEventSource = pName.Open
' Create a new feature layer and add the layer to the display.
Set pFLayer = New FeatureLayer
Set pFLayer.FeatureClass = pXYEventSource
pFLayer.Name = "XYEvent"
pMap.AddLayer pFLayer
End Sub
```

Part 3 opens *pXYEventSource* from *pXYEventSourceName*. An event source object can only be created from an event source name object. Finally, the code creates a new feature layer from *pXYEventSource* and adds the layer to the active map.

Box 6.4   XYEvents_GP

XYEvents_GP uses the MakeXYEventLayer tool in the Data Management toolbox to convert a text file, *events.txt*, with x-, y-coordinates into an event layer. In the macro, the tool is specified without the toolbox (Management) because the tool's name is unique. Run this macro in ArcMap and examine the output.

```
Private Sub XYEvents_GP()
    ' Run this macro in ArcMap so that the event layer can be displayed.
    ' Create the Geoprocessing object.
    Dim GP As Object
    Set GP = CreateObject("esriGeoprocessing.GpDispatch.1")
    ' MakeXYEventLayer <table> <in_x_field> <in_y_field> <out_layer> {spatial_reference}
    ' Execute the makexyeventlayer tool.
    GP.MakeXYEventLayer "c:\data\chap6\events.txt", "easting", "northing", "c:\data\chap6\XYEvent2", _
    esriSR_ProjCS_NAD1927UTM_11N
End Sub
```

CHAPTER **7**

Coordinate Systems

A coordinate system is a location reference system for spatial features on the Earth's surface. Because a geographic information system (GIS) works with geospatial data, the coordinate system plays a key role in a GIS project. There are two types of coordinate systems: geographic and projected. A geographic coordinate system consists of lines of longitude and latitude. A projected coordinate system is a plane coordinate system based on a map projection. For example, the Universal Transverse Mercator (UTM) coordinate system is based on the transverse Mercator projection.

A coordinate system, either geographic or projected, is defined by a set of parameters. The basic parameter of a geographic coordinate system is a datum, which is in turn based on a spheroid. Two commonly used datums in the United States are NAD27 (North American Datum of 1927) based on the Clarke 1866 spheroid and NAD83 (North American Datum of 1983) based on the GRS80 (Geodetic Reference System 1980) spheroid. The parameters of a projected coordinate system define not only the projection but also the geographic coordinate system that the projection is based on. For example, the NAD27UTM_11N coordinate system is based on NAD27 as well as the projection parameters of UTM_Zone_11 North, which includes the central meridian at 117°W, a scale factor of 0.9996, the latitude of projection origin at the equator, a false easting of 500,000 meters, and a false northing of 0.

ArcGIS saves the definition of a coordinate system in a spatial reference file (e.g., a prj file for a shapefile). ArcGIS considers a geographic dataset without the spatial reference information to have an unknown coordinate system. The presence of the spatial reference information is not only required for projecting a dataset from one coordinate system to another but also important for spatial analysis.

This chapter covers coordinate systems. Section 7.1 reviews commands for managing coordinate systems in ArcGIS. Section 7.2 discusses objects that are related to coordinate systems. Section 7.3 includes macros and a Geoprocessing (GP) macro for defining on-the-fly projection. Section 7.4 has macros and a GP macro for defining and importing a coordinate system. Section 7.5 includes macros for datum change. Section 7.6 discusses macros for projecting and reprojecting geographic datasets and a GP macro for projecting a shapefile. All macros start with the listing of key interfaces and key members (properties and methods) and the usage.

7.1 MANAGING COORDINATE SYSTEMS IN ARCGIS

ArcMap uses on-the-fly projection for data display. On-the-fly projection takes the spatial reference information of each dataset in a data frame and automatically converts these datasets to a common coordinate system. This common coordinate system is the coordinate system of the first layer added to the map. Alternatively, it can be set as a property of the data frame. On-the-fly projection does not actually change the original coordinate system of a dataset; it is designed for data display. If a dataset has an unknown coordinate system, ArcMap may use an assumed coordinate system. For example, NAD27 is the assumed geographic coordinate system.

On-the-fly projection cannot replace the projecting tasks. Projecting a dataset is recommended if the dataset is used frequently in a different coordinate system. Likewise, projecting the datasets to be used in spatial analysis to the same coordinate system is also recommended to obtain the most accurate results.

ArcToolbox has a suite of projection tools for working with coordinate systems. By function, these tools can be classified into the following three groups:

Defining coordinate systems
Performing geographic transformations
Projecting datasets

7.1.1 Defining Coordinate Systems

ArcToolbox has the Define Projection tools for defining coordinate systems. There are three basic methods for defining a coordinate system:

- *Select* lets the user choose a predefined coordinate system. The parameters of a predefined coordinate system are known and already coded in the software.
- *Import* lets the user copy the spatial reference information from one dataset to another.
- *New* is for a custom coordinate system. A custom geographic coordinate system requires a datum, an angular unit, and a prime meridian for its definition. A custom projected coordinate system requires the parameters of a geographic coordinate system as well as the parameters of the projected coordinate system for its definition.

7.1.2 Performing Geographic Transformations

Geographic transformation refers to the transformation or conversion from one geographic coordinate system to another. A common scenario for geographic transformation is datum change, such as changing from NAD27 to NAD83. Geographic transformation is included in the Project tools of ArcToolbox. Whenever a projection involves a datum change, a dialog appears with the datum to convert from, the datum to convert to, and a list of transformation methods. In North America, NADCON is a popular method for transformation between NAD27 and NAD83. The Project tools save new coordinates from a geographic transformation into the output dataset.

7.1.3 Projecting Datasets

ArcToolbox has the Project tools for feature and raster datasets. In both cases, the tool produces a new dataset based on the defined coordinate systems for the input and the output. Two common types of projections exist. The first type projects a dataset from a geographic to a projected coordinate system, such as from NAD27 geographic coordinates to UTM coordinates. The second type projects a dataset from one projected coordinate system to another, such as from UTM coordinates to State Plane coordinates. The second type of projection is sometimes called reprojection.

7.2 ARCOBJECTS FOR COORDINATE SYSTEMS

A primary ArcObjects component for managing coordinate systems is the *Spatial-Reference* abstract class. A spatial reference object implements *ISpatialReference*, which provides access to the spatial reference properties of a dataset, such as the name of its coordinate system.

The *SpatialReference* class is inherited by three coclasses: *GeographicCoordinate-System*, *ProjectedCoordinateSystem*, and *UnknownCoordinateSystem* (Figure 7.1). These coclasses manage geographic, projected, and unknown coordinate systems respectively.

A geographic coordinate system object implements *IGeographicCoordinateSys-tem* and *IGeographicCoordinateSystemEdit* (Figure 7.2). *IGeographicCoordinate-System* provides access to the properties of coordinate unit, datum, name, and prime meridian. *IGeographicCoordinateSystemEdit* has methods for defining the properties of a new geographic coordinate system.

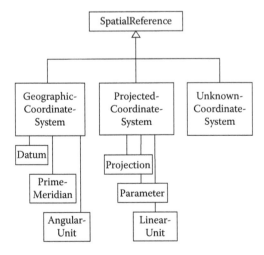

Figure 7.1   *SpatialReference* is an abstract class with three types of coordinate systems: geographic, projected, and unknown. Each geographic and projected coordinate system is associated with a set of parameters.

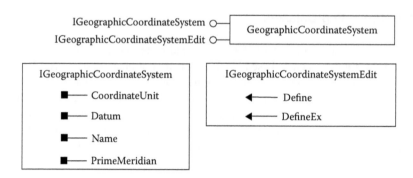

Figure 7.2   A *GeographicCoordinateSystem* object supports *IGeographicCoordinateSystem* and *IGeographicCoordinateSystemEdit*.

Likewise, a projected coordinate system object implements *IProjectedCoordinateSystem* and *IProjectedCoordinateSystemEdit* (Figure 7.3). *IProjectedCoordinateSystem* provides access to the properties of central meridian, central parallel, coordinate unit, false easting, false northing, geographic coordinate system, name, scale factor, first standard parallel, second standard parallel, and others. *IProjectedCoordinateSystemEdit* has the method to define the properties of a projected coordinate system.

Two other primary components for coordinate systems are *SpatialReferenceEnvironment* and *GeoTransformation*. The *SpatialReferenceEnvironment* coclass lets

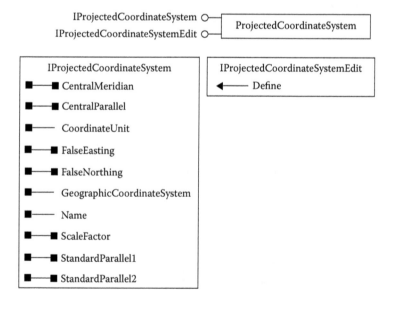

Figure 7.3   A *ProjectedCoordinateSystem* object supports *IProjectedCoordinateSystem* and *IProjectedCoordinateSystemEdit*.

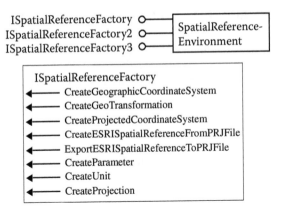

Figure 7.4   Methods on *ISpatialReferenceFactory* that a *SpatialReferenceEnvironment* object supports.

the user create coordinate systems by using predefined spatial reference objects. A spatial reference environment object implements three versions of *ISpatialReference-Factory* (Figure 7.4). *ISpatialReferenceFactory* has the methods for using predefined spatial reference objects such as NAD27 and UTM to define a coordinate system. *ISpatialReferenceFactory2* has additional methods for working with predefined geographic transformations and *ISpatialReferenceFactory3* has additional methods for a vertical coordinate system and a high precision spatial reference.

The *GeoTransformation* abstract class manages geographic transformations (Figure 7.5). Of particular interest to ArcGIS users in the United States is *GridTransformation*, a subclass of *GeoTransformation* that uses a grid-based method for converting geographic coordinates between the NAD27 and NAD83 datums.

Besides objects that are directly related to coordinate systems, macros for projection and reprojection must use other objects to create a new dataset. These additional objects deal mainly with data conversion, fields, and the geometry field. They are discussed in Section 7.5 and Section 7.6.

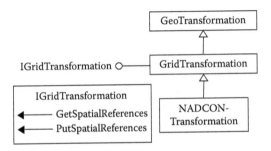

Figure 7.5   *NADCONTransformation* is a type of *GridTransformation*, which is in turn a type of *GeoTransformation*.

7.3 MANIPULATING ON-THE-FLY PROJECTION

This section offers two examples of altering the common coordinate system for on-the-fly projection. The first uses the UTM, a predefined coordinate system. The second uses the IDTM (Idaho Statewide Transverse Mercator) coordinate system, a custom coordinate system.

7.3.1 UTM_OnTheFly

UTM_OnTheFly adopts a UTM coordinate system for an active map. The macro performs the same function as using the map property to specify a UTM coordinate system for the active map. *UTM_OnTheFly* has two parts. Part 1 defines the active map, and Part 2 creates the NAD27UTM_11N coordinate system and assigns it to the active map.

> **Key Interfaces:** *ISpatialReferenceFactory, IProjectedCoordinateSystem*
> **Key Members:** *CreateProjectedCoordinateSystem, SpatialReference*
> **Usage:** Import *UTM_OnTheFly* to Visual Basic Editor in ArcMap. If necessary, insert a new data frame. Right-click the new data frame, select Properties, and make sure that the Coordinate System tab shows no projection for the current coordinate system. Run the macro. NAD_1927_UTM_Zone_11N should appear as the current coordinate system.

```
Private Sub UTM_OnTheFly()
    ' Part 1: Define the active map.
    Dim pMxDoc As IMxDocument
    Dim pMap As IMap
    Set pMxDoc = Application.Document
    Set pMap = pMxDoc.FocusMap
```

Part 1 sets *pMap* to be the focus map of the current document.

```
    ' Part 2: Create a Projected Coordinate System.
    Dim pSpatRefFact As ISpatialReferenceFactory
    Dim pPCS As IProjectedCoordinateSystem
    Set pSpatRefFact = New SpatialReferenceEnvironment
    Set pPCS = pSpatRefFact.CreateProjectedCoordinateSystem(esriSRProjCS_NAD1927UTM_11N)
    Set pMap.SpatialReference = pPCS
    ' Refresh the map.
    pMxDoc.ActiveView.Refresh
End Sub
```

Part 2 first creates *pSpatRefFact* as an instance of the *SpatialReferenceEnvironment* class. Next, the code uses the *CreateProjectedCoordinateSystem* method on *ISpatialReferenceFactory* to create the NAD27UTM_11N coordinate system, which is referenced by *pPCS*. Then the code sets *pPCS* to be the spatial reference of the active map before refreshing the active view of the map document.

7.3.2 IDTM_OnTheFly

IDTM_OnTheFly adopts IDTM as the common coordinate system for an active map. The macro performs the same function as using the map property to specify the IDTM coordinate system for the active map. ***IDTM_OnTheFly*** has three parts. Part 1 defines the projection, geographic coordinate system, and linear unit of IDTM; Part 2 stores IDTM's projection parameters in an array; and Part 3 creates IDTM and sets IDTM to be the active map's coordinate system.

> **Key Interfaces:** *ISpatialReferenceFactory2, IGeographicCoordinateSystem, IProjection, IUnit, ILinearUnit, IParameter, IProjectedCoordinateSystemEdit*
>
> **Key Members:** *CreateProjection, CreateGeographicCoordinateSystem, CreateUnit, CreateParameter, Define, SpatialReference*
>
> **Usage:** Import ***IDTM_OnTheFly*** to Visual Basic Editor in ArcMap. Run the macro. To make sure that the macro works, right-click the active map and select Properties. On the Coordinate System tab, UserDefinedPCS with the IDTM parameters should appear as the current coordinate system.

```
Private Sub IDTM_OnTheFly()
    ' Part 1: Define the active map and the IDTM coordinate system.
    Dim pMxDoc As IMxDocument
    Dim pMap As IMap
    Set pMxDoc = Application.Document
    Set pMap = pMxDoc.FocusMap
    ' Define the coordinate system's projection, geographic coordinate system (datum), and unit.
    Dim pSpatRefFact As ISpatialReferenceFactory2
    Dim pProjection As IProjection
    Dim pGCS As IGeographicCoordinateSystem
    Dim pUnit As IUnit
    Dim pLinearUnit As ILinearUnit
    Set pSpatRefFact = New SpatialReferenceEnvironment
    Set pProjection = pSpatRefFact.CreateProjection(esriSRProjection_Transverse Mercator)
    Set pGCS = pSpatRefFact.CreateGeographicCoordinateSystem(esriSRGeoCS_ NAD1983)
    Set pUnit = pSpatRefFact.CreateUnit(esriSRUnit_Meter)
    Set pLinearUnit = pUnit
```

Part 1 creates *pSpatRefFact* as an instance of the *SpatialReferenceEnvironment* class and uses methods on *ISpatialReferenceFactory2* to define the following: transverse Mercator for the projection, NAD83 for the geographic coordinate system, and meter for the linear unit. The *CreateUnit* method initially creates an *IUnit* object, but because *ILinearUnit* inherits *IUnit*, *pLinearUnit* can be set to be equal to *pUnit*.

```
    ' Part 2: Store the 5 known parameters of IDTM in an array.
    Dim aParamArray(5) As IParameter
    Set aParamArray(0) = pSpatRefFact.CreateParameter(esriSRParameter_FalseEasting)
    aParamArray(0).Value = 2500000
    Set aParamArray(1) = pSpatRefFact.CreateParameter(esriSRParameter_FalseNorthing)
    aParamArray(1).Value = 1200000
    Set aParamArray(2) = pSpatRefFact.CreateParameter(esriSRParameter_CentralMeridian)
    aParamArray(2).Value = -114
    Set aParamArray(3) = pSpatRefFact.CreateParameter(esriSRParameter_LatitudeOfOrigin)
    aParamArray(3).Value = 42
    Set aParamArray(4) = pSpatRefFact.CreateParameter(esriSRParameter_ScaleFactor)
    aParamArray(4).Value = 0.9996
```

Part 2 uses the *CreateParameter* method on *ISpatialReferenceFactory2* to store the five known parameters of IDTM in an array. Using the index and value properties of *IParameter*, the code stores the false easting of 2500000 as the first element of the array, the false northing of 1200000 as the second element, the central meridian of 114 as the third element, the latitude of origin of 42 as the fourth element, and the scale factor of 0.9996 at the central meridian as the fifth element.

```
' Part 3: Use the Define method of IProjectedCoordinateSystemEdit to create IDTM.
Dim pProjCoordSysEdit As IProjectedCoordinateSystemEdit
Set pProjCoordSysEdit = New ProjectedCoordinateSystem
pProjCoordSysEdit.Define "UserDefinedPCS", _
    "UserDefinedAlias", _
    "UsrDefAbbrv", _
    "Custom IDTM", _
    "Suitable for Idaho", _
    pGCS, _
    pLinearUnit, _
    pProjection, _
    aParamArray
' Set the map to use IDTM and refresh the map.
Set pMap.SpatialReference = pProjCoordSysEdit
pMxDoc.ActiveView.Refresh
End Sub
```

Part 3 first creates *pProjCoordSysEdit* as an instance of the *ProjectedCoordinateSystem* class. The code then uses the *Define* method on *IProjectedCoordinateSystemEdit* to define the coordinate system. The *Define* method requires nine object qualifiers and arguments, each of which is listed in a separate line for clarity. The first five arguments are the name, alias name, abbreviation, remarks, and usage of the coordinate system. The last four object qualifiers relate to the projection parameters from Part 1. After IDTM is created, the macro sets the active map to use IDTM as its spatial reference and refreshes the map.

7.4 DEFINING COORDINATE SYSTEMS

This section covers two scenarios in which the coordinate system of a geographic dataset is defined. The first uses a VBA (Visual Basic for Applications) macro to code the spatial reference information of a dataset. The information is presumably available in the dataset's metadata. The second defines the spatial reference information of a dataset by copying the information from another dataset.

A *GeoDataset* object that implements *IGeoDataset* and *IGeoDatasetSchemaEdit* is required to read or change a dataset's spatial reference (Figure 7.6). *IGeoDataset* has the read-only properties of *Extent* and *SpatialReference*. *IGeoDatasetSchemaEdit* has the members of *CanAlterSpatialReference* and *AlterSpatialReference* for editing a dataset's spatial reference. Both interfaces are used in this section.

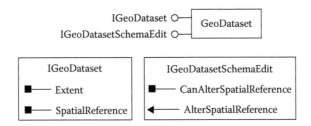

Figure 7.6   A *GeoDataset* object supports *IGeoDataset* and *IGeoDatasetSchemaEdit*.

7.4.1 *DefineGCS*

DefineGCS defines a shapefile's geographic coordinate system as NAD27. The macro performs the same function as using the Define Projection tool in ArcToolbox. *DefineGCS* has three parts. Part 1 creates the NAD27 coordinate system, Part 2 defines the input shapefile, and Part 3 alters the shapefile's spatial reference information and exports the information to a prj file.

> **Key Interfaces:** *ISpatialReferenceFactory, IGeographicCoordinateSystem, ISpatial-Reference, IGeoDataset, IGeoDatasetSchemaEdit*
>
> **Key Members:** *CreateGeographicCoordinateSystem, ExportESRISpatialReference-ToPRJFile, CanAlterSpatialReference, CreateESRISpatialReferenceFromPRJFile, AlterSpatialReference, SpatialReference*
>
> **Usage:** Add *idll.shp* to the active map in ArcMap. Without a spatial reference file, *idll* has an assumed NAD27 geographic coordinate system. Import *DefineGCS* to Visual Basic Editor in ArcMap. Run the macro. A message box verifies that GCS_North_American_1927 is the new spatial reference for *idll*. The macro also creates a prj file with the prefix of the shapefile name.

```
Private Sub DefineGCS()
    ' Part 1: Create the NAD27 coordinate system.
    Dim pSpatRefFact As ISpatialReferenceFactory
    Dim pGeogCS As IGeographicCoordinateSystem
    Dim pSpatialReference As ISpatialReference
    Set pSpatRefFact = New SpatialReferenceEnvironment
    Set pGeogCS = pSpatRefFact.CreateGeographicCoordinateSystem(esriSRGeoCS_NAD1927)
```

Part 1 creates *pSpatRefFact* as an instance of the *SpatialReferenceEnvironment* class and uses the *CreateGeographicCoordinateSystem* method to create a geographic coordinate system based on NAD27 and referenced by *pGeogCS*.

```
    ' Part 2: Define the input shapefile.
    Dim pMxDoc As IMxDocument
    Dim pMap As IMap
    Dim pFeatureLayer As IFeatureLayer
    Dim pFeatureClass As IFeatureClass
    Set pMxDoc = Application.Document
    Set pMap = pMxDoc.FocusMap
```

```
Set pFeatureLayer = pMap.Layer(0)
Set pFeatureClass = pFeatureLayer.FeatureClass
```

Part 2 defines *pFeatureLayer* to be the top layer in the active map and *pFeature-Class* to be its feature class.

```
' Part 3: Verify and alter the shapefile's spatial reference.
Dim pGeoDataset As IGeoDataset
Dim pGeoDatasetEdit As IGeoDatasetSchemaEdit
Set pGeoDatasetEdit = pFeatureClass
' Verify if the shapefile's spatial reference can be altered.
If (pGeoDatasetEdit.CanAlterSpatialReference = True) Then
     ' Alter the target layer's spatial reference.
     pGeoDatasetEdit.AlterSpatialReference pGeogCS
Else
     Exit Sub
End If
' Get the spatial reference information and export it to a prj file.
Set pGeoDataset = pFeatureClass
Set pSpatialReference = pGeoDataset.SpatialReference
pSpatRefFact.ExportESRISpatialReferenceToPRJFile "c:\data\chap7\idll", pSpatialReference
MsgBox pSpatialReference.Name
End Sub
```

Part 3 first performs a QueryInterface (QI) for the *IGeoDatasetSchemaEdit* interface and uses the *CanAlterSpatialReference* property to test whether the spatial reference of *pFeatureClass* can be altered. If it can be, the code uses the *AlterSpatialReference* method to alter the spatial reference with *pGeogCS*. If it cannot be, exit the sub. Next, the code uses the *IGeoDataset* interface to assign the spatial reference of *pFeatureClass* to *pSpatialReference*. The *ExportESRISpatialReferenceToPRJFile* method on *ISpatialReferenceFactory* then exports *pSpatialReference* to *idll.prj*. Finally, a message box displays the name of the updated spatial reference.

Box 7.1   DefineGCS_GP

DefineGCS_GP uses the DefineProjection tool in the Data Management toolbox to define the geographic coordinate system of *idll2.shp*, which has an assumed NAD27 geographic coordinate system. Run the macro in ArcCatalog and use the metadata tab to check the outcome.

```
Private Sub DefineGCS_GP()
    ' Create the Geoprocessing object.
    Dim GP As Object
    Set GP = CreateObject("esriGeoprocessing.GpDispatch.1")
    ' DefineProjection <in_dataset> <coordinate_system>
    ' Execute the defineprojection tool.
    GP.DefineProjection_management "c:\data\chap7\idll2. shp", esriSRGeoCS_NAD1927
End Sub
```

7.4.2 *CopySpatialReference*

CopySpatialReference copies the spatial reference from a source layer to a target layer. The macro performs the same function as using the Project tool in ArcToolbox to import the spatial reference from one dataset to another. *CopySpatialReference* assumes that the prj file of the source layer already exists on disk. If it does not exist, one can use a macro similar to *DefineGCS* to first create the prj file. The macro has two parts. Part 1 defines the target layer, and Part 2 verifies that the target layer's spatial reference is unknown and can be altered, before copying the prj file from the source layer.

> **Key Interfaces:** *IGeoDataset, ISpatialReference, ISpatialReferenceFactory, IGeo-DatasetSchemaEdit, IProjectedCoordinateSystem*
>
> **Key Members:** *SpatialReference, CanAlterSpatialReference, CreateESRISpatialReferenceFromPRJFile, AlterSpatialReference*
>
> **Usage:** Add *emidastrm.shp*, a stream shapefile with an unknown coordinate system, to an active map. Import *CopySpatialReference* to Visual Basic Editor in ArcMap. Run the macro. The macro creates *emidastrm.prj* by copying the prj file of *emidalat* on disk. *emidalat* is an elevation grid projected onto the NAD1927_UTM_Zone 11N coordinate system. Its prj file resides in the emidalat folder as *prj.adf*. Two messages appear during the execution of the macro: "unknown," the initial coordinate system for *emidastrm*; and "NAD_1927_UTM_11N," the new spatial reference for *emidastrm*.

```
Private Sub CopySpatialReference()
    ' Part 1: Define the target layer.
    Dim pMxDoc As IMxDocument
    Dim pMap As IMap
    Dim pTargetFL As IFeatureLayer
    Dim pTargetGD As IGeoDataset
    Set pMxDoc = ThisDocument
    Set pMap = pMxDoc.FocusMap
    Set pTargetFL = pMap.Layer(0)
    Set pTargetGD = pTargetFL.FeatureClass
```

Part 1 sets *pTargetFL* to be the top layer in the active map and *pTargetDG* to be its feature class.

```
    ' Part 2: Copy a prj file on disk to be the target layer's.
    Dim pTargetSR As ISpatialReference
    Dim pGeoDatasetEdit As IGeoDatasetSchemaEdit
    Dim pSpatRefFact As ISpatialReferenceFactory
    Dim pProjCoordSys As IProjectedCoordinateSystem
    Set pTargetSR = pTargetGD.SpatialReference
    MsgBox pTargetSR.Name
    ' Verify that the target layer has unknown spatial reference.
    If (pTargetSR.Name = "Unknown") Then
        Set pGeoDatasetEdit = pTargetGD
        ' If the spatial reference can be altered, create a prj file for the target layer.
        If (pGeoDatasetEdit.CanAlterSpatialReference = True) Then
```

```
          Set pSpatRefFact = New SpatialReferenceEnvironment
          Set pProjCoordSys = pSpatRefFact.CreateESRISpatialReferenceFromPRJFile _
          (c:\data\chap7\emidalat\prj.adf")
          ' Alter the target layer's spatial reference.
          pGeoDatasetEdit.AlterSpatialReference pProjCoordSys
          ' Get the updated spatial reference and report its name.
          Set pTargetSR = pTargetGD.SpatialReference
          MsgBox pTargetSR.Name
      Else
          Exit Sub
      End If
  Else
      Exit Sub
  End If
End Sub
```

Part 2 sets *pTargetSR* to be the spatial reference of *pTargetGD*. If the name of *pTargetSR* is unknown, the code switches to the *IGeoDatasetSchemaEdit* interface and verifies that its spatial reference can be altered. If it can be, the code uses the *CreateESRISpatialReferenceFromPRJFile* method on *ISpatialReferenceFactory* to create *pProjCoordSys* from *prj.adf*, the prj file of the source layer. The code then alters the spatial reference of *pTargetDG* with *pProjCoordSys*. A message box follows and reports the name of the target layer's new spatial reference.

7.5 PERFORMING GEOGRAPHIC TRANSFORMATIONS

A datum change may take place between two geographic coordinate systems, between a geographic and a projected coordinate system, or between two projected coordinate systems. Each scenario requires a geographic transformation. This section deals only with the geographic transformations between two geographic coordinate systems. The other two scenarios are treated as reprojection and are covered in Section 7.6.

7.5.1 *NAD27to83_Map*

NAD27to83_Map transforms the geographic coordinate system of an active map from NAD27 to NAD83. The macro performs the same function as using the map property to change the geographic coordinate system of an active map. Because on-the-fly projection includes datum transformation, NAD83 applies to existing as well as new layers in the active map. *NAD27to83_Map* has two parts. Part 1 performs the geographic transformation from NAD27 to NAD83, and Part 2 applies the transformation to the spatial reference property of the active map.

> **Key Interfaces:** *ISpatialReferenceFactory2, IGridTransformation, IGeographicCoordinateSystem*
> **Key Members:** *CreateGeoTransformation, GetSpatialReferences, SpatialReference*
> **Usage:** Right-click the active data frame in ArcMap, and select Properties. On the Coordinate System tab of the next dialog, click the Clear button to clear the current

coordinate system, if necessary. Add *idll27.shp* to the map. The map property shows GCS_North_American_1927 as the current coordinate system. Import *NAD27to83_Map* to Visual Basic Editor. Run the macro. The map's current coordinate system should now appear as GCS_North_American_1983.

```
Private Sub Nad27to83_Map()
    ' Part 1: Perform a geographic transformation from NAD27 to NAD83.
    Dim pSpatRefFact As ISpatialReferenceFactory2
    Dim pGeotransNAD27toNAD83 As IGridTransformation
    Dim pFromGCS As IGeographicCoordinateSystem
    Dim pToGCS As IGeographicCoordinateSystem
    Set pSpatRefFact = New SpatialReferenceEnvironment
    Set pGeotransNAD27toNAD83 = pSpatRefFact.CreateGeoTransformation _
    (esriSRGeoTransformation_NAD_1927_TO_NAD_1983_NADCON)
    pGeotransNAD27toNAD83.GetSpatialReferences pFromGCS, pToGCS
    MsgBox "The geographic coordinate system has been changed from " & pFromGCS.Name & "to " & _
    pToGCS.Name
```

Part 1 creates *pSpatRefFact* as an instance of the *SpatialReferenceEnvironment* class and uses the *CreateGeoTransformation* method on *ISpatialReferenceFactory2* to create *pGeotransNAD27toNAD83*, a reference to *IGridTransformation*. The transformation is grid based and uses the NADCON method. The code then uses the *GetSpatialReferences* method on *IGridTransformation* to verify the geographic coordinate systems involved in the transformation.

```
    ' Part 2: Set the spatial reference of the active map.
    Dim pMxDoc As IMxDocument
    Dim pMap As IMap
    Set pMxDoc = Application.Document
    Set pMap = pMxDoc.FocusMap
    Set pMap.SpatialReference = pToGCS
End Sub
```

Part 2 sets the spatial reference of the active map to be that of *pToGCS*, the output of the geographic transformation from Part 1.

7.5.2 *NAD27to83_Shapefile*

NAD27to83_Shapefile reprojects a shapefile based on NAD27 to a new shapefile based on NAD83. The macro performs the same function as using the Project tool in ArcToolbox to transform the shapefile from NAD27 to NAD83 coordinates. *NAD27to83_Shapefile* has six parts. Part 1 defines the input shapefile. Part 2 performs a geographic transformation from NAD27 to NAD83. Part 3 defines the output shapefile. Part 4 creates fields in the output based on the input fields. Part 5 finds the geometry field and defines the field's spatial reference and spatial index. Part 6 projects the input into the output and reports any processing errors.

Because *NAD27to83_Shapefile* creates a new shapefile from reprojection, the macro requires the use of objects that are not directly related to coordinate systems.

A *FeatureDataConverter* object implements *IFeatureDataConverter* and *IFeatureDataConverter2* (Figure 7.7). Both interfaces have methods for converting

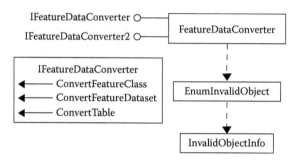

Figure 7.7   A *FeatureDataConverter* object creates an *EnumInvalidObject*, which in turn creates an *InvalidObjectInfo* object.

feature classes, feature datasets, and tables. *IFeatureDataConverter2* has the additional functionality of working with data subsets. Created by a feature data converter object, an *EnumInvalidObject* enumerator captures objects that have failed the conversion process. Created by an *EnumInvalidObject* enumerator, *InvalidObjectInfo* provides information about failed objects.

A *FieldChecker* object implements *IFieldChecker*, which has methods for creating a new set of fields from another set and for validating the fields (Figure 7.8). A field checker object creates an *EnumFieldError* enumerator and a *FieldError* object for fields that have caused problems in the conversion process.

Accessed from a field, a *GeometryDef* object defines the spatial properties of a feature class (Figure 7.9). A *GeometryDef* object implements *IGeometryDef* and *IGeometryDefEdit*. *IGeometryDef* has the read-only access to the spatial properties, whereas *IGeometryDefEdit* has the write-only access to define the spatial properties.

Key Interfaces: *ISpatialReferenceFactory2, IGridTransformation, IGeographicCoordinateSystem, IWorkspaceName, IFeatureClassName, IDatasetName, IName, IGeoDataset, IFields, IFieldChecker, IField, IGeometryDef, IGeometryDefEdit, IFeatureDataConverter*

Key Members: *CreateGeoTransformation, GetSpatialReferences, SpatialReference, WorkspaceFactoryProgID, PathName, Name, WorkspaceName, Fields, Validate, FieldCount, Field(), GeometryDef, ConvertFeatureClass*

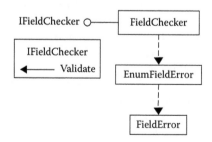

Figure 7.8   A *FieldChecker* object creates an *EnumFieldError* object, which in turn creates a *FieldError* object.

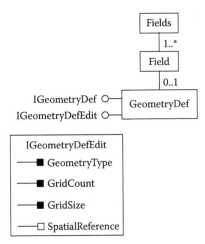

Figure 7.9   The relationship between the *Fields*, *Field*, and *GeometryDef* classes.

Usage: Import ***NAD27to83_Shapefile*** to Visual Basic Editor. Run the macro. The macro transforms *idll27.shp* in NAD27 geographic coordinates into *idll83.shp* in NAD83 geographic coordinates. Check the metadata of both shapefiles to verify the transformation. Also check the fields of *idll83* to make sure that they are identical to those of *idll27*.

```
Private Sub NAD27to83_Shapefile()
    ' Part 1: Define the shapefile.
    Dim plnWSName As IWorkspaceName
    Dim plnFCName As IFeatureClassName
    Dim plnDatasetName As IDatasetName
    Dim pName As IName
    Dim plnFC As IFeatureClass
    ' Set the input shapefile's workspace and path names.
    Set plnWSName = New WorkspaceName
    plnWSName.WorkspaceFactoryProgID = "esriCore.ShapefileWorkspaceFactory.1"
    plnWSName.PathName = "c:\data\chap7"
    ' Set the input feature class and dataset names.
    Set plnFCName = New FeatureClassName
    Set plnDatasetName = plnFCName
    plnDatasetName.Name = "idll27"
    Set plnDatasetName.WorkspaceName = plnWSName
    ' Set the input geodataset.
    Set pName = plnFCName
    Set plnFC = pName.Open
```

Part 1 uses the name objects to define the input shapefile's workspace, feature class, and name. The code creates *pInFCName* as an instance of the *FeatureClass-Name* class and performs a QI for the *IDatasetName* interface to define its name and workspace. Next, the code sets *pName*, a reference to *IName*, to be *pInFCName*, and uses the *Open* method on *IName* to open *pInFC*.

```
' Part 2: Perform a geographic transformation from NAD27 to NAD83.
Dim pSpatRefFact As ISpatialReferenceFactory2
Dim pGeotransNAD27toNAD83 As IGridTransformation
Dim pFromGCS As IGeographicCoordinateSystem
Dim pToGCS As IGeographicCoordinateSystem
Set pSpatRefFact = New SpatialReferenceEnvironment
Set pGeotransNAD27toNAD83 = pSpatRefFact.CreateGeoTransformation _
(esriSRGeoTransformation_NAD_1927_TO_NAD_1983_NADCON)
' Verify the From and To geographic coordinate systems.
pGeotransNAD27toNAD83.GetSpatialReferences pFromGCS, pToGCS
MsgBox "The geographic transformation is from " & pFromGCS.Name & " to " & pToGCS.Name
```

Part 2 performs a geographic transformation from NAD27 to NAD83. The code creates *pSpatRefFact* as an instance of the *SpatialReferenceEnvironment* class and uses the *CreateGeoTransformation* method on *ISpatialReferenceFactory2* to create *pGeotransNAD27toNAD83*. The code then uses the *GetSpatialReferences* method on *IGridTransformation* to return the input and output geographic coordinate systems, which are referenced by *pFromGCS* and *pToGCS* respectively.

```
' Part 3: Define the output shapefile.
Dim pOutWSName As IWorkspaceName
Dim pOutFCName As IFeatureClassName
Dim pOutDataSetName As IDatasetName
' Set the output shapefile's workspace and path names.
Set pOutWSName = New WorkspaceName
pOutWSName.WorkspaceFactoryProgID = "esriCore.ShapeFileWorkspaceFactory.1"
pOutWSName.PathName = "c:\data\chap7"
' Set the output feature class and dataset names.
Set pOutFCName = New FeatureClassName
Set pOutDataSetName = pOutFCName
Set pOutDataSetName.WorkspaceName = pOutWSName
pOutDataSetName.Name = "idll83"
```

Part 3 defines the output shapefile's workspace, feature class, and name in the same way as Part 1. Only the name objects are used.

```
' Part 4: Create the output fields based on the input's fields.
Dim pOutFCFields As IFields
Dim pInFCFields As IFields
Dim pFieldCheck As IFieldChecker
Dim i As Long
Set pInFCFields = pInFC.Fields
Set pFieldCheck = New FieldChecker
pFieldCheck.Validate pInFCFields, Nothing, pOutFCFields
```

Part 4 creates *pFieldCheck* as an instance of the *FieldChecker* class and uses the *Validate* method on *IFieldChecker* to create the output fields referenced by *pOut-FCFields* from the input fields referenced by *pInFCFields*. Because the input and output are both shapefiles, the code does not have to perform error checking using the *EnumFieldError* and *FieldError* objects.

```
' Part 5: Locate and define the geometry field.
Dim pGeoField As IField
Dim pOutFCGeoDef As IGeometryDef
Dim pOutFCGeoDefEdit As IGeometryDefEdit
' Find the index of the geometry field.
For i = 0 To pOutFCFields.FieldCount - 1
    If pOutFCFields.Field(i).Type = esriFieldTypeGeometry Then
        Set pGeoField = pOutFCFields.Field(i)
        Exit For
    End If
Next i
' Get the geometry field's geometry definition.
Set pOutFCGeoDef = pGeoField.GeometryDef
' Define the spatial index and spatial reference.
Set pOutFCGeoDefEdit = pOutFCGeoDef
pOutFCGeoDefEdit.GridCount = 1
pOutFCGeoDefEdit.GridSize(0) = 200
Set pOutFCGeoDefEdit.SpatialReference = pToGCS
```

Part 5 locates and defines the geometry field. The code loops through *pOut-FCFields*, checks for the geometry type, and saves the geometry field, referenced by *pGeoField*, by its index value. Next, the code sets *pOutFCGeoDef* to be the geometry definition of *pGeoField* and uses the *IGeometryDefEdit* interface to set its properties of *GridCount*, *GridSize*, and *SpatialReference*. Grid count and grid size make up the spatial index of the output shapefile, which is designed to improve the display, spatial query, and feature identification of the shapefile. In this case, the spatial index has one grid and the grid size of 200. The spatial reference assigned to the output shapefile is *pToGCS*, a reference to NAD83.

```
' Part 6: Perform data conversion and error checking.
Dim pFDConverter As IFeatureDataConverter
Dim pEnumErrors As IEnumInvalidObject
Dim pErrInfo As InvalidObjectInfo
' Perform feature data conversion.
Set pFDConverter = New FeatureDataConverter
Set pEnumErrors = pFDConverter.ConvertFeatureClass(pInFCName, Nothing, Nothing, _
pOutFCName, pOutFCGeoDef, pOutFCFields, "", 1000, 0)
' If an error exists, report an error message in the immediate window.
Set pErrInfo = pEnumErrors.Next
If Not pErrInfo Is Nothing Then
    Debug.Print "Conversion completed with errors"
Else
    Debug.Print "Conversion completed"
End If
End Sub
```

Part 6 performs data conversion and error checking. The code first creates *pFDConverter* as an instance of the *FeatureDataConverter* class, and uses the *ConvertFeatureClass* method to import data from the input feature class to the output. The method uses qualifying objects that have been previously defined in

Parts 4 and 5. Additionally, the code uses a flush interval of 1000, which controls the interval for committing data. (The flush interval is important for loading large amounts of data into a geodatabase.) Finally, the code checks for invalid features during the conversion process. If an invalid feature exists, the Immediate window displays an error message. If not, the window displays the message of "Conversion completed."

7.6 PROJECTING DATASETS

Similar to use of the Project tools in ArcToolbox, projection using a VBA macro also follows the sequence of defining the input's coordinate system and the output's coordinate system. Additionally, the code must be able to handle the creation of the output dataset, which includes copying fields from the input dataset and defining the geometry field.

7.6.1 *ProjectShapefile*

ProjectShapefile projects a shapefile from NAD27 geographic coordinates to NAD27UTM_11N projected coordinates. The module performs the same function as using the Project tool in ArcToolbox to project a shapefile. ***ProjectShapefile*** has four parts. Part 1 defines the input shapefile and its spatial reference. Part 2 creates the output fields based on the input fields. Part 3 finds the geometry field and defines the field's spatial reference and spatial index. Part 4 projects the input into the output and reports any processing errors. ***ProjectShapefile*** is similar to ***NAD27to83_Shapefile*** in terms of programming except that ***ProjectShapefile*** uses functions to define the name objects of workspace and dataset, one for the input and the other for the output.

> **Key Interfaces:** *IWorkspaceName, IFeatureClassName, IDatasetName, IName, IGeo-Dataset, IGeographicCoordinateSystem, ISpatialReferenceFactory, IFields, IField, IFieldChecker, IGeometryDef, IProjectedCoordinateSystem, IGeometryDefEdit, IFeatureDataConverter*
> **Key Members:** *WorkspaceFactoryProgID, PathName, Name, WorkspaceName, CreateGeographicCoordinateSystem, SpatialReference, Fields, Validate, FieldCount, Field(), GeometryDef, CreateProjectedCoordinateSystem, ConvertFeatureClass*
> **Usage:** Import ***ProjectShapefile*** to Visual Basic Editor. Run the module. The module projects *idll27.shp* in NAD27 geographic coordinates into *idutm27.shp* in NAD27UTM_11N projected coordinates. Check the metadata of the shapefiles to verify the result.

```
Private Sub ProjectShapefile()
    ' Part 1: Define the input shapefile.
    Dim pInFCName As IFeatureClassName
    Dim pSpatRefFact As ISpatialReferenceFactory
    Dim pName As IName
    Dim pInFC As IFeatureClass
    ' Use the InputName function to find the input shapefile.
    Set pInFCName = InputName("c:\data\chap7", "idll27")
```

```
Set pName = pInFCName
Set pInFC = pName.Open
```

Part 1 locates the input shapefile. The code passes the input shapefile's workspace and name to the ***InputName*** function, runs the function, and gets *pInFCName*, a reference to *IFeatureClassName*, from the function. The code then opens *pInFC-Name* to get *pInFC*.

```
' Part 2: Create the output fields based on the input's fields.
Dim pOutFCFields As IFields
Dim pInFCFields As IFields
Dim pFieldCheck As IFieldChecker
Dim i As Long
Set pInFCFields = pInFC.Fields
Set pFieldCheck = New FieldChecker
pFieldCheck.Validate pInFCFields, Nothing, pOutFCFields
```

Part 2 creates *pFieldCheck* as an instance of the *FieldChecker* class and uses the *Validate* method to create *pOutFCFields* from *pInFCFields*.

```
' Part 3: Locate and define the geometry field.
Dim pGeoField As IField
Dim pOutFCGeoDef As IGeometryDef
Dim pPCS As IProjectedCoordinateSystem
Dim pOutFCGeoDefEdit As IGeometryDefEdit
' Loop through the fields to locate the geometry field.
For i = 0 To pOutFCFields.FieldCount - 1
    If pOutFCFields.Field(i).Type = esriFieldTypeGeometry Then
        Set pGeoField = pOutFCFields.Field(i)
        Exit For
    End If
Next i
' Set the output's coordinate system.
Set pSpatRefFact = New SpatialReferenceEnvironment
Set pPCS = pSpatRefFact.CreateProjectedCoordinateSystem(esriSRProjCS_NAD1927UTM_11N)
' Get and edit the geometry field's geometry definition.
Set pOutFCGeoDef = pGeoField.GeometryDef
Set pOutFCGeoDefEdit = pOutFCGeoDef
pOutFCGeoDefEdit.GridCount = 1
pOutFCGeoDefEdit.GridSize(0) = 200
Set pOutFCGeoDefEdit.SpatialReference = pPCS
```

Part 3 locates the geometry field referenced by *pGeoField* by looping through the fields in *pOutFCFields*. Next, the code creates *pPCS* as a new projected coordinate system based on NAD1927UTM_11N. The code then sets *pOutFCGeoDef* to be the geometry definition of *pGeoField*, and performs a QI for the *IGeometry-DefEdit* interface to edit its spatial index and spatial reference.

```
' Part 4: Create the output, and report any errors in creating the output.
Dim pFDConverter As IFeatureDataConverter
```

```
    Dim pOutFCName As IFeatureClassName
    Dim pEnumErrors As IEnumInvalidObject
    Dim pErrInfo As IInvalidObjectInfo
    Set pFDConverter = New FeatureDataConverter
    Set pOutFCName = OutputName("c:\data\chap7", "idutm27")
    Set pEnumErrors = pFDConverter.ConvertFeatureClass(pInFCName, Nothing, Nothing, pOutFCName, _
    pOutFCGeoDef, pOutFCFields, "", 1000, 0)
    ' If an error exists, show an error message.
    Set pErrInfo = pEnumErrors.Next
    If Not pErrInfo Is Nothing Then
        Debug.Print "Conversion completed with errors"
    Else
        Debug.Print "Conversion completed"
    End If
End Sub
```

Part 4 creates the output shapefile and reports any errors in data conversion. The code first runs the *OutputName* function to define the output shapefile. Then the code uses the *ConvertFeatureClass* method on *IFeatureDataConverter* to create *pOutFCName*, a reference to *IFeatureClassName*. The Immediate window reports any invalid object during the conversion.

```
Private Function InputName(InWSPath As String, InDataset As String) As IFeatureClassName
    Dim pInWSName As IWorkspaceName
    Dim pInFCName As IFeatureClassName
    Dim pInDatasetName As IDatasetName
    ' Set the input shapefile's workspace and path names.
    Set pInWSName = New WorkspaceName
    pInWSName.WorkspaceFactoryProgID = "esriCore.ShapefileWorkspaceFactory.1"
    pInWSName.PathName = InWSPath
    ' Set the input feature class and name.
    Set pInFCName = New FeatureClassName
    Set pInDatasetName = pInFCName
    pInDatasetName.Name = InDataset
    Set pInDatasetName.WorkspaceName = pInWSName
    Set InputName = pInFCName
End Function
```

The *InputName* function receives the names of the input shapefile and its workspace path as strings, and returns *pInFCName*, a reference to an *IFeatureClassName* object, to *ProjectShapefile*.

```
Private Function OutputName(OutWSPath As String, OutDataset As String) As IFeatureClassName
    Dim pOutWSName As IWorkspaceName
    Dim pOutFCName As IFeatureClassName
    Dim pOutDatasetName As IDatasetName
    ' Set the output shapefile's workspace and path names.
    Set pOutWSName = New WorkspaceName
    pOutWSName.WorkspaceFactoryProgID = "esriCore.ShapeFileWorkspaceFactory.1"
    pOutWSName.PathName = OutWSPath
```

```
' Set the output feature class and name.
Set pOutFCName = New FeatureClassName
Set pOutDatasetName = pOutFCName
Set pOutDatasetName.WorkspaceName = pOutWSName
pOutDatasetName.Name = OutDataset
Set OutputName = pOutFCName
End Function
```

The ***OutputName*** function receives the names of the output shapefile and its workspace path, and returns *pOutFCName*, a reference to an *IFeatureClassName* object, to ***ProjectShapefile***.

Box 7.2   ProjectShapefile_GP

ProjectShapefile_GP uses the Project tool in the Data Management toolbox to project *idll27_2.shp* from NAD27 geographic coordinates to NAD27UTM_11N projected coordinates and save the output as *idutm27_2.shp*. Run the macro in ArcCatalog and verify the result by using the metadata tab.

```
Private Sub ProjectShapefile_GP()
    ' Create the Geoprocessing object.
    Dim GP As Object
    Set GP = CreateObject("esriGeoprocessing.GpDispatch.1")
    ' Project <in_dataset> <out_dataset> <out_coordinate_system>
    ' {transform_method:transform_method}
    ' Execute the project tool.
    GP.Project_management "c:\data\chap7\idll27.shp", "c:\data\chap7\idutm27_2.shp", _
    esriSRProjCS_NAD1927UTM_11N
End Sub
```

7.6.2   Use of a Different Datum

The only change needed for projecting a shapefile from NAD27 to NAD83UTM_11N coordinates is the line statement for setting the output file's spatial reference in Part 4 of ***ProjectShapefile***:

Set pPCS = pSpatRefFact.CreateProjectedCoordinateSystem(esriSRProjCS_NAD-1983UTM_11N)

7.6.3   *ReprojectShapefile*

ReprojectShapefile reprojects a shapefile from one projected coordinate system to another. Specifically, it reprojects a shapefile from NAD27UTM_11N to IDTM coordinates. The module performs the same function as using the Project tool in ArcToolbox to project a shapefile.

ReprojectShapefile has five parts. Additionally, the module uses a function to define IDTM. Part 1 defines the input shapefile and its spatial reference. Part 2 defines the output shapefile. Part 3 creates the output fields based on the input fields. Part 4 finds the geometry field and defines the field's spatial reference and spatial index. Part 5 projects the input into the output and reports any processing errors. Many line statements in **ReprojectShapefile** have been used in previous sample macros. Therefore, it does not require a detailed explanation of its code structure.

Key Interfaces: *IWorkspaceName, IFeatureClassName, IDatasetName, IProjected-CoordinateSystem, ISpatialReferenceFactory, IName, IGeoDataset, IFields, IField, IFieldChecker, IGeometryDef, IGeometryDefEdit, IFeatureDataConverter, IProjection, IGeographicCoordinateSystem, IProjectedCoordinateSystemEdit*

Key Members: *WorkspaceFactoryProgID, PathName, Name, WorkspaceName, SpatialReference, Fields, Validate, FieldCount, GeometryDef, Field( ), CreateProjected CoordinateSystem, ConvertFeatureClass, CreateProjection, CreateGeographicCoordinateSystem, CreateParameter*

Usage: Import **ReprojectShapefile** to Visual Basic Editor. Run the macro. The macro projects *idutm27.shp* in NAD27UTM_11N projected coordinates into *idtm.shp* in IDTM projected coordinates. (*Idutm27.shp* was created in Section 7.6.1.) Check the metadata of the shapefiles to verify the result.

```
Private Sub ReprojectShapefile()
    ' Part 1: Define the input shapefile and its spatial reference.
    Dim pInWSName As IWorkspaceName
    Dim pInFCName As IFeatureClassName
    Dim pInDatasetName As IDatasetName
    Dim pInCS As IProjectedCoordinateSystem
    Dim pSpatRefFact As ISpatialReferenceFactory
    Dim pName As IName
    Dim pInFC As IFeatureClass
    Dim pInGeoDataset As IGeoDataset
    ' Define the input shapefile's workspace and path.
    Set pInWSName = New WorkspaceName
    pInWSName.WorkspaceFactoryProgID = "esriCore.ShapefileWorkspaceFactory.1"
    pInWSName.PathName = "c:\data\chap7"
    ' Define the input feature class.
    Set pInFCName = New FeatureClassName
    Set pInDatasetName = pInFCName
    pInDatasetName.Name = "idutm27"
    Set pInDatasetName.WorkspaceName = pInWSName
    ' Define the input shapefile's projected coordinate system.
    Set pSpatRefFact = New SpatialReferenceEnvironment
    Set pInCS = pSpatRefFact.CreateProjectedCoordinateSystem(esriSRProjCS_NAD1927UTM_11N)
    ' Assign the spatial reference to the input shapefile.
    Set pName = pInFCName
    Set pInFC = pName.Open
    Set pInGeoDataset = pInFC
    Set pInCS = pInGeoDataset.SpatialReference
```

Part 1 sets *pInFC* to be the input shapefile and *pInCS* to be its spatial reference. The code defines *pInCS* as the NAD27UTM_11N projected coordinate system.

```
    ' Part 2: Define the output shapefile.
    Dim pOutWSName As IWorkspaceName
```

```
Dim pOutFCName As IFeatureClassName
Dim pOutDatasetName As IDatasetName
' Set the output shapefile's workspace and path names.
Set pOutWSName = New WorkspaceName
pOutWSName.WorkspaceFactoryProgID = "esriCore.ShapeFileWorkspaceFactory.1"
pOutWSName.PathName = "c:\data\chap7"
' Set the output feature class and dataset names.
Set pOutFCName = New FeatureClassName
Set pOutDatasetName = pOutFCName
Set pOutDatasetName.WorkspaceName = pOutWSName
pOutDatasetName.Name = "idtm"
```

Part 2 defines the workspace and name of the output shapefile.

```
' Part 3: Create the output fields based on the input's fields.
Dim pOutFCFields As IFields
Dim pInFCFields As IFields
Dim pFieldCheck As IFieldChecker
Dim i As Long
Set pInFCFields = pInFC.Fields
Set pFieldCheck = New FieldChecker
pFieldCheck.Validate pInFCFields, Nothing, pOutFCFields
```

Part 3 creates the fields for the output based on the fields of the input shapefile.

```
' Part 4: Locate and define the geometry field.
Dim pGeoField As IField
Dim pOutFCGeoDef As IGeometryDef
Dim pPCS As IProjectedCoordinateSystem
Dim pOutFCGeoDefEdit As IGeometryDefEdit
' Loop through the fields to locate the geometry field.
For i = 0 To pOutFCFields.FieldCount - 1
    If pOutFCFields.Field(i).Type = esriFieldTypeGeometry Then
        Set pGeoField = pOutFCFields.Field(i)
        Exit For
    End If
Next i
' Use the OutCS function to define the output's coordinate system.
Set pPCS = OutCS()
' Get and edit the geometry field's geometry definition.
Set pOutFCGeoDef = pGeoField.GeometryDef
Set pOutFCGeoDefEdit = pOutFCGeoDef
pOutFCGeoDefEdit.GridCount = 1
pOutFCGeoDefEdit.GridSize(0) = 200
Set pOutFCGeoDefEdit.SpatialReference = pPCS
```

Part 4 loops through the output fields, locates the geometry field, and defines the geometry definition of the field. Then the code calls the *OutCS* function to get the projected coordinate system for the output.

```
' Part 5: Create the output, and report any errors in creating the output.
Dim pFDConverter As IFeatureDataConverter
Dim pEnumErrors As IEnumInvalidObject
Dim pErrInfo As IInvalidObjectInfo
Set pFDConverter = New FeatureDataConverter
Set pEnumErrors = pFDConverter.ConvertFeatureClass(pInFCName, Nothing, Nothing, pOutFCName, _
pOutFCGeoDef, pOutFCFields, "", 1000, 0)
' If an error exists, show an error message.
' pEnumErrors.Reset
Set pErrInfo = pEnumErrors.Next
If Not pErrInfo Is Nothing Then
    Debug.Print "Conversion completed with errors"
Else
    Debug.Print "Conversion completed"
End If
End Sub
```

Part 5 creates the output shapefile and reports any errors in creating the output.

```
Private Function OutCS() As IProjectedCoordinateSystem
    Dim pSpatRefFact As ISpatialReferenceFactory2
    Dim pProjection As IProjection
    Dim pGCS As IGeographicCoordinateSystem
    Dim pUnit As IUnit
    Dim pLinearUnit As ILinearUnit
    Dim aParamArray(5) As IParameter
    Dim pProjCoordSysEdit As IProjectedCoordinateSystemEdit
    Dim pProjCoordSys As IProjectedCoordinateSystem
    ' Define the IDTM Coordinate System.
    Set pSpatRefFact = New SpatialReferenceEnvironment
    Set pProjection = pSpatRefFact.CreateProjection(esriSRProjection_TransverseMercator)
    Set pGCS = pSpatRefFact.CreateGeographicCoordinateSystem(esriSRGeoCS_NAD1983)
    Set pUnit = pSpatRefFact.CreateUnit(esriSRUnit_Meter)
    Set pLinearUnit = pUnit
    ' Store the 5 known parameters of IDTM in an array.
    Set aParamArray(0) = pSpatRefFact.CreateParameter(esriSRParameter_FalseEasting)
    aParamArray(0).Value = 2500000
    Set aParamArray(1) = pSpatRefFact.CreateParameter(esriSRParameter_FalseNorthing)
    aParamArray(1).Value = 1200000
    Set aParamArray(2) = pSpatRefFact.CreateParameter(esriSRParameter_CentralMeridian)
    aParamArray(2).Value = -114
    Set aParamArray(3) = pSpatRefFact.CreateParameter(esriSRParameter_LatitudeOfOrigin)
    aParamArray(3).Value = 42
    Set aParamArray(4) = pSpatRefFact.CreateParameter(esriSRParameter_ScaleFactor)
    aParamArray(4).Value = 0.9996
    ' Create IDTM by defining its properties.
    Set pProjCoordSysEdit = New ProjectedCoordinateSystem
    pProjCoordSysEdit.Define "UserDefinedPCS", _
        "UserDefinedAlias", _
        "UsrDefAbbrv", _
```

```
        "Custom IDTM", _
        "Suitable for Idaho", _
        pGCS, _
        pLinearUnit, _
        pProjection, _
        aParamArray
    Set OutCS = pProjCoordSysEdit
End Function
```

The **OutCS** function creates *pProjCoordSysEdit* as an instance of the *ProjectedCoordinateSystem* class and defines its properties, including the geographic coordinate system, the linear unit, and five known parameters for the projection. The function then returns the defined projected coordinate system to **ReprojectShapefile**.

Box 7.3   ReprojectShapefile_GP

ReProjectShapefile_GP uses the Project tool in the Data Management toolbox to project *idutm27_2.shp* from NAD27 UTM_11N coordinates to NAD83 UTM_11N coordinates and save the output as *idutm83_2.shp*. The macro uses NAD_1927_-to_NAD_1983_NADCON for the geographic transformation. Run the macro in ArcCatalog.

```
Private Sub ReprojectShapefile_GP()
    ' Create the Geoprocessing object.
    Dim GP As Object
    Set GP = CreateObject("esriGeoprocessing.GpDispatch.1")
    ' Project <in_dataset> <out_dataset> <out_coordinate_system>
    ' {transform_method:transform_method}
    ' Notice that the transform method is NAD_1927_to_NAD_1983_NADCON.
    ' Execute the project tool.
    GP.project_management "c:\data\chap7\idutm27_2.shp", "c\data\chap7\idutm83_2.shp", _
    "Coordinate Systems\Projected Coordinate Systems\Utm\Nad 1983\NAD 1983 UTM Zone 11N.prj", _
    "NAD_1927_to_NAD_1983_NADCON"
End Sub
```

Data Display

Spatial features are characterized by their locations and attributes. Data display involves choice of symbols to show attribute data at the locations of spatial features. Cartographers consider symbol types and visual variables for choice of symbols. Symbol types correspond to feature types: point symbols for point features, line symbols for line features, and area symbols for area features. Visual variables, which include color, size, texture, shape, and pattern, distinguish between symbols and communicate data characteristics to the viewer.

Color is a popular visual variable but is often misused. A color has the three visual dimensions of hue, value, and chroma. Hue is the quality that distinguishes one color from another, value is the lightness or darkness of a color, and chroma refers to the richness of a color. The use of color and its visual dimensions depends on the type of data to be displayed. Cartographic studies have shown that hue is a visual variable better suited for qualitative or categorical data, whereas value and chroma are better suited for quantitative data. Cartographic studies have also recommended a number of conventional color schemes for mapping quantitative data.

Layout design is part of map design. A map requires a title, a legend, a north arrow, a scale bar, and other elements to communicate the map information. The task of layout design is to arrange these various elements on a map so that the map would look balanced and organized to the viewer. Cartographers used to use thumbnail sketches to experiment with layout design. Now the experimentation can be easily performed on the computer monitor.

This chapter covers data display with emphasis on symbology, color, and layout. Section 8.1 reviews data display options in ArcGIS. Section 8.2 discusses objects that are related to various aspects of data display. Section 8.3 offers macros for displaying vector data. Section 8.4 includes macros for displaying raster data. Section 8.5 discusses a macro for making a layout page. All macros start with the listing of key interfaces and key members (properties and methods) and the usage.

8.1 DISPLAYING DATA IN ARCGIS

ArcMap provides many data display options on the Symbology tab of the Layer Properties dialog. These options depend on the data type. The primary data types are vector and raster data. At the secondary level, vector data include point, line, and area features and raster data include categorical and numeric data.

8.1.1 Displaying Vector Data

For vector data, the display options include Features, Categories, Quantities, Charts, and Multiple Attributes. The features option draws all features with a single symbol. The categories option displays unique values from a field or multiple fields. The quantities option offers graduated colors, graduated symbols, proportional symbols, and dot density. Charts include pie, bar/column, and stacked charts. The multiple attributes option uses more than one attribute for data display. ArcMap initially assigns default symbols to a data display option: point symbols for point features, line symbols for line features, and area symbols for area features. The user can alter these default symbols in terms of symbol type, color, size, pattern, and other visual variables.

8.1.2 Displaying Raster Data

For raster data, the display options include Unique Values, Classified, and Stretched. The unique values option displays unique cell values of a raster. The classified option displays classes of cell values. The stretched option stretches cell values to increase the visual contrast in data display. ArcMap initially assigns a default set of area (fill) symbols to a display option. The user can alter these symbols individually or as a group.

8.1.3 Use of Color Ramp and Classification Tools

Color ramp and classification are two tools on the Symbology tab for data display. A color ramp represents a range of distinctive or sequential colors. ArcMap offers a series of predefined color ramps, and the user can choose a color ramp graphically or by description (for example, yellow to dark red). Each color ramp has a properties dialog that allows the user to select the two end colors and an algorithm for producing the intermediate colors between them. The classification tool lets the user choose number of classes and method for subdividing a dataset into classes. Available classification methods are manual, equal interval, defined interval, quantile, natural breaks, and standard deviation. The natural breaks method, which classifies data values into natural groupings statistically, is the default method.

8.1.4 Designing a Layout

ArcMap offers two options for layout design. The first option is to use a layout template. Current layout templates are grouped into general, industry, USA, and world. Each group has a list of choices. For example, the layout templates for the United States include USA, conterminous USA, and five different regions of the country. The second option is to open a layout page and build on it one map element at a time. These map elements can be graphically manipulated for size change, repositioning, and other modifications on the layout page.

8.2 ARCOBJECTS FOR DATA DISPLAY

A data display macro, especially a layout macro, tends to involve more objects than other types of macros do. This section presents a summary of objects that are important to data display.

8.2.1 Renderer Objects

ArcObjects uses the term *renderer* to describe a set of symbols for displaying data values. A renderer is therefore like a legend. ArcObjects has two general (abstract) classes of renderers: *FeatureRenderer* for vector data and *RasterRenderer* for raster data.

A variety of feature renderers inherit the functionality of the *FeatureRenderer* class (Figure 8.1). Two of these renderers to be covered in the sample macros are *UniqueValueRenderer* and *ClassBreaksRenderer*. A unique value renderer uses a different symbol for each unique value, which may come from a field or a combination of fields. A class breaks renderer uses a different symbol for each class of data values.

ArcObjects organizes feature-based symbols into three general classes: *Marker-Symbol* for point features, *LineSymbol* for line features, and *FillSymbol* for area features (Figure 8.2). Each of these abstract symbol classes is inherited by a number of coclasses. Of these coclasses, *SimpleMarkerSymbol*, *SimpleLineSymbol*, and *SimpleFillSymbol* can generate the commonly used point, line, and area symbols respectively.

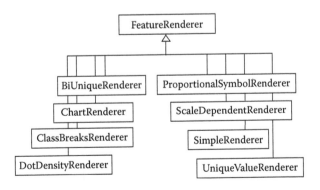

Figure 8.1 *FeatureRenderer* is an abstract class with many feature-based renderer types, each of which is a coclass.

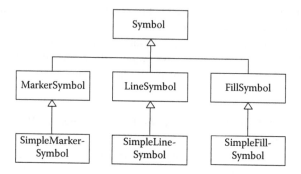

Figure 8.2 *SimpleMarkerSymbol* is a type of *MarkerSymbol*, which is in turn a type of *Symbol*. *SimpleMarkerSymbol* is a coclass, whereas both *MarkerSymbol* and *Symbol* are abstract classes. The same is true for the other two feature-based symbols.

The *RasterRenderer* abstract class is inherited by four coclasses: *RasterUniqueValueRenderer*, *RasterClassifyColorRampRenderer*, *RasterStretchColorRampRenderer*, and *RasterRGBRenderer*. As suggested by the name, the first three renderers are used for the display options of unique values, classified, and stretched respectively. A raster RGB (red, green, and blue) renderer object is designed for multiband data such as satellite images. The only symbol option for displaying cell-based raster data is the fill symbol.

8.2.2 Classification Objects

A macro for displaying classified data requires a classification object. ArcObjects offers five predefined classification objects (Figure 8.3):

- A *DefinedInterval* object uses a defined and precise interval such as 100 or 1,000.
- An *EqualInterval* object produces classes with an equal interval.
- A *NaturalBreaks* object uses a statistical method to create classes with natural breaks between them.

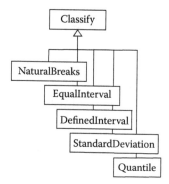

Figure 8.3 *Classify* is an abstract class with five classification types, each of which is a coclass.

- A *Quantile* object creates classes with an equal number of values in each class.
- A *StandardDeviation* object produces classes that are based on one whole or part of a standard deviation from the mean.

In addition to these predefined classification objects, one can also use a user-defined classification.

The use of a predefined classification object typically involves a *TableHistogram* object. A *TableHistogram* object implements *IHistogram* and *ITableHistogram*. These two interfaces have members to gather histogram data from a table, such as data values and frequencies, and to pass the histogram data to a classification object. The classification object can then use the histogram data to compute the class breaks.

8.2.3 Color Ramp and Color Objects

A color ramp is a collection of colors. ArcObjects has four coclasses of color ramps (Figure 8.4):

- A *RandomColorRamp* object creates a series of randomized colors.
- An *AlgorithmicColorRamp* object produces a sequential series of colors using two end colors and a defined algorithm.
- A *PresetColorRamp* object is a series of 13 specific colors.
- A *MultiPartColorRamp* object is a collection of color ramps.

The algorithms available for generating intermediate colors in an algorithmic color ramp are: esriHSVAlgorithm, esriCIELabAlgorithm, and esriLabLChAlgorithm. A color ramp produced by either the esriCIELab or esriLabLCh algorithm appears to blend the two end colors, whereas a color ramp produced by the esriHSV algorithm may contain additional hues.

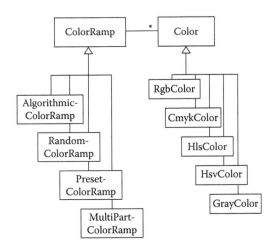

Figure 8.4   The relationship between the *ColorRamp* and *Color* classes, and the subtypes of each class.

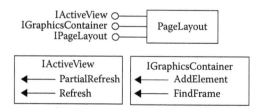

Figure 8.5 A *PageLayout* object supports *IActiveView*, *IGraphicsContainer*, and *IPageLayout*.

ArcObjects uses a color model to define colors in a color ramp (Figure 8.4). For example, a bright yellow color is represented by (255, 255, 0) in the RGB model. Besides RGB, ArcObjects has CMYK (cyan, magenta, yellow, and black), HLS (hue, lightness, and saturation), HSV (hue, saturation, and value), and Grayscale. Additionally, ArcObjects offers CIELAB as a device-independent color model and uses it to store colors internally.

8.2.4 Layout Objects

A layout consists of various map elements. Each element has its own display style and its own position on a layout page. Because of the complexity of the topic, this section limits the discussion of layout objects to only those that are used later in a sample module.

The primary component that works with a layout is the *PageLayout* coclass. A page layout object implements *IActiveView*, *IGraphicsContainer*, and *IPageLayout* (Figure 8.5). *IActiveView* has methods to refresh, or partially refresh, the layout view. *IGraphicsContainer* has methods for adding and finding elements in a layout. *IPageLayout* has members that are mainly important to the interaction between the user and a layout.

The *IGraphicsContainer* interface manages two types of elements in a layout: frame elements and text elements (Figure 8.6). Both types of elements implement *IElement*, which has access to the shape and the screen display of the element.

A frame element refers to an object that forms a border around other elements or objects. Two frame elements important to a layout design are *MapFrame* and *MapSurroundFrame* (Figure 8.7). A map frame object implements *IMapFrame*, which has access to a map object (for example, the focus map in the data view) and can create map surrounds. By definition, a map surround is an element that is

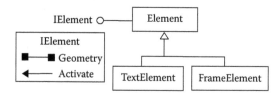

Figure 8.6 *TextElement* and *FrameElement* are types of the *Element* class.

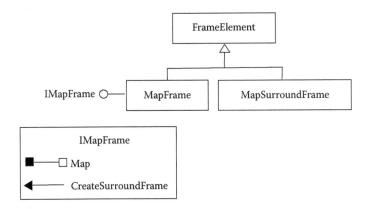

Figure 8.7   *MapFrame* and *MapSurroundFrame* are types of the *FrameElement* class. A map frame object has access to a map and can create map surround objects.

associated with a map, such as a legend, a north arrow, or a scale bar. A map surround frame object supports *IMapSurroundFrame*, which provides access to a map surround within the frame. In layout design, each map surround frame must be related to a map frame.

A text element may represent a title or a feature label on a layout page. A *TextElement* object supports *ITextElement* that can access the text string and symbol. A *TextSymbol* object implements *IFormattedTextSymbol* and *ITextSymbol*. Both interfaces let the user define text symbol properties such as color, font, horizontal alignment, size, and others. But *IFormattedTextSymbol* has more options than *ITextSymbol* does, especially in terms of character spacing and display properties. A *StdFont* object implements *IFontDisp*, which allows access to the font properties such as name, boldness, and size.

8.3 DISPLAYING VECTOR DATA

This section covers use of VBA (Visual Basic for Applications) macros for displaying vector data. These sample macros cover point, line, and area features; display both numeric and categorical data; and use both predefined and user-defined classification objects. Additionally, one sample macro uses a form to gather the necessary inputs from the user for data display.

8.3.1   *GraduatedColors*

GraduatedColors uses a sequential color ramp and a predefined classification object to display a choropleth map. A choropleth map shows data values based on administrative units. The macro performs the same function as using the Symbology/Quantities/Graduated colors command in the Layer Properties dialog in ArcMap.

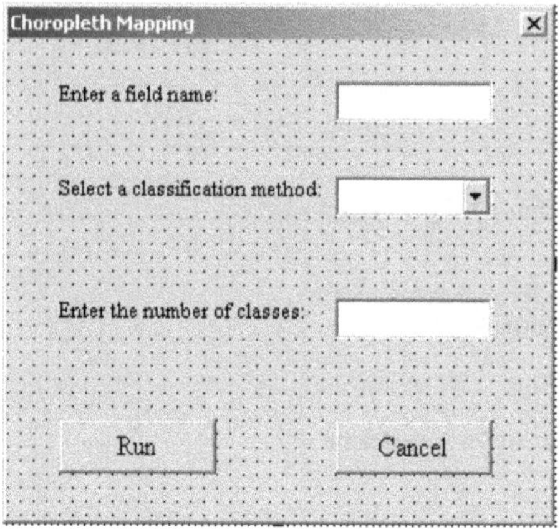

Figure 8.8 *GraduatedColors* uses the form to get a field name, a classification method, and a number of classes for data display.

GraduatedColors uses a user form to get a field, a classification method, and a number of classes for choropleth mapping (Figure 8.8). The form has two text boxes, a combo box, and two buttons:

txtField: A text box for entering the name of a numeric field.
cboMethod: A combo box with three predefined classification systems.
txtNumber: A text box for entering the number of classes.
cmdRun: A command button to execute data display.
cmdCancel: A command button to exit the form.

Associated with the *GraduatedColors* user form are three subs and one function:

UserForm_Initialize: Initialize the user form by populating the method dropdown list.
cmdRun_Click: Use the user inputs to display data.
GetRGBColor: Return a color based on the input RGB values.
cmdCancel_Click: Exit the user form.

Of the four procedures, *cmdRun_Click* involves most code writing and requires explanation. The other three are straightforward.

Key Interfaces: *ITable, ITableHistogram, IHistogram, IClassify, IClassBreaksRenderer, IAlgorithmicColorRamp, IEnumColors, IFillSymbol, IGeoFeatureLayer*

Key Members: *Field, Table, GetHistogram, SetHistogramData, Classify, ClassBreaks, BreakCount, MinimumBreak, Algorithm, ToColor, FromColor, Size, CreateRamp, Colors, Color, Break( ), Symbol( ), Label( ), Renderer, PartialRefresh, UpdateContents*

Usage: Add *idcounty.shp* to an active map. The shapefile has a field called change that shows the rate of population change between 1990 and 2000 in Idaho by county. Import *GraduatedColors.frm* to Visual Basic Editor. Right-click UserForm1 and select View Code. Select UserForm from the object list at the upper left of the Code window. Click Run Sub/UserForm. Enter "change" for the name of the field. Select a classification method from the method dropdown list. Enter the number of classes. Click on the Run button. The module uses a yellow-to-red color ramp to display the rate of population change data.

```
Private Sub UserForm_Initialize()
    cboMethod.AddItem "NaturalBreaks"
    cboMethod.AddItem "EqualInterval"
    cboMethod.AddItem "Quantile"
End Sub
```

UserForm_Initialize uses the *AddItem* method to add the three classification methods to the method combo box.

```
Private Sub cmdRun_Click()
    ' Part 1: Define the feature layer and derive its histogram data.
    Dim pMxDoc As IMxDocument
    Dim pMap As IMap
    Dim pLayer As IFeatureLayer
    Dim pTable As ITable
    Dim pTableHistogram As ITableHistogram
    Dim pHistogram As IHistogram
    Dim DataValues As Variant
    Dim DataFrequencies As Variant
    Set pMxDoc = ThisDocument
    Set pMap = pMxDoc.FocusMap
    Set pLayer = pMap.Layer(0)
    Set pTable = pLayer
    ' Define the table histogram.
    Set pTableHistogram = New TableHistogram
    pTableHistogram.Field = txtField.Value
    Set pTableHistogram.Table = pTable
    ' Derive the data values and frequencies from the histogram.
    Set pHistogram = pTableHistogram
    pHistogram.GetHistogram DataValues, DataFrequencies
```

Part 1 of *cmdRun_Click* derives from the feature layer the histogram data to be used for classification. The code first sets *pLayer* to be the top layer in the active map and *pTable*, a reference to *ITable*, to be the same as *pLayer*. Next, the code creates *pTableHistogram* as an instance of the *TableHistogram* class and defines the *Field* as *txtField.Value*, the field name entered in the user form, and the *Table* as *pTable* (Figure 8.9). The code then performs a QueryInterface (QI) for the *IHistogram* interface and uses the *GetHistogram* method to derive the histogram data from *pTableHistogram*. Of the derived histogram data, *DataValues* contains an array of data values, and *DataFrequencies* contains an array of frequencies corresponding to the data values.

```
    ' Part 2: Create a class breaks renderer.
    Dim pClassify As IClassify
    Dim Classes() As Double
```

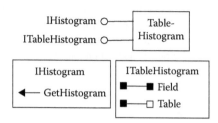

Figure 8.9   A *TableHistogram* object supports *IHistogram* and *ITableHistogram*. *ITableHistogram* has the properties to identify the table and the field. *IHistogram* has a method to get the histogram data.

```
Dim pClassBreaksRenderer As IClassBreaksRenderer
' Set up the classification method.
Select Case cboMethod.ListIndex
    Case 0
        Set pClassify = New NaturalBreaks
    Case 1
        Set pClassify = New EqualInterval
    Case 2
        Set pClassify = New Quantile
End Select
' Prepare a classify object.
pClassify.SetHistogramData DataValues, DataFrequencies
pClassify.Classify Val(txtNumber.Value)
' Create an array of class breaks.
Classes = pClassify.ClassBreaks
' Prepare a class breaks renderer.
Set pClassBreaksRenderer = New ClassBreaksRenderer
With pClassBreaksRenderer
    .Field = txtField.Value
    .BreakCount = Val(txtNumber.Value)
    .MinimumBreak = Classes(0)
End With
```

Part 2 of *cmdRun_Click* performs three tasks: using the selected classification method and the histogram data to prepare a classification object, deriving class breaks from the classification object, and linking the class breaks to a class breaks renderer. For the first task, the code creates *pClassify* as an instance of the *Natural-Breaks*, *EqualInterval*, or *Quantile* class, depending on the user's choice. Next, the code uses methods on *IClassify* to set the histogram data from Part 1 and to classify the histogram data into the number of classes entered by the user (i.e., *txtNumber.Value*) (Figure 8.10). For the second task, the code stores the class breaks into the array variable *Classes*. By default, the array is indexed from zero. For the third task, the code creates *pClassBreaksRenderer* as an instance of the *ClassBreaksRenderer* class and defines its properties of field, break count, and minimum break in a *With* block (Figure 8.11). In this case, the first class break is the minimum value

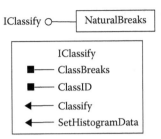

Figure 8.10 Properties and methods on the *IClassify* interface. All predefined classification classes including *NaturalBreaks* implement the *IClassify* interface.

in the dataset, and the subsequent breaks correspond to the upper class limits. The number of class breaks is therefore the number of classes plus one.

```
' Part 3: Create a color ramp.
Dim pAlgoRamp As IAlgorithmicColorRamp
Dim pColors As IEnumColors
' Prepare a color ramp.
Set pAlgoRamp = New AlgorithmicColorRamp
With pAlgoRamp
    .Algorithm = esriCIELabAlgorithm
    .ToColor = GetRGBColor(255, 0, 0)
    .FromColor = GetRGBColor(255, 255, 0)
    .Size = Val(txtNumber.Value)
    .CreateRamp (True)
End With
' Store the colors.
Set pColors = pAlgoRamp.Colors
```

Part 3 of *cmdRun_Click* creates a color ramp. The code creates *pAlgoRamp* as an instance of the *AlgorithmicColorRamp* class, and defines its properties of algorithm, end colors, and size before creating the ramp. The color ramp starts from a yellow color and ends with a red color. The code uses the ***GetRGBColor*** function

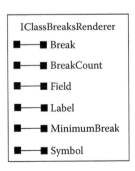

Figure 8.11 Properties on the *IClassBreaksRenderer* interface for defining graduated color symbols.

to define the end colors and the esriCIELabAlgorithm to generate the intermediate colors. After they are generated, these colors are stored as a sequence of colors referenced by *pColors*.

```
' Part 4: Assign a color symbol, break, and label to each of the classes.
Dim pFillSymbol As IFillSymbol
Dim I As Integer
For I = 0 To pClassBreaksRenderer.BreakCount - 1
    Set pFillSymbol = New SimpleFillSymbol
    pFillSymbol.Color = pColors.Next
    pClassBreaksRenderer.Symbol(I) = pFillSymbol
    pClassBreaksRenderer.Break(I) = Classes(I + 1)
    pClassBreaksRenderer.Label(I) = CSng(Classes(I)) & " - " & CSng(Classes(I + 1))
Next I
```

Part 4 of *cmdRun_Click* defines a color symbol, a break, and a label for each class in the class breaks renderer. The color symbols stored in *pColors* are initially assigned to *pFillSymbol*, which are in turn assigned as symbols to each class in *pClassBreaks-Renderer* in a *For...Next* loop. Two other properties of *pClassBreaksRenderer* are assigned within the same loop. The *Break(I)* property, which represents the upper bound of each class, is given the value of *Classes(I+1)*. The *Label(I)* property is given the value of *CSng(Classes(I)) & " - " & CSng(Classes(I + 1))*. The *CSng* (conversion to single) function truncates the long fractional part of numeric values.

```
' Part 5: Draw the graduated color map.
Dim pGeoFeatureLayer As IGeoFeatureLayer
' Assign the renderer to the feature layer.
Set pGeoFeatureLayer = pLayer
Set pGeoFeatureLayer.Renderer = pClassBreaksRenderer
' Refresh the map and its table of contents.
pMxDoc.ActiveView.PartialRefresh esriViewGeography, pLayer, Nothing
pMxDoc.UpdateContents
End Sub
```

Part 5 of *cmdRun_Click* accesses *IGeoFeatureLayer* and assigns *pClassBreaks-Renderer* to be the renderer. *IGeoFeatureLayer* controls the display of a feature layer. The code then refreshes the view and updates the document's table of contents.

```
Private Function GetRGBColor(R As Long, G As Long, B As Long)
    Dim pColor As IRgbColor
    Set pColor = New RgbColor
    pColor.Red = R
    pColor.Green = G
    pColor.Blue = B
    GetRGBColor = pColor
End Function
```

GetRGBColor uses the input values of R, G, and B from *cmdRun_Click* in Part 3 to create a new RGB color referenced by *pColor*. *GetRGBColor* then returns *pColor* to *cmdRun_Click*.

```
Private Sub cmdCancel_Click()
    End
End Sub
```

cmdCancel_Click exits the user form.

8.3.2   *GraduatedSymbols*

GraduatedSymbols uses different-sized circles to display different ranges of a field's values. The classification of the field values is user-defined rather than predefined. The macro performs the same function as using the Symbology/Quantities/Graduated symbols command in the Layer Properties dialog in ArcMap.

GraduatedSymbols has three parts. Part 1 creates a class breaks renderer. Part 2 prepares the symbol, break, and label for each class in the renderer. Part 3 assigns the renderer to the feature layer and refreshes the view.

> **Key Interfaces:** *IClassBreaksRenderer, ISimpleMarkerSymbol, IGeoFeatureLayer*
> **Key Members:** *Field, BreakCount, Color, Outline, OutlineColor, Size, Style, Symbol(), Break(), Label(), Renderer, PartialRefresh, UpdateContents*
> **Usage:** Add *idlcity.shp*, a shapefile that contains the ten largest cities in Idaho, to an active map. Import *GraduatedSymbols* to Visual Basic Editor. Run the macro. The macro produces a map showing the cities in graduated circles and the city names.

```
Private Sub GraduatedSymbols()
    ' Part 1: Create a class breaks renderer.
    Dim pClassBreaksRenderer As IClassBreaksRenderer
    Set pClassBreaksRenderer = New ClassBreaksRenderer
    With pClassBreaksRenderer
        .Field = "Population"
        .BreakCount = 3
    End With
```

Part 1 creates *pClassBreaksRenderer* as an instance of the *ClassBreaksRenderer* class and defines its properties of field and break count.

```
    ' Part 2: Set the symbol, break, and label for each class.
    Dim pMarkerSymbol As ISimpleMarkerSymbol
    ' Set the first class's symbol, break, and label.
    Set pMarkerSymbol = New SimpleMarkerSymbol
    With pMarkerSymbol
        .Color = GetRGBColor(255, 255, 0)
        .Outline = True
        .OutlineColor = GetRGBColor(0, 0, 0)
        .Size = 12
        .Style = esriSMSCircle
    End With
    pClassBreaksRenderer.Symbol(0) = pMarkerSymbol
    pClassBreaksRenderer.Break(0) = 20000
    pClassBreaksRenderer.Label(0) = "14300 - 20000"
    ' Set the second class's symbol, break, and label.
```

```
Set pMarkerSymbol = New SimpleMarkerSymbol
With pMarkerSymbol
    .Color = GetRGBColor(255, 125, 0)
    .Outline = True
    .OutlineColor = GetRGBColor(0, 0, 0)
    .Size = 18
    .Style = esriSMSCircle
End With
pClassBreaksRenderer.Symbol(1) = pMarkerSymbol
pClassBreaksRenderer.Break(1) = 30000
pClassBreaksRenderer.Label(1) = "20001 - 30000"
' Set the third class's symbol, break, and label.
Set pMarkerSymbol = New SimpleMarkerSymbol
With pMarkerSymbol
    .Color = GetRGBColor(255, 0, 0)
    .Outline = True
    .OutlineColor = GetRGBColor(0, 0, 0)
    .Size = 24
    .Style = esriSMSCircle
End With
pClassBreaksRenderer.Symbol(2) = pMarkerSymbol
pClassBreaksRenderer.Break(2) = 125660
pClassBreaksRenderer.Label(2) = "30001 - 125660"
```

Part 2 sets the symbol, break, and label for each class of the renderer. The code creates *pMarkerSymbol* as an instance of the *SimpleMarkerSymbol* class and defines the following five symbol properties: *Color* is the symbol color, *Outline* indicates whether or not to draw the outline, *OutlineColor* is the symbol outline color, *Size* is the symbol size measured in points, and *Style* is the symbol style (e.g., circle, square, or diamond). The ***GetRGBColor*** function provides the colors for the symbol and the symbol outline, and then the code assigns *pMarkerSymbol* to be the symbol of the first class, followed by assigning the class's upper limit and label. The same procedure for the first class applies to the other two classes.

```
' Part 3: Draw the feature layer and refresh the table of contents.
Dim pMxDoc As IMxDocument
Dim pMap As IMap
Dim pLayer As ILayer
Dim pGeoFeatureLayer As IGeoFeatureLayer
Set pMxDoc = ThisDocument
Set pMap = pMxDoc.FocusMap
Set pLayer = pMap.Layer(0)
' Set the renderer and annotation for drawing.
Set pGeoFeatureLayer = pLayer
Set pGeoFeatureLayer.Renderer = pClassBreaksRenderer
pGeoFeatureLayer.DisplayField = "City_Name"
pGeoFeatureLayer.DisplayAnnotation = True
pMxDoc.ActiveView.PartialRefresh esriViewGeography, pLayer, Nothing
pMxDoc.UpdateContents
End Sub
```

Part 3 first accesses *IGeoFeatureLayer* to assign the renderer and to display the field City_Name as annotation. The code then refreshes the map and updates the table of contents.

8.3.3 *UniqueSymbols*

UniqueSymbols uses a set of symbols to display each unique value of a field. These unique values represent categorical data such as different road types. The macro performs the same function as using the Symbology/Categories/Unique values command in the Layer Properties dialog in ArcMap.

UniqueSymbols has two parts. Part 1 creates a unique value renderer and populates the renderer with symbols for each unique value of a specified field, and Part 2 assigns the renderer to the feature layer and refreshes the view.

Key Interfaces: *IUniqueValueRenderer, ILineSymbol, IGeoFeatureLayer*

Key Members: *FieldCount, Field( ), Color, Width, AddValue, Renderer, PartialRefresh, UpdateContents*

Usage: Add *idroads.shp* to an active map. Import *UniqueSymbols* to Visual Basic Editor. Run the macro. The macro produces a map showing the interstate, U.S., and state highways in three different line symbols.

```
Private Sub UniqueSymbols()
    ' Part 1: Prepare a unique value renderer.
    Dim pUniqueValueRenderer As IUniqueValueRenderer
    Dim pSym1 As ILineSymbol
    Dim pSym2 As ILineSymbol
    Dim pSym3 As ILineSymbol
    ' Define the renderer.
    Set pUniqueValueRenderer = New UniqueValueRenderer
    pUniqueValueRenderer.FieldCount = 1
    pUniqueValueRenderer.Field(0) = "Route_Desc"
    ' Add the first symbol to the renderer.
    Set pSym1 = New SimpleLineSymbol
    pSym1.Color = GetRGBColor(255, 0, 0)
    pSym1.Width = 3
    pUniqueValueRenderer.AddValue "Interstate", "", pSym1
    ' Add the second symbol to the renderer.
    Set pSym2 = New SimpleLineSymbol
    pSym2.Color = GetRGBColor(255, 100, 0)
    pSym2.Width = 2
    pUniqueValueRenderer.AddValue "U.S.", "", pSym2
    ' Add the third symbol to the renderer.
    Set pSym3 = New SimpleLineSymbol
    pSym3.Color = GetRGBColor(255, 150, 0)
    pSym3.Width = 1
    pUniqueValueRenderer.AddValue "State", "", pSym3
```

Part 1 creates a unique value renderer with three symbols. The code creates *pUniqueValueRenderer* as an instance of the *UniqueValueRenderer* class and defines its

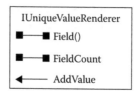

Figure 8.12 Members on the *IUniqueValueRenderer* interface for defining unique symbols.

field count as one and its field as Route_Desc (Figure 8.12). These properties are necessary because a unique value renderer can apply to two or more fields. Next the code creates *pSym1* as an instance of the *SimpleLineSymbol* class and defines its color and width properties. The color is generated by the **GetRGBColor** function, and the width is specified in points. Then the code uses the *AddValue* method on *IUniqueValueRenderer* to add the unique value (i.e., Interstate) and corresponding symbol (i.e., *pSym1*) to the renderer. The same procedure is repeated for the second and third unique values.

```
' Part 2: Assign the renderer to the feature layer and refresh the map and its table of contents.
Dim pMxDoc As IMxDocument
Dim pMap As IMap
Dim pLayer As IFeatureLayer
Dim pGeoFeatureLayer As IGeoFeatureLayer
Set pMxDoc = ThisDocument
Set pMap = pMxDoc.FocusMap
Set pLayer = pMap.Layer(0)
Set pGeoFeatureLayer = pLayer
Set pGeoFeatureLayer.Renderer = pUniqueValueRenderer
pMxDoc.ActiveView.PartialRefresh esriViewGeography, pLayer, Nothing
pMxDoc.UpdateContents
End Sub
```

Part 2 sets *pLayer* to be the top layer in the active map. Next, the code uses the *IGeoFeatureLayer* interface to assign the renderer. Finally, the code refreshes the view and updates the table of contents.

8.4 DISPLAYING RASTER DATA

Unlike vector data, which may be represented by point, line, or area (fill) symbols, raster data are represented only by fill symbols. Raster data can be categorical or numeric, however. Therefore, fill symbols for displaying raster data can be unique symbols or graduated color symbols.

8.4.1 *RasterUniqueSymbols*

RasterUniqueSymbols uses a symbol for each unique cell value to draw a raster layer. The macro performs the same function as using the Symbology/Unique values command in the Layer Properties dialog in ArcMap.

RasterUniqueSymbols has four parts. Part 1 defines the raster dataset to draw. Part 2 creates a raster unique value renderer and connects the renderer to the raster dataset. Part 3 assigns colors generated from a random color ramp and labels to each unique value. Part 4 applies the renderer to the raster layer and refreshes the view.

Key Interfaces: *ITable, IRasterBand, IRasterBandCollection, IRasterUniqueValue-Renderer, IRasterRenderer, IRandomColorRamp, ISimpleFillSymbol*

Key Members: *Raster, Item( ), AttributeTable, RowCount, FindField, Size, Create-Ramp, GetRow, Value, Update, Renderer, Refresh, UpdateContents*

Usage: Add *hucgd*, a raster containing major watersheds in Idaho, to an active map. Import *RasterUniqueSymbols* to Visual Basic Editor. Run the macro. The macro displays each watershed in *hucgd* with a unique symbol. These unique symbols differ from those initial symbols used by ArcMap for displaying *hucgd*.

```
Private Sub RasterUniqueSymbols()
    ' Part 1: Define the raster dataset to draw.
    Dim pMxDoc As IMxDocument
    Dim pMap As IMap
    Dim pRLayer As IRasterLayer
    Dim pRaster As IRaster
    Dim pTable As ITable
    Dim pBand As IRasterBand
    Dim pBandCol As IRasterBandCollection
    Dim TableExist As Boolean
    Dim NumOfValues As Integer
    Dim FieldIndex As Integer
    Dim FieldName As String
    Set pMxDoc = ThisDocument
    Set pMap = pMxDoc.FocusMap
    Set pRLayer = pMap.Layer(0)
    ' Work with the first band of the raster.
    Set pRaster = pRLayer.Raster
    Set pBandCol = pRaster
    Set pBand = pBandCol.Item(0)
    ' Make sure the band has an attribute table. If not, exit sub.
    pBand.HasTable TableExist
    If Not TableExist Then Exit Sub
    Set pTable = pBand.AttributeTable
    ' Get the row count.
    NumOfValues = pTable.RowCount(Nothing)
    ' Find the field Value.
    FieldName = "Value"
    FieldIndex = pTable.FindField(FieldName)
```

Part 1 defines the raster dataset and the field to draw. The code sets *pRaster* to be the raster of the top layer in the active map. Next, the code accesses the *IRasterBandCollection* interface to set *pBand* as the first band of *pRaster*. If *pBand* has a table, then the code assigns the attribute table to *pTable*. If not, the sub stops. Using *pTable* as the source, the code assigns the number of rows to the *NumOfValues* variable and the index of the field Value to the *FieldIndex* variable.

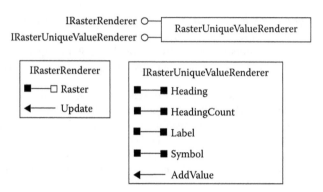

Figure 8.13 A *RasterUniqueValueRenderer* object supports *IRasterRenderer* and *IRaster-UniqueValueRenderer*. The interfaces provide access to the unique symbols and raster for data display.

```
' Part 2: Define a raster unique value renderer.
Dim pUVRen As IRasterUniqueValueRenderer
Dim pRasRen As IRasterRenderer
Set pUVRen = New RasterUniqueValueRenderer
Set pRasRen = pUVRen
Set pRasRen.Raster = pRaster
pRasRen.Update
```

Part 2 defines the renderer for displaying *pRaster*. The code creates *pUVRen* as an instance of the *RasterUniqueValueRenderer* class. A raster unique value renderer object supports *IRasterRenderer* and *IRasterUniqueValueRenderer* (Figure 8.13). Next, the code performs a QI for the *IRasterRenderer* interface to define *pRaster* as the raster and to update the renderer. A raster renderer must be updated for any changes that have been made.

```
' Part 3: Assign symbol and label to each unique value in the renderer.
Dim pRamp As IRandomColorRamp
Dim I As Long
Dim pRow As IRow
Dim UniqValue As Variant
Dim pFSymbol As ISimpleFillSymbol
' Create a random color ramp.
Set pRamp = New RandomColorRamp
With pRamp
    .Size = NumOfValues
    .CreateRamp (True)
End With
' Add unique values, labels, and symbols to the renderer.
For I = 0 To NumOfValues - 1
    Set pRow = pTable.GetRow(I)
    ' Add a unique value.
    UniqValue = pRow.Value(FieldIndex)
    pUVRen.AddValue 0, I, UniqValue
```

```
' Add corresponding label.
pUVRen.Label(0, I) = CStr(UniqValue)
' Add corresponding symbol.
Set pFSymbol = New SimpleFillSymbol
pFSymbol.Color = pRamp.Color(I)
pUVRen.Symbol(0, I) = pFSymbol
Next I
```

Part 3 populates the symbol and label in the renderer. The code first creates *pRamp* as an instance of the *RandomColorRamp* class, and sets the number of colors to be the same as the number of unique values before creating the ramp. Then the code uses a *For...Next* loop to add the unique values and their corresponding labels and symbols to *pUVRen*. The unique value is the value at the row defined by *I* and the column defined by *FieldIndex*. The label is the string converted from the unique value, and the symbol is a simple fill symbol with a color generated from *pRamp*. The members of *AddValue*, *Label*, and *Symbol* on *IRasterUniqueValueRenderer* all use two indices: the first represents the *iHeading* and the second represents the *IClass*. Headings are used for organizing unique values. One can create a new heading, for example, by combining two or more unique values (e.g., two or more unique values from a multiband raster). Because this macro works with individual unique values, the code specifies *iHeading* as zero (i.e., the first and only heading) and *IClass* as *I* (i.e., the row number).

```
' Part 4: Update the renderer, draw the layer, and refresh the view.
pRasRen.Update
Set pRLayer.Renderer = pUVRen
pMxDoc.ActiveView.Refresh
pMxDoc.UpdateContents
End Sub
```

Part 4 updates *pRasRen* before assigning it to be the renderer for *pRLayer*. Finally, the code refreshes the view and updates the table of contents.

8.4.2 *RasterClassifyColorRamp*

RasterClassifyColorRamp displays a raster layer by using a classification object, a raster classify renderer, and a user-defined color ramp. The macro performs the same function as using the Symbology/Classified command in the Layer Properties dialog in ArcMap. *RasterClassifyColorRamp* has three parts. Part 1 derives the histogram data from the raster for display, Part 2 prepares a raster renderer, and Part 3 defines the properties of the renderer and uses the renderer to draw the raster layer.

> **Key Interfaces:** *IRaster, ITable, IRasterBandCollection, IRasterBand, ITableHistogram, IHistogram, IClassify, IRasterClassifyColorRampRenderer, IRasterClassifyUIProperties, IRasterRenderer, IAlgorithmicColorRamp, IEnumColors, IFillSymbol*
>
> **Key Members:** *Raster, Item( ), AttributeTable, Field, Table, GetHistogram, SetHitogramData, Classify, ClassBreaks, ClassID, ClassCount, ClassField, ClassificationMethod,*

Update, Algorithm, FromColor, ToColor, Size, CreateRamp, Color, Symbol( ), Break( ), Label( ), Refresh, UpdateContents

Usage: Add *intemida*, an integer elevation raster, to an active map. Import **RasterClassifyColorRamp** to Visual Basic Editor. Run the macro. The macro displays *intemida* using ten equal interval classes and a color ramp from red to blue.

```
Private Sub RasterClassifyColorRamp()
    ' Part 1: Prepare the raster dataset for display.
    Dim pMxDoc As IMxDocument
    Dim pMap As IMap
    Dim pRLayer As IRasterLayer
    Dim pRaster As IRaster
    Dim pTable As ITable
    Dim pBandCol As IRasterBandCollection
    Dim pRasBand As IRasterBand
    Dim TestTable As Boolean
    Dim pTableHist As ITableHistogram
    Dim pHist As IHistogram
    Dim vValues As Variant
    Dim vFrequencies As Variant
    ' Define the raster dataset and verify it has a table.
    Set pMxDoc = ThisDocument
    Set pMap = pMxDoc.FocusMap
    Set pRLayer = pMap.Layer(0)
    Set pRaster = pRLayer.Raster
    Set pBandCol = pRaster
    Set pRasBand = pBandCol.Item(0)
    pRasBand.HasTable TestTable
    If TestTable = False Then Exit Sub
    Set pTable = pRasBand.AttributeTable
    ' Derive the histogram data from the table.
    Set pTableHist = New TableHistogram
    pTableHist.Field = "Value"
    Set pTableHist.Table = pTable
    Set pHist = pTableHist
    pHist.GetHistogram vValues, vFrequencies
```

Part 1 verifies the raster has a table and derives the histogram data from the table. The code first defines the following objects: *pRaster* to be the raster of the first layer in the active map, *pBandCol* to be the same as *pRaster*, and *pRasBand* to be the first band of the band collection. Next, the code verifies that *pRasBand* has a table and sets *pTable* to be the attribute table of *pRasBand*. The rest of Part 1 derives the histogram data from *pTable*. The code creates *pTableHist* as an instance of the *TableHistogram* class and defines its field and table properties, and then the code switches to the *IHistogram* interface and uses the *GetHistogram* method to derive the data values and frequencies from *pTable*.

```
    ' Part 2: Prepare the raster renderer.
    Dim pClassify As IClassify
    Dim ClassBreak As Variant
```

```
Dim pUID As UID
Dim pClassRen As IRasterClassifyColorRampRenderer
Dim pClassProp As IRasterClassifyUIProperties
Dim pRasRen As IRasterRenderer
' Prepare an equal interval classification object with 10 classes.
Set pClassify = New EqualInterval
pClassify.SetHistogramData vValues, vFrequencies
pClassify.Classify 10
' Store the class breaks.
ClassBreak = pClassify.ClassBreaks
' Obtain the classification UID.
Set pUID = pClassify.ClassID
' Prepare a raster classify renderer.
Set pClassRen = New RasterClassifyColorRampRenderer
pClassRen.ClassCount = 10
pClassRen.ClassField = "Value"
' Define the classification method.
Set pClassProp = pClassRen
Set pClassProp.ClassificationMethod = pUID
' Define the raster and update the renderer.
Set pRasRen = pClassRen
Set pRasRen.Raster = pRaster
pRasRen.Update
```

Part 2 prepares a classification object and a raster renderer. This preparation requires several steps. First, the code creates *pClassify* as an instance of the *Equal-Interval* class. The code then uses members on *IClassify* to get the histogram data for a classification with ten classes, to save the class breaks into an array referenced by *ClassBreak*, and to obtain the classification's unique identifier (UID). Second, the code creates *pClassRen* as an instance of the *RasterClassifyColorRampRenderer* class and defines its properties of class count and class field. Third, the code accesses *IRasterClassifyUIProperties* to specify *pUID* for the classification method. Finally, the code uses the *IRasterRenderer* interface to define the raster and to update the raster. Figure 8.14 shows the interfaces that a raster classify color ramp renderer object supports.

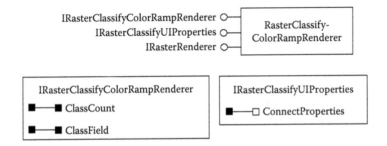

Figure 8.14 A *RasterClassifyColorRampRenderer* object supports *IRasterClassifyColorRampRenderer*, *IRasterClassifyUIProperties*, and *IRasterRenderer*. The first two interfaces can define the classification of raster data and the third, as shown in Figure 8.13, connects the raster and its renderer.

```
' Part 3: Define the properties of the renderer, and use the renderer to draw the map.
Dim pAlgoRamp As IAlgorithmicColorRamp
Dim pColors As IEnumColors
Dim pFillSymbol As IFillSymbol
Dim I As Integer
' Prepare a color ramp.
Set pAlgoRamp = New AlgorithmicColorRamp
With pAlgoRamp
    .Algorithm = esriHSVAlgorithm
    .FromColor = GetRGBColor(255, 0, 0)
    .ToColor = GetRGBColor(0, 0, 255)
    .Size = 10
    .CreateRamp (True)
End With
Set pColors = pAlgoRamp.Colors
' Define the symbol, break, and label for each class.
For I = 0 To pClassRen.ClassCount - 1
    Set pFillSymbol = New SimpleFillSymbol
    pFillSymbol.Color = pColors.Next
    pClassRen.Symbol(I) = pFillSymbol
    pClassRen.Break(I) = ClassBreak(I)
    pClassRen.Label(I) = ClassBreak(I) & " - " & ClassBreak(I + 1)
Next I
' Assign the raster renderer to the layer and refresh the map.
pRasRen.Update
Set pRLayer.Renderer = pRasRen
pMxDoc.ActiveView.Refresh
pMxDoc.UpdateContents
End Sub
```

Part 3 provides the symbol, break, and label to the renderer and uses the renderer to draw the raster layer. The code first creates *pAlgoRamp* as an instance of the *AlgorithmicColorRamp* class, and defines its properties. The **GetRGBColor** function is used to define the two end colors of the color ramp. The colors from *pAlgoRamp* are stored in an enumerator referenced by *pColors*. Next, the code uses a *For...Next* loop to assign the symbol, break, and label to each class in *pClassRen*. Finally, the code updates *pRasRen* with the changes made through *pClassRen* and uses *pRasRen* as the renderer to draw the raster layer.

8.4.3   *RasterUserDefinedColorRamp*

RasterUserDefinedColorRamp displays a classified raster layer by using user-defined class breaks and color ramp. The macro performs the same function as using the Symbology/Classified command in the Layer Properties dialog in ArcMap. *RasterUserDefinedColorRamp* has three parts. Part 1 defines the raster and prepares a raster renderer. Part 2 creates a color ramp, followed by specifying the label, break, and symbol of each class for the renderer. Part 3 assigns the renderer to the raster and refreshes the map.

Key Interfaces: *IRaster, IRasterClassifyColorRampRenderer, IRasterRenderer, IAlgorithmicColorRamp, IFillSymbol*

Key Members: *Raster, ClassCount, Update, Algorithm, FromColor, ToColor, Size, CreateRamp, Break( ), Label( ), Color, Symbol( ), Renderer, Refresh, UpdateContents*

Usage: Add *emidalat*, an elevation raster, to an active map. Import ***RasterUserDefinedColorRamp*** to Visual Basic Editor. Run the macro. The macro redraws *emidalat* in three classes using symbols from a color ramp.

```
Private Sub RasterUserDefinedColorRamp()
    ' Part 1: Define the raster dataset and a raster renderer.
    Dim pMxDoc As IMxDocument
    Dim pMap As IMap
    Dim pRLayer As IRasterLayer
    Dim pRaster As IRaster
    Dim pClassRen As IRasterClassifyColorRampRenderer
    Dim pRasRen As IRasterRenderer
    ' Define the raster dataset.
    Set pMxDoc = ThisDocument
    Set pMap = pMxDoc.FocusMap
    Set pRLayer = pMap.Layer(0)
    Set pRaster = pRLayer.Raster
    ' Define a raster classify color ramp renderer.
    Set pClassRen = New RasterClassifyColorRampRenderer
    pClassRen.ClassCount = 3
    Set pRasRen = pClassRen
    Set pRasRen.Raster = pRaster
    pRasRen.Update
```

Part 1 defines the raster and a raster renderer. The code sets *pRaster* to be the raster of the layer to draw. Next, the code creates *pClassRen* as an instance of the *RasterClassifyColorRampRenderer* class and specifies three for the class count. Then the code uses the *IRasterRenderer* interface to define the raster and to update the renderer.

```
    ' Part 2: Specify the properties for the renderer.
    Dim pRamp As IAlgorithmicColorRamp
    Dim pFSymbol As IFillSymbol
    ' Create a color ramp.
    Set pRamp = New AlgorithmicColorRamp
    With pRamp
        .Algorithm = esriCIELabAlgorithm
        .FromColor = GetRGBColor(0, 255, 255)
        .ToColor = GetRGBColor(0, 0, 255)
        .Size = 3
        .CreateRamp True
    End With
    ' Specify the label, break, and symbol for each class in the renderer.
    Set pFSymbol = New SimpleFillSymbol
    pFSymbol.Color = pRamp.Color(0)
    pClassRen.Symbol(0) = pFSymbol
```

```
pClassRen.Break(0) = 855
pClassRen.Label(0) = "855 - 1000"
pFSymbol.Color = pRamp.Color(1)
pClassRen.Symbol(1) = pFSymbol
pClassRen.Break(1) = 1000
pClassRen.Label(1) = "1000 - 1200"
pFSymbol.Color = pRamp.Color(2)
pClassRen.Symbol(2) = pFSymbol
pClassRen.Break(2) = 1200
pClassRen.Label(2) = "1200 - 1350"
```

Part 2 specifies the properties of the renderer. The code first creates *pRamp* as an instance of the *AlgorithmicColorRamp* class and defines its properties. Then the code uses colors from *pRamp* and hard-coded class breaks and labels to define the properties for each class in *pClassRen*.

```
' Part 3: Assign the renderer to the layer and refresh the map.
pRasRen.Update
Set pRLayer.Renderer = pRasRen
pMxDoc.ActiveView.Refresh
pMxDoc.UpdateContents
End Sub
```

After the renderer is set up in Part 2, Part 3 assigns the updated *pRasRen* to the raster layer, refreshes the map, and updates the document's table of contents.

8.5 MAKING A PAGE LAYOUT

A page layout may contain one or more maps and various map elements. A module for making a page layout typically involves more objects and code lines than other types of applications.

8.5.1 *Layout*

Layout prepares a layout of a thematic map showing the rate of population change by county in Idaho between 1990 and 2000. The layout includes the map body, a title, a subtitle, a legend, a north arrow, and a scale bar. The module performs the same function as using the Insert menu in the Layout View to add different map elements to the layout. *Layout* is organized into five subs. *Start* is the startup sub, which calls the other subs. *AddTitle* adds the title and subtitle, *AddLegend* adds the legend, *AddNorthArrow* adds the north arrow, and *AddScaleBar* adds the scale bar to the layout.

> **Key Interfaces:** *IGraphicsContainer, IMapFrame, IElement, IActiveView, IPageLayout, ITextElement, IFormattedTextSymbol, IFontDisp, IPoint, IEnvelope, IMapSurroundFrame, IMapSurround, IMarkerNorthArrow, ICharacterMarkerSymbol*
> **Key Members:** *PageLayout, FindFrame, Text, Name, Size, Bold, Font, Case, HorizontalAlignment, Symbol, Geometry, AddElement, Value, PutCoords, CreateSurroundFrame, Activate, MarkerSymbol, CharacterIndex*

Usage: Add *idcounty.lyr* to a new map (if necessary, set the layer file's data source to be *idcounty.shp*). Make sure that no other maps are in the map document. The layer file shows the rate of population change by county between 1990 and 2000 in Idaho. In ArcMap's table of contents, double-click and delete *idcounty* (the layer name) and Change (the field name). (The layer name and the field name are confusing to the viewer.) Change from Data View to Layout View. The layout page size for this macro is 8.5 by 11 inches, and all point measurements are in inches. Therefore, select Page and Print Setup from the File menu and, in the next dialog, uncheck Use Printer Paper Settings and set the Width of the page to be 8.5 inches and the Height to be 11 inches. Now use the handle of the map frame to resize the map so that it fills the page. Import *Layout* to Visual Basic Editor. Run the module. The module adds the title, subtitle, legend, north arrow, and scale bar to the layout. Because the map and the map elements are graphic elements, they can be reduced, enlarged, and moved on the page layout if necessary.

```
Public Sub Start()
    ' Set the variables and run the subs.
    Dim pMxDoc As IMxDocument
    Dim pPageLayout As IPageLayout
    Dim pGraphicsContainer As IGraphicsContainer
    Dim pActiveView As IActiveView
    Dim pMapFrame As IMapFrame
    Dim pElement As IElement
    ' Set the layout view.
    Set pMxDoc = Application.Document
    Set pPageLayout = pMxDoc.PageLayout
    Set pGraphicsContainer = pPageLayout
    Set pActiveView = pPageLayout
    ' Set the map frame.
    Set pMapFrame = pGraphicsContainer.FindFrame(pMxDoc.FocusMap)
    Set pElement = pMapFrame
    ' Call subs to add map elements.
    Call AddTitleSubtitle
    Call AddLegend(pElement)
    Call AddNorthArrow(pElement)
    Call AddScalebar(pElement)
    ' Refresh the layout.
    pActiveView.PartialRefresh esriViewGraphics, Nothing, Nothing
End Sub
```

Start defines the page layout and runs the subs to add the title, subtitle, legend, north arrow, and scale bar to the layout. The code sets *pPageLayout* to be the page layout of the map document, and sets both *pGraphicsContainer* and *pActiveView* to be the same as *pPageLayout*. Next, *Start* sets *pMapFrame* to be the map frame for the focus map of the map document. *IMapFrame* provides access to the map within the frame and has methods to create map surrounds, such as the legend, north arrow, and scale bar, that are associated with the map. Therefore, *pElement*, which is set to be *pMapFrame*, must be passed as an argument to the subs that add the legend, north arrow, and scale bar to the layout. After all the elements are added to the layout, the code refreshes the layout view.

```
Private Sub AddTitleSubtitle()
    ' Part 1: Set the page layout.
    Dim pMxDoc As IMxDocument
    Dim pPageLayout As IPageLayout
    Dim pGraphicsContainer As IGraphicsContainer
    Set pMxDoc = Application.Document
    Set pPageLayout = pMxDoc.PageLayout
    Set pGraphicsContainer = pPageLayout
```

Part 1 of *AddTitleSubtitle* defines the page layout.

```
    ' Part 2: Add the title.
    Dim pTextElement As ITextElement
    Dim pTextSymbol As IFormattedTextSymbol
    Dim pTextFont As IFontDisp
    Dim pElement As IElement
    Dim pPoint As IPoint
    ' Define the text font.
    Set pTextFont = New StdFont
    With pTextFont
        .Name = "Times New Roman"
        .Size = 24
        .Bold = True
    End With
    ' Define the text symbol.
    Set pTextSymbol = New TextSymbol
    With pTextSymbol
        .Font = pTextFont
        .Case = esriTCAllCaps
        .HorizontalAlignment = esriTHACenter
    End With
    ' Define the title as a text element.
    Set pTextElement = New TextElement
    pTextElement.Text = "Idaho 1990-2000"
    pTextElement.Symbol = pTextSymbol
    ' Define the position to plot the title.
    Set pElement = pTextElement
    Set pPoint = New Point
    pPoint.X = 5#
    pPoint.Y = 10#
    pElement.Geometry = pPoint
    ' Add the title to the graphics container.
    pGraphicsContainer.AddElement pTextElement, 0
```

Part 2 of *AddTitleSubtitle* adds a title to the layout. The code performs two tasks: it defines the title as a text element and locates the title on the layout. The definition of a text element includes a text symbol, the definition of which in turn includes a text font (Figure 8.15). Therefore, the code first creates *pTextFont* as an instance of the *StdFont* class and defines its properties of name, size, and boldness in a *With* block. Next, the code creates *pTextSymbol* as an instance of the *TextSymbol* class

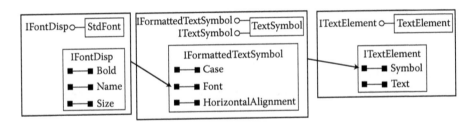

Figure 8.15 The diagram shows text-related objects and their properties for adding the title and subtitle to the layout. *IFontDisp* defines the font for the text symbol, and *IFormattedTextSymbol* defines the symbol for the text element.

and defines its properties of font, case, and horizontal alignment. To complete the first task of defining the title, the code creates *pTextElement* as an instance of the *TextElement* class and defines its properties of text and symbol. The code starts the second task by creating *pPoint* as an instance of the *Point* class and defines the point's *X* and *Y* properties. Both *X* and *Y* values are measured in inches, with the origin at the lower left corner of the page layout. The code then accesses the *IElement* interface and assigns *pPoint* to be the geometry of *pTextElement*. Because the horizontal alignment of the text symbol is center justified (i.e., esriTHACenter), *pPoint* is at the center point of the title. After the definition is complete, the code uses the *AddElement* method on *IGraphicsContainer* to add *pTextElement* to the graphics container.

```
' Part 3: Add the subtitle.
' Define the text font.
Set pTextFont = New StdFont
With pTextFont
      .Name = "Times New Roman"
      .Size = 14
End With
' Define the text symbol.
Set pTextSymbol = New TextSymbol
With pTextSymbol
      .Font = pTextFont
      .Case = esriTCAllCaps
      .HorizontalAlignment = esriTHACenter
End With
' Define the subtitle as a text element.
Set pTextElement = New TextElement
pTextElement.Text = "Rate of Population Change by County"
pTextElement.Symbol = pTextSymbol
' Define the position to plot the subtitle.
Set pElement = pTextElement
Set pPoint = New Point
pPoint.X = 5#
pPoint.Y = 9.5
pElement.Geometry = pPoint
```

```
   ' Add the subtitle to the graphic container.
   pGraphicsContainer.AddElement pTextElement, 0
End Sub
```

Part 3 of ***AddTitleSubtitle*** follows the same procedure as in Part 2 to add a subtitle to the layout. The subtitle should appear half an inch below the title in a smaller text size.

```
Private Sub AddLegend(pElement As IElement)
   ' Part 1: Define the page layout.
   Dim pMxDoc As IMxDocument
   Dim pPageLayout As IPageLayout
   Dim pGraphicsContainer As IGraphicsContainer
   Dim pActiveView As IActiveView
   Set pMxDoc = Application.Document
   Set pPageLayout = pMxDoc.PageLayout
   Set pGraphicsContainer = pPageLayout
   Set pActiveView = pPageLayout
```

Part 1 of ***AddLegend*** defines the page layout.

```
   ' Part 2: Create a legend map surround frame.
   Dim pMapFrame As IMapFrame
   Dim pID As New UID
   Dim pMapSurroundFrame As IMapSurroundFrame
   Dim pMapSurround As IMapSurround
   ' Get the map frame.
   Set pMapFrame = pElement
   ' Create a legend map surround frame.
   pID.Value = "esriCore.Legend"
   Set pMapSurroundFrame = pMapFrame.CreateSurroundFrame(pID, pMapSurround)
```

Part 2 of ***AddLegend*** creates a map surround frame for the legend. The code first sets *pMapFrame* to be *pElement*, an argument passed from ***Layout***. Next, the code sets the value of *pID*, a new unique identifier (*UID*) object, to be "esriCore.Legend." Then the code uses the *CreateSurroundFrame* method on *IMapFrame* to create a legend within a frame. The legend is referenced by *pMapSurround*, and the frame is referenced by *pMapSurroundFrame*.

```
   ' Part 3: Create the legend and add the legend to the graphics container.
   Dim pFrameElement As IElement
   Dim pEnvelope As IEnvelope
   ' Define the geometry of the legend.
   Set pEnvelope = New Envelope
   pEnvelope.PutCoords 5.5, 7, 6.5, 8.4
   Set pFrameElement = pMapSurroundFrame
   pFrameElement.Geometry = pEnvelope
   ' Activate the screen display of the legend.
   pFrameElement.Activate pActiveView.ScreenDisplay
```

```
    ' Add the legend to the graphics container.
    pGraphicsContainer.AddElement pFrameElement, 0
End Sub
```

Part 3 of ***AddLegend*** creates the legend on the layout. The code first creates *pEnvelope* as an instance of the *Envelope* class and uses the *PutCoords* method to define the envelope's xmin, ymin, xmax, and ymax values. An envelope object represents a rectangular shape, and in this case, it controls the size and position of the legend. Next, the code uses the *IElement* interface to assign *pEnvelope* to be the geometry of the map surround frame. Finally, the code activates the screen display of the legend frame and adds the frame to the graphics container.

```
Private Sub AddNorthArrow(pElement As IElement)
    ' Part 1: Define the page layout.
    Dim pMxDoc As IMxDocument
    Dim pPageLayout As IPageLayout
    Dim pGraphicsContainer As IGraphicsContainer
    Dim pActiveView As IActiveView
    Set pMxDoc = Application.Document
    Set pPageLayout = pMxDoc.PageLayout
    Set pGraphicsContainer = pPageLayout
    Set pActiveView = pPageLayout
```

Part 1 of ***AddNorthArrow*** defines the page layout.

```
    ' Part 2: Create a north arrow map surround frame.
    Dim pMapFrame As IMapFrame
    Dim pID As New UID
    Dim pMapSurroundFrame As IMapSurroundFrame
    Dim pFrameElement As IElement
    Dim pMapSurround As IMapSurround
    Dim pMarkerNorthArrow As IMarkerNorthArrow
    Dim pCharacterMarkerSymbol As ICharacterMarkerSymbol
    ' Get the map frame.
    Set pMapFrame = pElement
    ' Create a north arrow map surround frame.
    pID.Value = "esriCore.MarkerNorthArrow"
    ' Choose a north arrow design other than the default.
    Set pMarkerNorthArrow = New MarkerNorthArrow
    Set pCharacterMarkerSymbol = pMarkerNorthArrow.MarkerSymbol
    pCharacterMarkerSymbol.CharacterIndex = 176
    pMarkerNorthArrow.MarkerSymbol = pCharacterMarkerSymbol
    Set pMapSurround = pMarkerNorthArrow
    Set pMapSurroundFrame = pMapFrame.CreateSurroundFrame(pID, pMapSurround)
```

Part 2 of ***AddNorthArrow*** creates a map surround frame for the north arrow. The code first sets *pMapFrame* to be *pElement*, an argument passed from ***Layout***. Next, the code sets the value of *pID* to be "esriCore.MarkerNorthArrow." ArcObjects offers a wide variety of north arrow objects. The default option is a rather fancy north arrow, but one can use additional line statements to specify a simpler north arrow.

The code creates *pMarkerNorthArrow* as an instance of the *MarkerNorthArrow* class, and initially sets *pCharacterMarkerSymbol* to be its marker symbol. Then the code specifies *pCharacterMarkerSymbol* to be the character at index 176 and assigns the symbol to be the marker symbol for *pMarkerNorthArrow*. In this case, the north arrow, a simpler symbol than the default, is a character marker symbol at the index of 176. Finally, the code uses the chosen symbol as an object qualifier to create a north arrow map surround frame. (To see other character marker symbols for the north arrow, do the following in the Layout View of ArcMap: select North Arrow from the Insert menu, click Properties in the North Arrow Selector dialog, and then click the Character dropdown arrow in the North Arrow dialog. The index value of the character symbol shows up in the ToolTip message box.)

```
' Part 3: Create the north arrow and add it to the graphics container.
Dim pEnvelope As IEnvelope
' Create a envelope for the north arrow.
Set pEnvelope = New Envelope
pEnvelope.PutCoords 5.7, 6.2, 5.9, 6.4
Set pFrameElement = pMapSurroundFrame
pFrameElement.Geometry = pEnvelope
pFrameElement.Activate pActiveView.ScreenDisplay
' Add the north arrow to the graphics container.
pGraphicsContainer.AddElement pFrameElement, 0
End Sub
```

Part 3 of *AddNorthArrow* creates a new envelope and assigns the envelope to be the geometry of the map surround frame for the north arrow. Then the code activates the screen display of the frame element and adds the element to the graphics container.

```
Private Sub AddScalebar(pElement As IElement)
    ' Part 1: Define the page layout.
    Dim pMxDoc As IMxDocument
    Dim pPageLayout As IPageLayout
    Dim pGraphicsContainer As IGraphicsContainer
    Dim pActiveView As IActiveView
    Set pMxDoc = Application.Document
    Set pPageLayout = pMxDoc.PageLayout
    Set pGraphicsContainer = pPageLayout
    Set pActiveView = pPageLayout
```

Part 1 of *AddScaleBar* defines the page layout.

```
    ' Part 2: Create a scalebar map surround frame.
    Dim pMapFrame As IMapFrame
    Dim pID As New UID
    Dim pMapSurroundFrame As IMapSurroundFrame
    Dim pFrameElement As IElement
    Dim pMapSurround As IMapSurround
    Dim pScaleMarks As IScaleMarks
```

```
' Get the map frame.
Set pMapFrame = pElement
' Create a scale bar map surround frame.
pID.Value = "esriCore.Scalebar"
Set pMapSurround = New AlternatingScaleBar
Set pScaleMarks = pMapSurround
pScaleMarks.MarkFrequency = esriScaleBarMajorDivisions
Set pMapSurroundFrame = pMapFrame.CreateSurroundFrame(pID, pMapSurround)
```

Part 2 of ***AddScalebar*** creates a scale bar map surround frame. The code first sets *pMapFrame* to be *pElement*, an argument passed from ***Layout***. Next, the code defines the value of *pID* to be "esriCore.Scalebar." Then the code creates *pMapSurround* as an instance of the *AlternatingScaleBar* class and uses this scale bar option to create a map surround frame. An *AlternatingScaleBar* object uses two symbols such as black and white to create a scale bar. The *MarkFrequency* property of *IScaleMarks* stipulates that the markings are to be added to the major divisions.

```
' Part 3: Create the scalebar and add it to the graphics container.
Dim pEnvelope As IEnvelope
' Create an envelope for the scale bar.
Set pEnvelope = New Envelope
pEnvelope.PutCoords 5.2, 5.4, 7.2, 6#
Set pFrameElement = pMapSurroundFrame
pFrameElement.Geometry = pEnvelope
pFrameElement.Activate pActiveView.ScreenDisplay
' Add the scale bar to the graphics container.
pGraphicsContainer.AddElement pFrameElement, 0
End Sub
```

Part 3 of ***AddScalebar*** creates a new envelope and assigns the envelope to be the geometry of the map surround frame for the scale bar. The code then activates the screen display of the frame element and adds the element to the graphics container.

Data Exploration

Data exploration involves data-centered query and analysis. It allows users to examine the general trends in the data, to take a close look at data subsets, and to focus on possible relationships between datasets. For some users of geographic information systems (GIS), data exploration serves as a starting point in formulating research questions and hypotheses. For others, data exploration constitutes their routine tasks.

Data exploration in ArcGIS can be based on an attribute query or spatial query. An attribute query selects records by using a SQL (Structured Query Language) statement and attribute data in a feature class or an attribute table. A spatial query selects features by using a cursor, a graphic object, or a spatial relationship. The use of a cursor or a graphic object (for example, a circle) for selecting features is straightforward. The use of a spatial relationship, on the other hand, requires a statement that can link "features to select" and "features to be selected" spatially. Common spatial relationships are containment, intersect, and proximity. The result from an attribute or spatial query is a data subset, which can be highlighted in a table, a feature layer, or both.

A data exploration task may use both attribute and spatial queries. This happens when the selection of a data subset involves both attribute and spatial conditions. One of the sample macros in this chapter selects thermal wells and springs that are within a certain distance of an interstate highway (a spatial condition) and have water temperatures above a certain reading (an attribute condition). Therefore the macro must perform both attribute and spatial queries.

Data exploration in GIS may also involve derivation of descriptive statistics such as minimum, maximum, range, mean, and standard deviation from numeric data. These descriptive statistics are derived on a field using either all records or a data subset in a table.

This chapter covers data exploration. Section 9.1 reviews data exploration options in ArcGIS. Section 9.2 discusses objects that are important to data exploration. Section 9.3 offers macros and a Geoprocessing (GP) macro for attribute queries. Section 9.4 offers macros and a GP macro for spatial queries. Section 9.5 includes macros for combining attribute and spatial queries. Section 9.6 has macros and a GP macro for deriving and reporting descriptive statistics. All macros start with the listing of key interfaces and key members (that is, properties and methods) and the usage.

9.1 EXPLORING DATA IN ARCGIS

ArcMap has the Select By Attributes command for attribute queries. SQL is the query language. The basic syntax of a SQL statement is

select <attribute list>
from <table>
where <condition>

The *select* keyword selects field(s) from a database, the *from* keyword selects table(s) from a database, and the *where* keyword specifies the condition for data query. SQL is integrated with the user interface in ArcMap: the table to be selected from is the activated table, and fields to be queried are available in a dropdown list. The user only has to prepare the *where* clause (the query expression) in the Select By Attributes dialog.

ArcMap has Select Features, Select By Graphics, and Select By Location for spatial queries. These three commands correspond to selecting features by using a cursor, a graphic, and a spatial relationship respectively. Select By Location is the only command that uses a dialog. The dialog requires the user to specify one or more layers whose features will be selected, and a layer whose features will be used for selection. The dialog offers 11 spatial relationships to connect features to be selected and features used for selection. These relationships are: "are completely within," "completely contain," "have their center in," "contain," "are contained by," "intersect," "are crossed by the outline of," "are within a distance of," "share a line segment with," "touch the boundary of," and "are identical to." The spatial concepts of containment, intersect, and proximity are the basis for these relationships.

Both Select By Attributes and Select By Location offer the selection methods of "create a new selection," "add to current selection," "remove from current selection," and "select from current selection." Therefore, after selecting a feature subset, we may add features to, remove features from, or select features from the subset or we may select a new subset.

ArcMap offers two ways to derive descriptive statistics on a field. The context menu of a field in a feature attribute table has the Statistics command. When selected, the command displays the field's statistics using all or selected records. The Statistics command is also available in the Selection menu. This command derives statistics for selected features.

9.2 ARCOBJECTS FOR DATA EXPLORATION

The primary ArcObjects components for data query are *QueryFilter* and *SpatialFilter* (Figure 9.1). A query filter object can retrieve a data subset from a feature layer, a feature class, or a table by using the condition expressed in the *WhereClause* property.

The *SpatialFilter* class is a type of the *QueryFilter* class. This means that a spatial filter object can include both spatial and attribute constraints in performing a data query. Two important properties of a spatial filter object are *Geometry* and

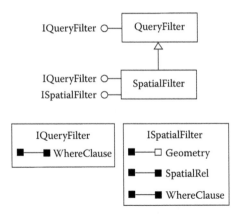

Figure 9.1 *SpatialFilter* is a type of *QueryFilter*. Notice that a spatial filter object has its own properties in addition to the *WhereClause* property that it inherits from *QueryFilter*.

SpatialRel. The *Geometry* property defines the query geometry to filter results. The query geometry may be the shape of a selected polygon in a feature layer, a rectangle based on the extent drawn by the user, or a buffer zone around a highway segment. The *SpatialRel* property defines the spatial relationship for filtering. ArcObjects has ten predefined spatial relationships (Figure 9.2).

9.2.1   Use of a Query Filter

A query filter object can be used on a feature layer or a feature class. When used on a feature layer, a query filter produces a set of selected features for visual inspection. A *FeatureLayer* object implements *IFeatureSelection*, and the *Select-Features* method on *IFeatureSelection* can use a query filter to select features (Figure 9.3). *IFeatureSelection* also has properties to set the color and symbol for displaying the selected features.

When used on a feature class, a query filter produces a data subset for data manipulation or presentation. Figure 9.4 shows that a *SelectionSet* object can be

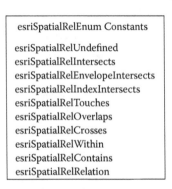

Figure 9.2 Predefined spatial relationships in ArcObjects.

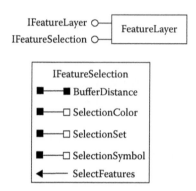

Figure 9.3   Properties and methods on *IFeatureSelection*.

created from a table object and a query filter object. A selection set object supports *ISelectionSet*, which has methods to manage the selection set such as making it permanent or combining it with another selection set. A *TableWindow* object implements *ITableWindow*, which has members to set up a table for data presentation and to highlight the selected subset (Figure 9.5).

Because the relationship between selected features and a selection set is like that between a feature layer and its feature class, there are ways to connect selected features and a selection set. For example, the *SelectionSet* property on *IFeatureSelection* can produce a selection set from the selected features.

9.2.2   Cursor

A *cursor* is a data-access object that allows the programmer to step through each row of a table for such purposes as counting and editing. Figure 9.6 shows two ways

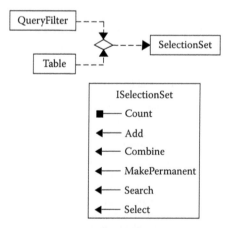

Figure 9.4   A *QueryFilter* object and a *Table* object together can create a *SelectionSet* object. *ISelectionSet* has members to manage the selection set.

ITableWindow

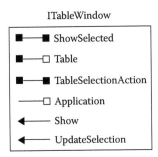

Figure 9.5   *ITableWindow* has members to set up a table for view.

for creating a cursor: combining a query filter and a table or a feature class or a feature layer, or combining a query filter and a selection set.

A cursor object created from a feature class or a feature layer is a feature cursor. A feature cursor object supports *IFeatureCursor* (Figure 9.7). Two important methods on *IFeatureCursor* are *NextFeature* and *UpdateFeature*. The *NextFeature* method advances the position of the cursor by one and returns the feature at that position. The method is therefore useful for setting up a loop to step through each feature in the cursor. The *UpdateFeature* method can update the feature corresponding to the current position of the cursor.

9.2.3 Data Statistics

The primary component for deriving descriptive statistics is *DataStatistics*. A data statistics object implements *IDataStatistics*, which has properties that let the user set a field to gather statistics on and a cursor to use either all records or the selected records (Figure 9.8).

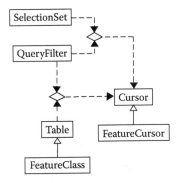

Figure 9.6   Two ways for creating a *Cursor* object.

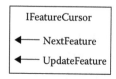

Figure 9.7 Methods on *IFeatureCursor.*

9.3 PERFORMING ATTRIBUTE QUERY

This section includes three sample macros for attribute query from a feature layer and from a table.

9.3.1 *SelectFeatures*

SelectFeatures selects from a feature layer features that meet a query expression and highlights the selected features. The macro performs the same function as using the Select By Attributes command in ArcMap. *SelectFeatures* has two parts. Part 1 defines the feature layer, and Part 2 prepares a new query filter, uses the query filter to select features, and highlights the selected features.

> **Key Interfaces:** *IActiveView, IFeatureLayer, IFeatureSelection, IQueryFilter*
> **Key Members:** *WhereClause, SelectFeatures, PartialRefresh*
> **Usage:** Add *idcities.shp* to the focus map as the top layer. Import *SelectFeatures* to Visual Basic Editor in ArcMap. Run the macro. Those cities that have population > 10000 are selected and highlighted.

```
Private Sub SelectFeatures()
    ' Part 1: Define the feature layer.
    Dim pDoc As IMxDocument
    Dim pMap As IMap
    Dim pActiveView As IActiveView
    Dim pFeatureLayer As IFeatureLayer
    Set pDoc = ThisDocument
    Set pMap = pDoc.FocusMap
    Set pActiveView = pMap
    Set pFeatureLayer = pMap.Layer(0)
```

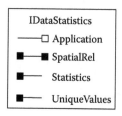

Figure 9.8 *IDataStatistics* has properties for accessing statistics and unique values on a field.

Part 1 sets *pFeatureLayer* to be the top layer in the active map.

```
' Part 2: Select features.
Dim pQueryFilter As IQueryFilter
Dim pFeatureSelection As IFeatureSelection
' Prepare a query filter.
Set pQueryFilter = New QueryFilter
pQueryFilter.WhereClause = "Population > 10000"
' Refresh the old selection if any to erase it.
pActiveView.PartialRefresh esriViewGeoSelection, Nothing, Nothing
' Select features.
Set pFeatureSelection = pFeatureLayer
pFeatureSelection.SelectFeatures pQueryFilter, esriSelectionResultNew, False
' Refresh again to draw the new selection.
pActiveView.PartialRefresh esriViewGeoSelection, Nothing, Nothing
End Sub
```

Part 2 first creates *pQueryFilter* as an instance of the *QueryFilter* class and defines its *WhereClause* property. Next, the code refreshes the active view to erase any previous selection. Then the code performs a QueryInterface (QI) for *IFeature-Selection* and uses the *SelectFeatures* method to create a new selection of features. Finally, the code refreshes the active view and draws the new selection. The *esri-ViewGeoSelection* option limits the refresh to the selected features only.

Box 9.1   SelectFeatures_GP

SelectFeatures_GP uses the SelectLayerByAttribute tool in the Data Management toolbox to select from *idcities.shp* those cities with population > 10,000. Add *idci-ties.shp* as the top layer in an active map. Run the macro. Selected cities are highlighted.

```
Private Sub SelectFeatures_GP()
    ' Run this macro in ArcMap with idcities.shp as the top layer in the active map.
    ' Create the Geoprocessing object.
    Dim GP As Object
    Set GP = CreateObject("esriGeoprocessing.GpDispatch.1")
    ' SelectLayerByAttribute <in_layer_or_view> {NEW_SELECTION | ADD_TO_SELECTION |
    ' CLEAR_SELECTION} {where_clause}
    'Execute the selectlayerbyattribute tool.
    GP.SelectLayerByAttribute "idcities", "NEW_SELECTION", "POPULATION > 10000"
End Sub
```

9.3.2 *SelectRecords*

SelectRecords selects records from the attribute table of a shapefile, displays the selected records in a table window, and displays the feature layer with selected features highlighted. *SelectRecords* has four parts. Part 1 defines the feature class. Part 2 prepares a query filter and uses it to select records that meet the query expression. Part 3 prepares a table window, shows the table, and highlights the selected records. Part 4 displays a new feature layer based on the feature class and highlights features of the selected records.

> **Key Interfaces:** *IWorkspaceName, IDatasetName, IName, ITable, IQueryFilter, IScratchWorkspaceFactory, ISelectionSet, ITableWindow, IFeatureLayer, IFeature-Selection, IActiveView*
>
> **Key Members:** *WorkspaceFactoryProgID, PathName, WorkspaceName, Name, Open, WhereClause, Select, Table, TableSelectionAction, UpdateSelection, Application, Show, FeatureClass, PartialRefresh, SelectionSet*
>
> **Usage:** Import ***SelectRecords*** to Visual Basic Editor in ArcMap. Run the macro. The macro opens the attribute table of *idcounty*, highlights the selected records, adds *idcounty* to the active map, and highlights features of the selected records.

```
Private Sub SelectRecords()
    ' Part 1: Define the table.
    Dim pWSName As IWorkspaceName
    Dim pDatasetName As IDatasetName
    Dim pName As IName
    Dim pTable As ITable
    ' Get the dbf file.
    Set pWSName = New WorkspaceName
    pWSName.WorkspaceFactoryProgID = "esriCore.ShapefileWorkspaceFactory.1"
    pWSName.PathName = "c:\data\chap9"
    Set pDatasetName = New TableName
    pDatasetName.Name = "idcounty.dbf"
    Set pDatasetName.WorkspaceName = pWSName
    Set pName = pDatasetName
    ' Open the dbf table.
    Set pTable = pName.Open
```

Part 1 creates *pDatasetName* as an instance of the *TableName* class and defines its workspace and name. Next, the code accesses *IName* and uses the *Open* method to open *pTable*, a reference to the actual table.

```
    ' Part 2: Perform selection.
    Dim pQFilter As IQueryFilter
    Dim pScratchWS As IWorkspace
    Dim pScratchWSFactory As IScratchWorkspaceFactory
    Dim pSelectionSet As ISelectionSet
    ' Prepare a query filter.
    Set pQFilter = New QueryFilter
    pQFilter.WhereClause = "Change > 30"
    ' Select records and save the selection in a scratch workspace.
    Set pScratchWSFactory = New ScratchWorkspaceFactory
    Set pScratchWS = pScratchWSFactory.DefaultScratchWorkspace
    Set pSelectionSet = pTable.Select (pQFilter, esriSelectionTypeHybrid, esriSelectionOptionNormal, pScratchWS)
```

Part 2 creates *pQFilter* as an instance of the *QueryFilter* class and defines its *WhereClause* property. Next, the code creates a scratch workspace referenced by *pScratchWS*. The code then uses the *Select* method on *ITable* to create a selection set referenced by *pSelectionSet*. The *Select* method requires two selection constants in addition to the object qualifiers of *pQFilter* and *pScratchWS*.

```
' Part 3: Show the selection in a table window.
Dim pTableWindow As ITableWindow
' Create a table window and specify its properties and methods.
Set pTableWindow = New TableWindow
With pTableWindow
    Set .Table = pTable
    .TableSelectionAction = esriSelectFeatures
    .ShowSelected = False
    Set .Application = Application
    .Show True
    .UpdateSelection pSelectionSet
End With
```

Part 3 creates *pTableWindow* as an instance of the *TableWindow* class and defines its display properties as follows: *pTable* for *Table*, esriSelectFeatures for *TableSelectionAction*, False for *ShowSelected*, Application for *Application*, True for *Show*, and *pSelectionSet* for *UpdateSelection*. A false value for *ShowSelected* means that all records are displayed. A true value for *ShowSelected*, on the other hand, means that only the selected records are displayed. The entry of Application for *Application* is important because it specifies ArcMap (Application) as the window for displaying *pTableWindow*. If this property is not specified, the macro will crash.

```
' Part 4: Show selected features in the feature layer.
Dim pMxDoc As IMxDocument
Dim pMap As IMap
Dim pFeatureLayer As IFeatureLayer
Dim pFeatSelection As IFeatureSelection
Dim pActiveView As IActiveView
Set pMxDoc = ThisDocument
Set pMap = pMxDoc.FocusMap
' Use the dBASE table to create a new feature layer.
Set pFeatureLayer = New FeatureLayer
Set pFeatureLayer.FeatureClass = pTable
pFeatureLayer.Name = "idcounty"
' Add the new layer to the active map.
pMap.AddLayer pFeatureLayer
' Refresh the view to draw selected features.
Set pActiveView = pMap
Set pFeatSelection = pFeatureLayer
Set pFeatSelection.SelectionSet = pSelectionSet
pActiveView.PartialRefresh esriViewGeoSelection, Nothing, Nothing
End Sub
```

Part 4 creates *pFeatureLayer* as an instance of the *FeatureLayer* class, sets *pTable* to be its feature class, and sets idcounty to be its name. Next, the code adds *pFeatureLayer* to the active map. Finally, the code refreshes the view to draw the selected features in *pSelectionSet*.

9.4 PERFORMING SPATIAL QUERY

A VBA (Visual Basic for Applications) macro for a spatial query requires the use of a spatial filter object. A spatial filter in turn requires a query object whose geometry will be used for selecting features and a spatial relationship by which features of a target layer will be selected. Because a spatial filter object can have both spatial and attribute constraints, it is more versatile than a query filter object for custom applications.

9.4.1 *SpatialQuery*

SpatialQuery uses the shape of a preselected county and the spatial relationship of containment to select cities that the county contains. The macro performs the same function as using the Select By Location command in ArcMap to select features from the city layer that are completely within the selected feature in the county layer.

SpatialQuery has three parts. Part 1 defines the selected feature on the layer for selection (the county layer), Part 2 sets up a spatial filter, and Part 3 selects features and uses a cursor to count the number of selected features (selected cities).

> **Key Interfaces:** *IFeatureSelection, ISelectionSet, IFeatureCursor, IFeature, ISpatial-Filter, IFeatureClass*
>
> **Key Members:** *SelectionSet, Search, NextFeature, Geometry, Shape, SpatialRel, FeatureClass*
>
> **Usage:** Add *idcounty.shp* and *idcities.shp* to the active map. *idcities* must be on top of *idcounty*. Use the Select Features tool to select a county. Import *SpatialQuery* to Visual Basic Editor in ArcMap. Run the macro. A message box reports the number of cities selected.

```
Private Sub SpatialQuery()
    ' Part 1: Define the selected feature on the layer for selection.
    Dim pMxDoc As IMxDocument
    Dim pCountyLayer As IFeatureSelection
    Dim pCountySelection As ISelectionSet
    Dim pCountyCursor As IFeatureCursor
    Dim pCounty As IFeature
    Set pMxDoc = ThisDocument
    Set pCountyLayer = pMxDoc.FocusMap.Layer(1)
    Set pCountySelection = pCountyLayer.SelectionSet
    ' Create a cursor from the selected feature in Layer (1).
    pCountySelection.Search Nothing, True, pCountyCursor
    ' Return the selected feature from the cursor.
    Set pCounty = pCountyCursor.NextFeature
    ' Exit the sub if no feature has been selected.
    If pCounty Is Nothing Then
        MsgBox "Please select a county"
        Exit Sub
    End If
```

Part 1 locates and verifies the selected feature on the layer for selection. The code sets *pCountyLayer* to be the second layer in the active map and *pCountySelection* to be the selection set of the layer. Next, the code uses the *Search* method on *ISelectionSet* to create a feature cursor referenced by *pCountyCursor*. Because the macro is designed to work with a selected county, the feature cursor should contain a single county. The *NextFeature* method on *IFeatureCursor* returns the county and assigns it to *pCounty*. If *pCountyCursor* contains no selected feature, the macro terminates.

```
' Part 2: Prepare a spatial filter.
Dim pSpatialFilter As ISpatialFilter
Set pSpatialFilter = New SpatialFilter
Set pSpatialFilter.Geometry = pCounty.Shape
pSpatialFilter.SpatialRel = esriSpatialRelContains
```

Part 2 creates *pSpatialFilter* as an instance of the *SpatialFilter* class and defines its query geometry and spatial relationship. In this case, the query geometry is the shape of *pCounty* and the spatial relationship is *esriSpatialRelContains*, which stipulates that the query geometry contains the target geometry (that is, cities).

```
' Part 3: Select features and report number of selected features.
Dim pCityLayer As IFeatureLayer
Dim pCityFClass As IFeatureClass
Dim pCityCursor As IFeatureCursor
Dim pCity As IFeature
Dim intCount As Integer
Set pCityLayer = pMxDoc.FocusMap.Layer(0)
' Select features from Layer (0) and save them to a cursor.
Set pCityFClass = pCityLayer.FeatureClass
Set pCityCursor = pCityFClass.Search(pSpatialFilter, False)
' Loop through the cursor and count number of selected features.
Set pCity = pCityCursor.NextFeature
Do Until pCity Is Nothing
    intCount = intCount + 1
    Set pCity = pCityCursor.NextFeature
Loop
MsgBox "This county has " & intCount & " cities"
End Sub
```

Part 3 sets *pCityLayer* to be the top layer in the active map and *pCityFClass* to be its feature class. Next, the code uses the *Search* method on *IFeatureClass* to create a feature cursor referenced by *pCityCursor*. (An alternative is to use the *Search* method on *IFeatureLayer* to create the cursor.) The code then steps through each feature in *pCityCursor* and counts the number of features using the *intCount* variable. Finally, the code reports *intCount* in a message box.

Box 9.2   SpatialQueryByName_GP

SpatialQueryByName_GP uses the SelectLayerByAttribute tool and a name given by the user, such as Ada, to select a county and then the SelectLayerByLocation tool to select cities that are within the county. Add *idcities.shp* and *idcounty.shp* to the active map in ArcMap. The macro selects and highlights the selected county and cities.

```
Private Sub SpatialQueryByName_GP()
    ' Run this macro in ArcMap with idcounty.shp and idcities.shp in the active map.
    ' idcities must be on top of idcounty.
    'Create the Geoprocessing object
    Dim GP As Object
    Set GP = CreateObject("esriGeoprocessing.GpDispatch.1")
    ' SelectLayerByAttribute <in_layer_or_view> {NEW_SELECTION | ADD_TO_SELECTION |
    ' CLEAR_SELECTION} {where_clause}
    'Execute the selectlayerbyattribute tool.
    Dim name As String
    name = InputBox("Enter the name of a county")
    GP.SelectLayerByAttribute "idcounty", "NEW_SELECTION", "CO_NAME = '" & name & "'"
    ' SelectLayerByLocation <in_layer> {INTERSECT | WITHIN_A_DISTANCE | COMPLETELY_CONTAINS |
    ' COMPLETELY_WITHIN | HAVE_THEIR_CENTER_IN | SHARE_A_LINE_SEGMENT_WITH |
    ' BOUNDARY_TOUCHES | ARE_IDENTICAL_TO | CROSSED_BY_THE_OUTLINE_OF |
    ' CONTAINS | CONTAINED_BY}{select_features} {search_distance} {NEW_SELECTION |
    ' ADD_TO_SELECTION | REMOVE_FROM_SELECTION | SUBSET_SELECTION | SWITCH_SELECTION}
    ' Execute the selectlayerbylocation tool.
    GP.SelectLayerByLocation "idcities", "COMPLETELY_WITHIN", "idcounty"
End Sub
```

9.4.2 *SpatialQueryByName*

Instead of using a preselected county, ***SpatialQueryByName*** uses an input box to get the name of a county from the user. The code then selects and highlights the county before selecting cities that the county contains. The only difference between ***SpatialQuery*** and ***SpatialQueryByName*** is Part 1. Parts 2 and 3 are the same and are not listed below. (SpatialQueryByName.txt on the companion CD has all three parts.)

To use ***SpatialQueryByName***, first add *idcounty.shp* and *idcities.shp* to the active map. *idcities* must be on top of *idcounty*. Import ***SpatialQueryByName*** to Visual Basic Editor in ArcMap. Run the macro. Enter the name of a county, such as Ada. (The evaluation of the county name is case sensitive.) A message box reports the number of cities selected within Ada County.

```
Private Sub SpatialQueryByName()
    ' Part 1: Define the selected feature on the layer for selection.
    Dim pMxDoc As IMxDocument
    Dim pActiveview As IActiveview
    Dim pCountyLayer As IFeatureLayer
    Dim pFeatureSelection As IFeatureSelection
    Dim name As String
    Dim pQueryFilter As IQueryFilter
    Dim pCountySelection As ISelectionSet
    Dim pCountyCursor As IFeatureCursor
```

```
Dim pCounty As IFeature
Set pMxDoc = ThisDocument
Set pActiveview = pMxDoc.FocusMap
Set pCountyLayer = pMxDoc.FocusMap.Layer(1)
' Get a county name from the user.
name = InputBox("Enter a county name: ", "")
' Prepare a new query filter.
Set pQueryFilter = New QueryFilter
pQueryFilter.WhereClause = "CO_NAME = '" & name & "'"
' Refresh or erase any previous selection.
pActiveview.PartialRefresh esriviewGeoSelection, Nothing, Nothing
' Select features.
Set pFeatureSelection = pCountyLayer
' Refresh again to draw the new selection.
pFeatureSelection.SelectFeatures pQueryFilter, esriSelectionResultNew, False
pActiveview.PartialRefresh esriviewGeoSelection, Nothing, Nothing
' Create a selection set from the selected feature.
Set pCountySelection = pFeatureSelection.SelectionSet
' Create a cursor from the selection set.
pCountySelection.Search Nothing, True, pCountyCursor
' Return the selected feature.
Set pCounty = pCountyCursor.NextFeature
```

Part 1 uses an input box to get a county name from the user and assigns it to the string variable *name*. Next, the code performs a QI for the *IFeatureSelection* interface and uses the *SelectFeatures* method to select the county. The code then highlights the selected county in the map. The rest of Part 1 processes the selected county before using it for selecting cities. The code sets *pCountySelection* to be the selection set of *pFeatureSelection*. Then the code creates a feature cursor and uses the *NextFeature* method to return the selected county.

9.4.3 *MultipleSpatialQueries*

MultipleSpatialQueries selects cities that are within two or more selected counties. The code steps through each county and adds selected cities to those that are already in a cursor. *MultipleSpatialQueries* has three parts. Part 1 defines the layer for selection (county layer) and the selected features on the layer. Part 2 loops through each selected county, uses a spatial filter to select cities that are within the county, and adds the selected cities to a cursor. Part 3 draws all selected features including counties and cities and reports the number of selected cities.

Key Interfaces: *IFeatureSelection, ISelectionSet, IFeatureCursor, IFeature, ISpatial-Filter, IFeatureSelection, IActiveView*
Key Members: *SelectionSet, Search, Geometry, SpatialRel, SelectFeatures, NextFeature*
Usage: Add *idcounty.shp* and *idcities.shp* to the active map. *idcities* must be on top of *idcounty*. Use the Select Features tool to select two or more counties. Import *MultipleSpatialQueries* to Visual Basic Editor in ArcMap. Run the macro. The macro draws all selected counties and cities, and a message box reports the number of cities selected.

```
Private Sub MultipleSpatialQueries()
    ' Part 1: Define the selected counties.
    Dim pMxDoc As IMxDocument
    Dim pCountyLayer As IFeatureSelection
    Dim pCountySelection As ISelectionSet
    Dim pCountyCursor As IFeatureCursor
    Dim pCounty As IFeature
    Set pMxDoc = ThisDocument
    Set pCountyLayer = pMxDoc.FocusMap.Layer(1)
    Set pCountySelection = pCountyLayer.SelectionSet
    ' Create a cursor from the selected counties.
    pCountySelection.Search Nothing, True, pCountyCursor
```

Part 1 defines *pCountyLayer* to be the second layer in the active map and creates *pCountyCursor* as a feature cursor that contains the selection set in *pCountyLayer*.

```
    ' Part 2: Loop through each selected county to select cities.
    Dim intCount As Integer
    Dim pCityLayer As IFeatureLayer
    Dim pSpatialFilter As ISpatialFilter
    Dim pCitySelection As IFeatureSelection
    intCount = 0
    Set pCityLayer = pMxDoc.FocusMap.Layer(0)
    Set pCitySelection = pCityLayer
    ' Step through each county to select cities.
    Set pCounty = pCountyCursor.NextFeature
    Do Until pCounty Is Nothing
        ' Prepare a spatial filter.
        Set pSpatialFilter = New SpatialFilter
        Set pSpatialFilter.Geometry = pCounty.Shape
        pSpatialFilter.SpatialRel = esriSpatialRelContains
        ' Select cities and add them to those already selected.
        pCitySelection.SelectFeatures pSpatialFilter, esriSelectionResultAdd, False
        Set pCounty = pCountyCursor.NextFeature
    Loop
```

Part 2 first sets *pCityLayer* to be the top layer in the active map. Next, the code steps through each selected feature in *pCountyCursor*. Within the loop, the code prepares a spatial filter and defines its properties of geometry and spatial relationship. The code then uses the *SelectFeatures* method on *IFeatureSelection* to select cities that are within the county. The argument for the selection method is esriSelection-ResultAdd, which adds selected features to those already selected.

```
    ' Part 3: Draw all selected features and report number of selected cities.
    Dim pActiveView As IActiveView
    Dim pSelCity As ISelectionSet
    Dim pCityCursor As IFeatureCursor
    Dim pCity As IFeature
    Set pActiveView = pMxDoc.FocusMap
    ' Refresh selected features.
```

```
pActiveView.PartialRefresh esriviewGeoSelection, Nothing, Nothing
' Prepare a cursor for selected cities.
Set pSelCity = pCitySelection.SelectionSet
pSelCity.Search Nothing, True, pCityCursor
' Count number of cities in the cursor.
Set pCity = pCityCursor.NextFeature
Do Until pCity Is Nothing
    intCount = intCount + 1
    Set pCity = pCityCursor.NextFeature
Loop
MsgBox "These counties have " & intCount & " cities."
End Sub
```

Part 3 draws all selected features by using the *PartialRefresh* method on *IActive-View*. The code then creates a cursor for the selection set of *pCitySelection* from Part 2 and counts the number of selected cities in the cursor.

9.4.4 *SelectByShape*

SelectByShape uses a rectangle element drawn by the user as the query geometry to select cities that meet an attribute query expression. The macro performs the same function as using the Select Features tool, followed by Select By Location, in ArcMap to select features. *SelectByShape* is organized into three parts. Part 1 uses the extent drawn by the user to prepare a rectangle element. Part 2 prepares a fill symbol, shown only with an outline in red, for the rectangle. Part 3 prepares a spatial filter, selects cities within the rectangle, and counts the number of cities selected.

> **Key Interfaces:** *IGraphicsContainer, IEnvelope, IRubberBand, IElement, IFillShape-Element, IFillSymbol, IColor, ILineSymbol, ISpatialFilter, IFeatureCursor, IFeature*
> **Key Members:** *TrackNew, Geometry, Symbol, Color, Outline, Transparency, Width, AddElement, WhereClause, SpatialRel, Search, NextFeature*
> **Usage:** Add *idcities.shp* to the active map. Follow the steps below to attach *SelectByShape* to a tool so that the user can use the tool to draw a rectangle and to run the macro.

1. Double-click the empty space to the right of a toolbar in ArcMap to open the Customize dialog.
2. Click New in the Customize dialog. In the New Toolbar dialog, use the dropdown menu to select Untitled to save in. Click OK. Custom Toolbar 1 is added as a new toolbar to ArcMap.
3. Click the Commands tab. Select UIControls from the list of categories. Click New UIControl to open its dialog. Check the option for UIToolControl and then Create. Drag Project UIToolControl1 from the Commands list to the new toolbar.
4. Right-click UIToolControl1 and, if desired, select Change Button Image to change the tool icon. Right-click UIToolControl1 again and select View Source to open Visual Basic Editor. Make sure that the object list on the upper left of the Code window shows UIToolControl1 and the procedure list on the upper right shows MouseDown rather than the default option of Select. Copy and paste *SelectBy-Shape* to the Code window. Check the code lines and make sure that there are no errors due to misaligned code lines. Close Visual Basic Editor.

5. Click on UIToolControl1 and draw a rectangle on *idcities*. The rectangle is shown in a red outline symbol and a message box reports how many cities meet the selection criteria.

6. The rectangle is a graphic element, which can be deleted from the display by first using the Select Elements tool to select it.

```
Private Sub UIToolControl1_MouseDown(ByVal button As Long, ByVal shift As Long,ByVal x As Long, ByVal y As Long)
    ' Part 1: Get a rectangle element drawn by the user.
    Dim pEnv As IEnvelope
    Dim pRubberEnv As IRubberBand
    Dim pMxDoc As IMxDocument
    Dim pElem As IElement
    ' Create a new rubber envelope.
    Set pRubberEnv = New RubberEnvelope
    ' Return a new envelope from the tracker object.
    Set pMxDoc = ThisDocument
    Set pEnv = pRubberEnv.TrackNew(pMxDoc.ActiveView.ScreenDisplay, Nothing)
    ' Create a new envelope element.
    Set pElem = New RectangleElement
    pElem.Geometry = pEnv
```

Part 1 makes a rectangle element based on the extent drawn by the user. The code first creates *pRubberEnv* as an instance of the *RubberEnvelope* class, which implements *IRubberBand* (Figure 9.9). Next, the code uses the *TrackNew* method on *IRubberBand* to track (rubberband) the shape on the screen and assigns the shape to an *IEnvelope* object referenced by *pEnv*. The code then creates *pElem* as an instance of the *RectangleElement* class and assigns *pEnv* to be its geometry (Figure 9.10).

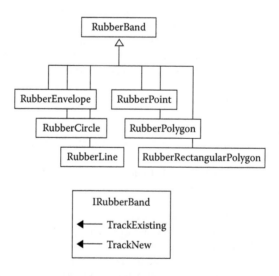

Figure 9.9 *RubberEnvelope* is one of six types of *RubberBand*. A rubber envelope object shares the methods on *IRubberBand*.

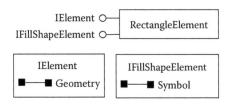

Figure 9.10   A *RectangleElement* object supports *IElement* and *IFillShapeElement*. The interfaces provide access to the geometry and symbol of the object.

```
' Part 2: Add the rectangle with a red outline to view.
Dim pFillShapeElement As IFillShapeElement
Dim pFillSymbol As IFillSymbol
Dim pColor As IColor
Dim pLineSymbol As ILineSymbol
Dim pGContainer As IGraphicsContainer
Set pFillShapeElement = pElem
' Set the fill symbol.
Set pFillSymbol = pFillShapeElement.Symbol
Set pColor = pFillSymbol.Color
pColor.Transparency = 0
pFillSymbol.Color = pColor
' Set the outline symbol.
Set pLineSymbol = pFillSymbol.Outline
pColor.Transparency = 255
pColor.RGB = RGB(255, 0, 0)
pLineSymbol.Width = 0.1
pFillSymbol.Outline = pLineSymbol
' Assign the symbol to the rectangle.
pFillShapeElement.Symbol = pFillSymbol
' Add the rectangle to the graphics container, and refresh the display.
Set pGContainer = pMxDoc.ActiveView
pGContainer.AddElement pElem, 0
pMxDoc.ActiveView.Refresh
```

Part 2 adds the rectangle in a red outline symbol to view. The code first accesses
IFillShapeElement to define a fill symbol for *pElem* (Figure 9.10). The symbol
consists of a fill color and an outline symbol. In this case, the fill color has a
transparency of zero, meaning that the color is transparent. The outline symbol is a
thin line symbol in solid red. The red, green, blue (*RGB*) property of *pColor* defines
the color of the outline symbol, which is derived by using Visual Basic's RGB
function. The *Width* property of *pLineSymbol* defines the width of the outline symbol.
After the fill symbol is defined, the code assigns *pFillSymbol* to be the symbol of
pElem. Finally, the code adds *pElem* to the graphics container, which is set to be
the active view of the document, and refreshes the view.

```
' Part 3: Use the rectangle to search features in the top layer.
Dim pSpatialFilter As ISpatialFilter
Dim pCityLayer As IFeatureLayer
```

```
Dim pCityCursor As IFeatureCursor
Dim pCity As IFeature
Dim intCount As Integer
' Prepare a spatial filter.
Set pSpatialFilter = New SpatialFilter
pSpatialFilter.WhereClause = "Population > 5000"
Set pSpatialFilter.Geometry = pEnv
pSpatialFilter.SpatialRel = esriSpatialRelContains
' Create a cursor by searching the top layer's feature class.
Set pCityLayer = pMxDoc.FocusMap.Layer(0)
Set pCityCursor = pCityLayer.Search(pSpatialFilter, False)
' Count number of features in the cursor.
Set pCity = pCityCursor.NextFeature
Do Until pCity Is Nothing
    intCount = intCount + 1
    Set pCity = pCityCursor.NextFeature
Loop
MsgBox "There are " & intCount & " cities over 5000 within the rectangular area."
End Sub
```

Part 3 creates *pSpatialFilter* as an instance of the *SpatialFilter* class and defines its properties of *WhereClause*, geometry, and spatial relationship. Next, the code sets *pCityLayer* to be the top layer in the active map and uses the *Search* method on *IFeatureLayer* to create a feature cursor. Finally, the code steps through each feature in the cursor and reports the number of selected cities in a message box.

9.5 COMBINING SPATIAL AND ATTRIBUTE QUERIES

Because a spatial filter object can use both spatial and attribute constraints, it seems redundant to have a separate section on combining spatial and attribute queries. In many scenarios, however, a macro needs to perform an attribute query or a spatial query or both more than once; this creates a more challenging task to solve than the sample macros covered so far in this chapter.

9.5.1 *BufferSelect*

BufferSelect selects thermal springs and wells that have water temperatures higher than 60°C and are within 8000 meters of an interstate in Idaho. The macro has three parts. Part 1 defines a road layer, selects interstates by attributes, and saves the interstates into a feature cursor. Part 2 loops through each interstate segment in the cursor, buffers it with a distance of 8000 meters, and uses a spatial filter object to select thermal springs and wells that the buffer contains. Part 3 draws all selected features and reports the total number of thermal springs and wells selected.

BufferSelect uses *ITopologicalOperator* to create the buffer zone around the interstates. Implemented by a point, polyline, or polygon object, *ITopologicalOperator* has methods such as buffer, clip, intersect, and union to create new geometries (Figure 9.11).

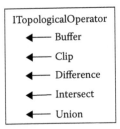

Figure 9.11 Methods on *ITopologicalOperator*.

Key Interfaces: *IFeatureSelection, IQueryFilter, ISelectionSet, IFeatureCursor, ISpatialFilter, IElement, IFeature, ITopologicalOperator, IActiveView*

Key Members: *WhereClause, SelectFeatures, BufferDistance, SelectionSet, Search, NextFeature, Shape, Buffer, Geometry, SpatialRel, PartialRefresh*

Usage: Add *idroads.shp* and *thermal.shp* to the active map. *idroads* must be on top of *thermal*. Import ***BufferSelect*** to Visual Basic Editor in ArcMap. Run the macro. A message box reports the number of thermal springs and wells that meet the selection criteria. The interstates and the selected thermal springs and wells are highlighted on the map.

```
Private Sub BufferSelect()
    ' Part 1: Create a cursor of interstates.
    Dim pMxDoc As IMxDocument
    Dim pMap As IMap
    Dim pRoadLayer As IFeatureLayer
    Dim pFeatSelection As IFeatureSelection
    Dim pQueryFilter As IQueryFilter
    Dim pRoadSelSet As ISelectionSet
    Dim pRoadCursor As IFeatureCursor
    Set pMxDoc = ThisDocument
    Set pMap = pMxDoc.FocusMap
    Set pRoadLayer = pMap.Layer(0)
    Set pFeatSelection = pRoadLayer
    ' Select interstates.
    Set pQueryFilter = New QueryFilter
    pQueryFilter.WhereClause = "Route_Desc = 'Interstate'"
    pFeatSelection.SelectFeatures pQueryFilter, esriSelectionResultNew, False
    ' Display an 8000-meter buffer around the interstates.
    pFeatSelection.BufferDistance = 8000
    ' Create a feature cursor of selected interstates.
    Set pRoadSelSet = pFeatSelection.SelectionSet
    pRoadSelSet.Search Nothing, False, pRoadCursor
```

Part 1 first defines *pRoadLayer* as the top layer in the active map. To select the interstates, the code creates a query filter object, switches to the *IFeatureSelection* interface, and uses the *SelectFeatures* method to create a feature selection referenced by *pFeatSelection*. The *BufferDistance* property of *pFeatSelection* is set to display a buffer distance of 8000 meters. Then the code assigns the selection set of *pFeatSelection* to *pRoadSelSet* and creates from *pRoadSelSet* a feature cursor referenced by *pRoadCursor*. Because the creation of the cursor uses no query filter, all selected interstates are included in the cursor.

```
' Part 2: Buffer each interstate segment, and select thermals.
Dim pSpatialFilter As ISpatialFilter
Dim pThermalLayer As IFeatureLayer
Dim pElement As IElement
Dim pThermalSelection As IFeatureSelection
Dim pRoad As IFeature
Dim pTopoOperator As ITopologicalOperator
' Prepare a spatial filter.
Set pSpatialFilter = New SpatialFilter
pSpatialFilter.WhereClause = "temp > 60"
pSpatialFilter.SpatialRel = esriSpatialRelContains
' Define the thermal layer whose features will be selected.
Set pThermalLayer = pMxDoc.FocusMap.Layer(1)
Set pThermalSelection = pThermalLayer
' Step through each interstate segment and select thermals.
Set pRoad = pRoadCursor.NextFeature
Do Until pRoad Is Nothing
    ' Create an 8000-meter buffer from the road.
    Set pTopoOperator = pRoad.Shape
    Set pElement = New PolygonElement
    pElement.Geometry = pTopoOperator.Buffer(8000)
    ' Define the geometry of the spatial filter.
    Set pSpatialFilter.Geometry = pElement.Geometry
    ' Select thermals and add them to the selection set.
    pThermalSelection.SelectFeatures pSpatialFilter, esriSelectionResultAdd, False
    Set pRoad = pRoadCursor.NextFeature
Loop
```

Part 2 selects thermal springs and wells that meet the following two criteria: within 8000 meters of an interstate, and having water temperatures above 60°C. The code initially creates a spatial filter object referenced by *pSpatialFilter* and defines its attribute constraint and spatial relationship. Next, the code sets *pThermalLayer* to be the second layer in the active map and accesses the *IFeatureSelection* interface. Then the code steps through each selected interstate segment. Within the loop, two tasks are performed. First, the code uses the *Buffer* method on *ITopologicalOperator* to create an 8000-meter buffer polygon around the shape of each line segment. This buffer polygon is then assigned to be the geometry of *pSpatialFilter*. Second, the code uses the *SelectFeatures* method on *IFeatureSelection* to create a feature selection. The argument used for the selection method is esriSelectionResultAdd, which adds selected thermal springs and wells to the current selection. The loop continues until all interstates are exhausted.

```
' Part 3: Draw all selected features, and report number of thermals selected.
Dim pActiveView As IActiveView
Dim pThermalSelSet As ISelectionSet
```

```
Set pActiveView = pMxDoc.FocusMap
pActiveView.PartialRefresh esriViewGeoSelection, Nothing, Nothing
Set pThermalSelSet = pThermalSelection.SelectionSet
MsgBox "There are " & pThermalSelSet.Count & " thermals selected"
End Sub
```

Part 3 draws all selected interstates and thermal springs and wells. Finally, a message box reports the number of thermal springs and wells that meet the selection criteria.

9.5.2 *IntersectSelect*

IntersectSelect selects counties in Idaho that intersect the interstate highways and have a rate of population increase over 15% between 1990 and 2000. Like *Buffer-Select*, *IntersectSelect* uses both attribute and spatial queries. One major difference is that *IntersectSelect* applies the intersect relationship in selecting counties.

IntersectSelect has three parts. Part 1 selects interstates from the road layer and creates a feature cursor of the selected features. Part 2 uses a spatial filter object to select counties that intersect an interstate and have a high rate of population growth. Part 3 shows the selected features and reports the number of counties selected.

> **Key Interfaces:** *IFeatureSelection, IQueryFilter, ISelectionSet, IFeatureCursor, ISpatialFilter, IElement, IFeature, IActiveView*
>
> **Key Members:** *WhereClause, SelectFeatures, SelectionSet, Search, NextFeature, Shape, Geometry, SpatialRel, PartialRefresh*
>
> **Usage:** Add *idroads.shp* and *idcounty.shp* to the active map. *idroads* must be on top of *idcounty*. (If necessary, clear selected features of *idroads*.) Import *IntersectSelect* to Visual Basic Editor in ArcMap. Run the macro. A message box reports the number of counties that meet the selection criteria. The interstates and the selected counties are highlighted on the map.

```
Private Sub IntersectSelect()
    ' Part 1: Create a cursor of interstates.
    Dim pMxDoc As IMxDocument
    Dim pMap As IMap
    Dim pRoadLayer As IFeatureLayer
    Dim pFeatSelection As IFeatureSelection
    Dim pQueryFilter As IQueryFilter
    Dim pRoadSelSet As ISelectionSet
    Dim pRoadCursor As IFeatureCursor
    Set pMxDoc = ThisDocument
    Set pMap = pMxDoc.FocusMap
    Set pRoadLayer = pMap.Layer(0)
    Set pFeatSelection = pRoadLayer
    ' Select interstates.
    Set pQueryFilter = New QueryFilter
    pQueryFilter.WhereClause = "Route_Desc = 'Interstate'"
    pFeatSelection.SelectFeatures pQueryFilter, esriSelectionResultNew, False
    ' Create a feature cursor of selected interstates.
    Set pRoadSelSet = pFeatSelection.SelectionSet
    pRoadSelSet.Search Nothing, False, pRoadCursor
```

Part 1 uses a query filter to select interstates from the road layer, before creating a feature cursor from the selected interstates.

```
' Part 2: Select high-growth counties that intersect an interstate.
Dim pCountyLayer As IFeatureLayer
Dim pElement As IElement
Dim pCountySelection As IFeatureSelection
Dim pRoad As IFeature
Dim pSpatialFilter As ISpatialFilter
Set pCountyLayer = pMxDoc.FocusMap.Layer(1)
Set pCountySelection = pCountyLayer
' Prepare a spatial filter.
Set pSpatialFilter = New SpatialFilter
pSpatialFilter.WhereClause = "change > 15"
pSpatialFilter.SpatialRel = esriSpatialRelIntersects
' Step through each interstate and select counties.
Set pRoad = pRoadCursor.NextFeature
Do Until pRoad Is Nothing
     ' Define the geometry of the spatial filter.
     Set pSpatialFilter.Geometry = pRoad.Shape
     ' Select counties and add them to the selection set.
     pCountySelection.SelectFeatures pSpatialFilter, esriSelectionResultAdd, False
     Set pRoad = pRoadCursor.NextFeature
Loop
```

Part 2 first creates a spatial filter object and defines its attribute constraint and spatial relationship. Next, the code steps through each interstate in a loop. Within the loop, the code defines the query geometry as the shape of the interstate and selects counties that intersect the interstate. As the loop continues, selected counties are added to the current selection set.

```
' Part 3: Draw all selected features and report number of counties selected.
Dim pActiveView As IActiveView
Dim pCountySelSet As ISelectionSet
Set pActiveView = pMxDoc.FocusMap
' Draw all selected features.
pActiveView.PartialRefresh esriViewGeoSelection, Nothing, Nothing
Set pCountySelSet = pCountySelection.SelectionSet
MsgBox "There are " & pCountySelSet.Count & " counties selected"
End Sub
```

Part 3 draws all selected interstates and counties. A message box then reports the number of selected counties.

9.6 DERIVING DESCRIPTIVE STATISTICS

This section discusses how to program ArcObjects to derive descriptive statistics on a field of a feature layer. These statistics can be based on all records or a subset of records.

9.6.1 *DataStatistics*

DataStatistics derives and reports descriptive statistics on a field using all records of a feature layer. The macro performs the same function as using the Statistics command in a field's context menu in ArcMap. *DataStatistics* has three parts. Part 1 defines the feature layer and creates a cursor from the layer. Part 2 creates a data statistics object and uses the object to derive the descriptive statistics of a field. Part 3 displays the statistics in a message box.

> **Key Interfaces:** *ICursor, IDataStatistics, IStatisticsResults*
> **Key Members:** *Search, Field, Cursor, Statistics, Maximum, Minimum, Mean*
> **Usage:** Add *idcounty.shp* to an active map. Import *DataStatistics* to Visual Basic Editor. Run the macro. The macro uses a message box to display the maximum, minimum, and mean values of the field change.

```
Private Sub DataStatistics()
    ' Part 1: Define the feature layer and cursor.
    Dim pMxDoc As IMxDocument
    Dim pFLayer As IFeatureLayer
    Dim pCursor As ICursor
    Set pMxDoc = ThisDocument
    Set pFLayer = pMxDoc.FocusMap.Layer(0)
    Set pCursor = pFLayer.Search(Nothing, False)
```

Part 1 Sets *pFLayer* to be the top layer in the active map and creates *pCursor* that includes every feature in *pFLayer*.

```
    ' Part 2: Derive statistics on the field change.
    Dim pData As IDataStatistics
    Dim pStatResults As IStatisticsResults
    Dim pChangeMax As Double
    Dim pChangeMin As Double
    Dim pChangeMean As Double
    ' Create a data statistics object.
    Set pData = New DataStatistics
    pData.Field = "change"
    Set pData.Cursor = pCursor
    Set pStatResults = pData.Statistics
    ' Get the maximum, minimum, and mean values.
    pChangeMax = pStatResults.Maximum
    pChangeMin = pStatResults.Minimum
    pChangeMean = pStatResults.Mean
```

Part 2 creates *pData* as an instance of the *DataStatistics* class and defines the following properties for the object: *Field* to be change, *Cursor* to be *pCursor*, and *Statistics* to be *pStatResults*. The code then uses the properties of *IStatisticsResults* to save the maximum, minimum, and mean statistics into the proper variables.

```
' Part 3: Display the statistics in a message box.
Dim sMsg As String
sMsg = "Statistics of the field change are:" & vbCrLf
sMsg = sMsg & "===================" & vbCrLf
sMsg = sMsg & "Maximum: " & pChangeMax & vbCrLf
sMsg = sMsg & "Minimum: " & pChangeMin & vbCrLf
sMsg = sMsg & "Mean: " & pChangeMean & vbCrLf
sMsg = sMsg & "==================="
MsgBox sMsg
End Sub
```

Part 3 uses a message box to display the statistics derived from Part 2. The constant vbCrLf creates a new line.

Box 9.3 DataStatistics_GP

DataStatistics_GP uses the Statistics tool in the Analysis toolbox to derive from *idcounty.shp* the minimum, maximum, and mean on the field Change and save the statistics to a dbf file. Run the macro in ArcCatalog, and open the dbf file to view the results.

```
Private Sub DataStatistics_GP()
    ' Create the Geoprocessing object.
    Dim GP As Object
    Set GP = CreateObject("esriGeoprocessing.GpDispatch.1")
    ' Statistics <in_table> <out_table> <field{Statistic Type}; field{Statistic Type}...> {case_field}
    ' Define the third parameter for the command.
    Dim Parameter3 As String
    Parameter3 = "CHANGE MIN;CHANGE MAX;CHANGE MEAN;CHANGE STD"
    ' Execute the statistics tool.
    GP.Statistics "c:\data\chap9\idcounty.shp", "c:\data\chap9\stats2.dbf", Parameter3
End Sub
```

9.6.2 *DataSubsetStatistics*

DataSubsetStatistics report statistics on a set of selected records. The macro performs the same function as using the Statistics command on a field of selected records in ArcMap. *DataSubsetStatistics* has three parts. Part 1 defines the feature layer and creates a cursor from the layer, Part 2 creates a data statistics object and uses the object to derive a field's descriptive statistics, and Part 3 displays the statistics in a message box.

> **Key Interfaces:** *IQueryFilter, ICursor, IDataStatistics, IStatisticsResults*
> **Key Members:** *WhereClause, Search, Field, Cursor, Statistics, Maximum, Minimum, Mean*
> **Usage:** Add *idcounty.shp* to an active map. Import *DataSubsetStatistics* to Visual Basic Editor. Run the macro. The macro uses a message box to display the maximum, minimum, and mean values of the field change for those counties that have change > 20.

```
Private Sub DataSubsetStatistics()
    ' Part 1: Get a handle on the feature layer and cursor.
    Dim pMxDoc As IMxDocument
    Dim pFLayer As IFeatureLayer
    Dim pQueryFilter As IQueryFilter
    Dim pCursor As ICursor
    Set pMxDoc = ThisDocument
    Set pFLayer = pMxDoc.FocusMap.Layer(0)
    Set pQueryFilter = New QueryFilter
    pQueryFilter.WhereClause = "change > 20"
    Set pCursor = pFLayer.Search(pQueryFilter, False)
```

Part 1 creates a query filter object and defines its *WhereClause* property. The code then creates a cursor that contains only those counties that have change > 20.

```
    ' Part 2: Derive statistics on the field change.
    Dim pData As IDataStatistics
    Dim pStatResults As IStatisticsResults
    Dim pChangeMax As Double
    Dim pChangeMin As Double
    Dim pChangeMean As Double
    ' Define an IDataStatistics object.
    Set pData = New DataStatistics
    pData.Field = "change"
    Set pData.Cursor = pCursor
    Set pStatResults = pData.Statistics
    ' Get the maximum, minimum, and mean values.
    pChangeMax = pStatResults.Maximum
    pChangeMin = pStatResults.Minimum
    pChangeMean = pStatResults.Mean
```

Part 2 creates a data statistics object, and derives its maximum, minimum, and mean.

```
    ' Part 3: Display the statistics in a message box.
    Dim sMsg As String
    sMsg = "Statistics of the field change are:" & vbCrLf
    sMsg = sMsg & "===================" & vbCrLf
    sMsg = sMsg & "Maximum: " & pChangeMax & vbCrLf
    sMsg = sMsg & "Minimum: " & pChangeMin & vbCrLf
    sMsg = sMsg & "Mean: " & pChangeMean & vbCrLf
    sMsg = sMsg & "==================="
    MsgBox sMsg
End Sub
```

Part 3 reports the statistics in a message box.

Vector Data Operations

Vector data analysis is based on the geometric objects of point, line, and polygon. Vector-based operations typically involve the shape of spatial features. Some operations also involve feature attributes. Four common types of vector data analyses are buffering, overlay, spatial join, and feature manipulation.

Buffering creates a buffer zone that is within a specified distance of each spatial feature in the input layer. If an input layer has 20 line segments, the first step in buffering is to create 20 buffer zones. These buffer zones are often dissolved to remove the overlapped areas between them. An important requirement for buffering is that the input layer must have the spatial reference information, which is needed for distance measurements.

Overlay combines the shapes and attributes of two layers to create the output. One of the two layers is the input layer and the other, the overlay layer. Each feature in the output contains a combination of attributes from both layers, and this combination separates the feature from its neighbors. Two common overlay methods are union and intersect. Union preserves all features from both layers, whereas intersect preserves only those features that fall within the overlapped area between the layers.

Spatial join joins attribute data from two layers by using a spatial relationship such as nearest neighbor, containment, or intersection. If nearest neighbor is applied, the distance measures between features of the two layers can also be included in the output.

Feature manipulation refers to various methods for manipulating features in a layer or between two layers. Unlike overlay, feature manipulation does not combine the shapes and attributes of the two layers into the output. Common manipulations include clip, dissolve, and merge.

This chapter covers vector data operations. Section 10.1 reviews vector data analysis using ArcGIS. Section 10.2 discusses objects that are related to vector data analysis. Section 10.3 offers a macro and a Geoprocessing (GP) macro for buffering. Section 10.4 offers a macro and a GP macro for an overlay operation. Section 10.5 has a macro for a spatial join operation. Section 10.6 includes macros and a GP macro for feature manipulation operations. All macros start with the listing of key interfaces and key members (properties and methods) and the usage.

10.1 ANALYZING VECTOR DATA IN ARCGIS

ArcGIS offers feature-based analysis tools through ArcToolbox. Analysis Tools includes clip, select, and split in the Extract toolset; erase, identity, intersect, symmetrical difference, union, and update in the Overlay toolset; and buffer, near, and point distance in the Proximity toolset. Data Management Tools has append and merge in the General toolset and dissolve and eliminate in the Generalization toolset.

Spatial Join in ArcMap can join attributes from two feature attribute tables based on a spatial relationship between features. The spatial relationship can be proximity, so that the joining of attributes takes place between features of a layer and their closest features in another layer. Alternatively, the spatial relationship can be containment, so that the joining of attributes takes place between features of a layer and the features of another layer that contain them. Spatial join can use a maximum search distance as a constraint, and the result of spatial join can include distance measures or aggregate statistics.

10.2 ARCOBJECTS FOR VECTOR DATA ANALYSIS

The primary ArcObjects components for vector data analysis are *FeatureCursor-Buffer*, *BasicGeoprocessor*, and *SpatialJoin*. A feature cursor buffer object supports *IFeatureCursorBuffer*, which has different methods and properties for the buffering operation (Figure 10.1). As examples, the *Dissolve* property determines whether overlapping buffered features should be dissolved, the *RingDistance* property sets the number of buffer rings, and the *ValueDistance* property specifies the constant buffer distance. *IFeatureCursorBuffer2* has additional properties that specify the spatial reference systems of the data frame, the source data, the target, and the buffering. A feature cursor buffer object also implements *IBufferProcessingParameter*, which has access to members that set and retrieve parameters for the buffering process.

A basic geoprocessor object implements *IBasicGeoprocessor*. *IBasicGeoprocessor* offers methods for intersect, union, clip, dissolve, and merge (Figure 10.2). All methods except for merge allow use of selected features as inputs.

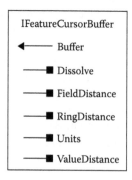

Figure 10.1 Properties on *IFeatureCursorBuffer* can define the buffering parameters.

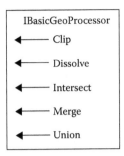

Figure 10.2 Methods on *IBasicGeoProcessor.*

A spatial join object supports *ISpatialJoin*, which provides methods for joining attributes of features based on a spatial relationship between features (Figure 10.3). The *JoinAggregate* method joins features using aggregate statistics, the *JoinNearest* method joins features using the nearest relationship, and the *JoinWithin* method joins features using the containment relationship.

Because vector data analysis deals with discrete features of points, lines, and areas, the properties and methods of these discrete features may become important for some applications. One example is the updating of the area and perimeter values of polygons on the output shapefile of an overlay operation. Another example is the derivation of centroids of a polygon feature class. Figure 10.4 summarizes properties and methods on *IPoint, ICurve,* and *IArea* that are important for vector data analysis.

When developing VBA (Visual Basic for Applications) macros for vector data analysis, we must be careful in separating vector data analysis from spatial data query (Chapter 9). A vector data analysis produces an output layer, whereas a spatial query produces a data subset that meets the selection criteria. Terms used in vector data analysis also appear for spatial data query, however. Among the spatial relationships that can be queried are constants such as esriSpatialRelIntersects and esriSpatialRelEnvelopeIntersects. For creating geometries for spatial query, the *ITopologicalOperator* interface actually provides methods for buffer, clip, intersect, and union. Therefore, it is important to know the context in which buffer, intersect, or union is used.

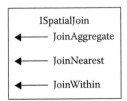

Figure 10.3 Methods on *ISpatialJoin.*

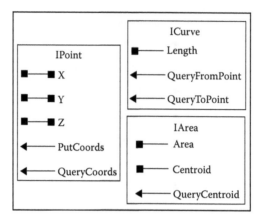

Figure 10.4 *IPoint, ICurve*, and *IArea* have members that can be important for vector data analysis.

10.3 BUFFERING

Writing VBA macros for buffering involves two important considerations. First, because the distance units used in buffering may or may not be the same as the map units of the input layer, a macro must specify the spatial reference of the input layer as well as the output layer. Second, because buffering is performed for each point, each line, or each polygon, a macro must set up a feature cursor for stepping through each feature for buffering.

10.3.1 *Buffer*

Buffer creates a buffer zone around features of a line shapefile. *Buffer* has four parts. Part 1 creates a cursor of all features in the input dataset and gets the spatial reference of the active map. Part 2 defines the workspace and dataset name for the output. Part 3 performs buffering. Part 4 creates a feature layer from the output and adds the layer to the active map.

> **Key Interfaces:** *IFeatureClass, IFeatureCursor, ISpatialReference, IFeatureCursor-Buffer2, IWorkspaceName, IDatasetName, IFeatureClassName, IName*
>
> **Key Members:** *FeatureClass,* Search, *SpatialReference, FeatureCursor, Dissolve, ValueDistance, BufferSpatialReference, DataFrameSpatialReference, SourceSpatialReference, TargetSpatialReference, WorkspaceFactoryProgID, PathName, Name, WorkspaceName, Buffer*
>
> **Usage:** Add *sewers.shp* to an active map. *sewers.shp* is measured in UTM coordinates and has the spatial reference information. Import ***Buffer*** to Visual Basic Editor. Run the macro. The macro creates a new shapefile named *Buffer_Result* and adds the shapefile as a feature layer to the active map.

```
Private Sub Buffer()
    ' Part 1: Get the map's spatial reference, and prepare a feature cursor.
    Dim pMxDoc As IMxDocument
```

```
Dim pMap As IMap
Dim pFeatureLayer As IFeatureLayer
Dim pFCursor As IFeatureCursor
Dim pSpatialReference As ISpatialReference
Set pMxDoc = ThisDocument
Set pMap = pMxDoc.FocusMap
' Set the spatial reference.
Set pSpatialReference = pMap.SpatialReference
' Prepare a feature cursor of all features in the top layer.
Set pFeatureLayer = pMap.Layer(0)
Set pFCursor = pFeatureLayer.Search(Nothing, False)
```

Part 1 sets *pSpatialReference* to be the spatial reference of the active map. Next, the code uses the *Search* method on *IFeatureLayer* to create a feature cursor referenced by *pFCursor*. Because the *Search* method does not use a query filter object, *pFCursor* contains all features of the feature layer.

```
' Part 2: Define the output.
Dim pBufWSName As IWorkspaceName
Dim pBufDatasetName As IDatasetName
Dim pBufFCName As IFeatureClassName
' Define the output's workspace and name.
Set pBufFCName = New FeatureClassName
Set pBufDatasetName = pBufFCName
Set pBufWSName = New WorkspaceName
pBufWSName.WorkspaceFactoryProgID = "esriCore.ShapeFileWorkspaceFactory.1"
pBufWSName.PathName = "c:\data\chap10"
Set pBufDatasetName.WorkspaceName = pBufWSName
pBufDatasetName.Name = "Buffer_result"
```

Part 2 creates *pBufFCName* as an instance of the *FeatureClassName* class. Then the code performs a QueryInterface (QI) for *IDatasetName* to define the workspace and name of *pBufFCName*.

```
' Part 3: Perform buffering.
Dim pFeatureCursorBuffer2 As IFeatureCursorBuffer2
' Define a feature cursor buffer object.
Set pFeatureCursorBuffer2 = New FeatureCursorBuffer
With pFeatureCursorBuffer2
    Set .FeatureCursor = pFCursor
    .Dissolve = True
    .ValueDistance = 300
    Set .BufferSpatialReference = pSpatialReference
    Set .DataFrameSpatialReference = pSpatialReference
    Set .SourceSpatialReference = pSpatialReference
    Set .TargetSpatialReference = pSpatialReference
End With
' Use the buffer method.
pFeatureCursorBuffer2.buffer pBufFCName
```

Part 3 creates *pFeatureCursorBuffer2* as an instance of the *FeatureCursorBuffer* class and defines its properties in a *With* block. The code specifies *pFCursor* for the feature cursor, True for dissolving overlapped areas, 300 (meters) for the constant buffer distance, and *pSpatialReference* for the spatial reference for buffering, the data frame, the source dataset, and the target dataset. The code then uses the *Buffer* method on *IFeatureCursorBuffer2* to buffer features in *pFCursor* and creates the output referenced by *pBufFCName*.

```
' Part 4: Create the output layer and add it to the active map.
Dim pName As IName
Dim pBufFC As IFeatureClass
Dim pBufFL As IFeatureLayer
Set pName = pBufFCName
Set pBufFC = pName.Open
Set pBufFL = New FeatureLayer
Set pBufFL.FeatureClass = pBufFC
pBufFL.Name = "Buffer_Result"
pMap.AddLayer pBufFL
End Sub
```

Part 4 first accesses the *IName* interface and uses the *Open* method to open a feature class object referenced by *pBufFC*. The code then creates a feature layer from *pBufFC* and adds the layer to the active map.

Box 10.1 Buffer_GP

Buffer_GP uses the Buffer tool in the Analysis toolbox to buffer *sewers.shp* with a constant buffer distance of 300 meters. The macro also specifies that the buffer zones be dissolved. Run the macro in ArcCatalog and examine the output shapefile (*Buffer_result2.shp*) in the Catalog tree.

```
Private Sub Buffer_GP()
    ' Create the Geoprocessing object.
    Dim GP As Object
    Set GP = CreateObject("esriGeoprocessing.GpDispatch.1")
    ' Buffer <in_features> <out_feature_class> <buffer_distance_or_field> {FULL I LEFT I RIGHT}
    ' {ROUND I FLAT} {NONE I ALL I LIST} {dissolve_field;dissolve_field...}
    ' Execute the buffer tool.
    GP.Buffer_analysis "c:\data\chap10\sewers.shp", "c:\data\chap10\Buffer_result2.shp", "300", "", "", "ALL"
End Sub
```

10.3.2 Buffer Options

In Section 10.3.1, **Buffer** buffers line features with a constant distance. A variety of buffer options are available through the properties of a feature cursor buffer object. The following summarizes these options by referring to statements in **Buffer**.

- To change from a constant buffer distance to varying buffer distances, change the statement .ValueDistance = 300 to .FieldDistance = "DistanceField" where DistanceField is the field that specifies the buffer distances.

- To create multiple buffer zones, change the statement `.ValueDistance = 300` to `.RingDistance(3) = 100`, where 3 means three rings and 100 means a 100-meter buffer distance for each ring.
- To create undissolved or discrete buffer zones, change the statement `.Dissolve = True` to `.Dissolve = False`.
- To use different buffer units (e.g., feet) than map units (e.g., meters), add the statement `Units(esriMeters) = esriFeet` as an additional property of *pFeatureCursorBuffer2*.

10.4 PERFORMING OVERLAY

To perform an overlay, a VBA macro must identify the input layer and the overlay layer. Among the methods offered by *IBasicGeoprocessor* are intersect, union, and clip. All three methods have exactly the same syntax. Therefore, the following macro on intersect can also be used for union and clip by simply changing the method on *IBasicGeoprocessor*.

10.4.1 *Intersect*

Intersect uses the intersect method to create an overlay output. *Intersect* has four parts. Part 1 defines the datasets for the intersect operation. Part 2 defines the output including its workspace and name. Part 3 sets up a basic geoprocessor object and runs the intersect method. Part 4 creates a feature layer from the output and adds the layer to the active map.

> **Key Interfaces:** *IFeatureClass, IWorkspaceName, IFeatureClassName, IDataset-Name, IBasicGeoprocessor*
> **Key Members:** *FeatureClass, WorkspaceFactoryProgID, PathName, Name, WorkspaceName, Intersect*
> **Usage:** Add *landsoil.shp* and *sewerbuf.shp* to an active map. *landsoil* will be used as the input layer, and *sewerbuf* as the overlay layer. *landsoil* must be on top of *sewerbuf* in the table of contents. Import *Intersect* to Visual Basic Editor. Run the macro. The macro creates *Intersect_result.shp* and adds the shapefile as a feature layer to the active map.

```
Private Sub Intersect()
    ' Part 1: Define the datasets for intersect.
    Dim pMxDoc As IMxDocument
    Dim pMap As IMap
    Dim pInputLayer As IFeatureLayer
    Dim pOverlayLayer As IFeatureLayer
    Dim pInputFC As IFeatureClass
    Dim pOverlayFC As IFeatureClass
    Set pMxDoc = ThisDocument
    Set pMap = pMxDoc.FocusMap
    ' Define the input feature class.
    Set pInputLayer = pMap.Layer(0)
    Set pInputFC = pInputLayer.FeatureClass
```

```
' Define the overlay table.
Set pOverlayLayer = pMap.Layer(1)
Set pOverlayFC = pOverlayLayer.FeatureClass
```

Part 1 sets the input layer to be the top layer, and the overlay layer the second layer, in the active map. The code then sets *pInputFC* to be the feature class of the input layer and *pOverlayFC* to be the feature class of the overlay layer.

```
' Part 2: Define the output dataset.
Dim pNewWSName As IWorkspaceName
Dim pFeatClassName As IFeatureClassName
Dim pDatasetName As IDatasetName
Set pFeatClassName = New FeatureClassName
Set pDatasetName = pFeatClassName
Set pNewWSName = New WorkspaceName
pNewWSName.WorkspaceFactoryProgID = "esriCore.ShapeFileWorkspaceFactory"
pNewWSName.PathName = "c:\data\chap10"
Set pDatasetName.WorkspaceName = pNewWSName
pDatasetName.name = "Intersect_result"
```

Part 2 creates *pFeatClassName* as an instance of the *FeatureClassName* class. The code then uses the *IDatasetName* interface to define the workspace and name of *pFCName*.

```
' Part 3: Perform intersect.
Dim pBGP As IBasicGeoprocessor
Dim tol As Double
Dim pOutputFC As IFeatureClass
' Define a basic geoprocessor object.
Set pBGP = New BasicGeoprocessor
' Use the default tolerance.
tol = 0#
' Run intersect.
Set pOutputFC = pBGP.Intersect(pInputTable, False, pOverlayTable, False, tol, pFeatClassName)
```

Part 3 creates *pBGP* as an instance of the *BasicGeoprocessor* class and uses the *Intersect* method on *IBasicGeoprocessor* to create the output referenced by *pOutputFC*. *Intersect* uses six object qualifiers and arguments. The four object qualifiers have been previously defined. The two arguments determine if only subsets of *pInputFC* and/or *pOverlayFC* are used in the overlay operation. If they are set to be false, the overlay operation ignores any selected subsets.

```
' Part 4: Create the output layer and add it to the active map.
Dim pOutputFeatLayer As IFeatureLayer
Set pOutputFeatLayer = New FeatureLayer
Set pOutputFeatLayer.FeatureClass = pOutputFC
pOutputFeatLayer.name = pOutputFeatClass.AliasName
pMap.AddLayer pOutputFeatLayer
End Sub
```

Part 4 creates a feature layer from *pOutputFC*, and adds the layer to the active map.

Box 10.2 Intersect_GP

Intersect_GP uses the Intersect tool in the Analysis toolbox to perform an overlay operation between *sewerbuf.shp* and *soils.shp*. Run the macro in ArcCatalog and view the output shapefile (*sewerbuf_soils.shp*) in the Catalog tree.

```
Private Sub intersect_GP()
    ' Create the Geoprocessing object and define its workspace.
    Dim GP As Object
    Set GP = CreateObject("esriGeoprocessing.GpDispatch.1")
    Dim filepath As String
    filepath = "c:\data\chap10"
    GP.Workspace = filepath
    ' Intersect <features{Ranks};features{Ranks}> <out_feature_class>
    ' {ALL | NO_FID | ONLY_FID} {cluster_tolerance} {INPUT | LINE | POINT}
    ' Define parameter 1 and parameter 2 for the command.
    Dim parameter1 As String
    parameter1 = "sewerbuf.shp;soils.shp"
    Dim parameter2 As String
    parameter2 = "sewerbuf_soils.shp"
    ' Execute the intersect tool.
    GP.Intersect_analysis parameter1, parameter2
End Sub
```

10.4.2 Updating Area and Perimeter of a Shapefile

A basic geoprocessor object does not automatically update the area and perimeter of an output shapefile from an overlay operation. The attribute table of the output from *Intersect* in fact contains two sets of area and perimeter from the input and overlay layers. However, neither set has been updated.

If geodatabase feature classes are used as inputs to an overlay operation, the output feature class will have the fields of shape_area and shape_length with correct area and perimeter values. (Chapter 14 has a sample macro that uses geodatabase feature classes in an overlay operation.) But if shapefiles are used as inputs, the area and perimeter of the output shapefile can be updated in two ways. The first method is to convert the shapefile to a geodatabase feature class by using a conversion macro such as *ShapefileToAccess* in Chapter 6. The geodatabase feature class will have the fields of shape_area and shape_length with the updated values. The second method is to add the following code fragment to *Intersect* as Part 5. (Intersect_Update.txt on the companion CD is a complete macro. The macro produces *Intersect_result2* with the updated area and perimeter values.)

```
' Part 5: Update area and perimeter of the output shapefile.
    Dim pField1 As IFieldEdit
    Dim pField2 As IFieldEdit
    Dim pUpdateCursor As IFeatureCursor
    Dim intArea As Integer
    Dim intLeng As Integer
    Dim pArea As IArea
    Dim pCurve As ICurve
    Dim pFeature As IFeature
    ' Define the two fields to be added.
    Set pField1 = New Field
    pField1.Name = "Shape_area"
    pField1.Type = esriFieldTypeDouble
    Set pField2 = New Field
    pField2.Name = "Shape_leng"
    pField2.Type = esriFieldTypeDouble
    ' Add the two new fields.
    pOutputFC.AddField pField1
    pOutputFC.AddField pField2
    ' Create a feature cursor to update each feature's new fields.
    Set pUpdateCursor = pOutputFC.Update(Nothing, False)
    ' Locate the new fields.
    intArea = pUpdateCursor.FindField("Shape_area")
    intLeng = pUpdateCursor.FindField("Shape_leng")
    Set pFeature = pUpdateCursor.NextFeature
    ' Loop through each feature.
    Do Until pFeature Is Nothing
        ' Update the area field.
        Set pArea = pFeature.Shape
        pFeature.Value(intArea) = pArea.Area
        ' Update the length field.
        Set pCurve = pFeature.Shape
        pFeature.Value(intLeng) = pCurve.Length
        ' Update the feature.
        pUpdateCursor.UpdateFeature pFeature
        Set pFeature = pUpdateCursor.NextFeature
    Loop
```

Part 5 first defines two new fields and adds them to *pOutputFC*. Next, the code uses the *Update* method on *IFeatureClass* to create a feature cursor referenced by *pUpdateCursor*, finds the index of the new field, and sets up a loop to step through each feature in *pUpdateCursor*. Within the loop, the Shape_area field is given the value of the area of the feature (*pArea.Area*) and the Shape_leng is given the value of the perimeter of the feature (*pCurve.Length*). Then the *UpdateFeature* method on *IFeatureCursor* updates the feature in the database.

When **Intersect** is run with the additional Part 5, the attribute table of the output shapefile will have the additional fields of Shape_area and Shape_leng. The values in those two fields are the updated values of area and perimeter.

10.5 JOINING DATA BY LOCATION

Spatial join joins attributes of two layers based on a spatial relationship between features. A common application of spatial join is to derive the distance between features using the nearest relationship.

10.5.1 *JoinByLocation*

JoinByLocation joins line features to point features using the nearest relationship. The macro performs the same function as using the option of "Join data from another layer based on spatial location" of Join Data in ArcMap. *JoinByLocation* has four parts. Part 1 defines the source table and the join table. Part 2 prepares the name objects of the output's workspace, feature class, and dataset. Part 3 defines a spatial join object and runs the nearest method. Part 4 creates a feature layer from the output and adds the layer to the active map.

> **Key Interfaces:** *IFeatureClass, IWorkspaceName, IFeatureClassName, IDataset-Name, ISpatialJoin*
>
> **Key Members:** *FeatureClass, WorkspaceFactoryProgID, PathName, Name, WorkspaceName, ShowProcess, LeftOuterJoin, SourceTable, JoinTable, JoinNearest*
>
> **Usage:** Add *deer.shp* and *edge.shp* to an active map. *deer* contains deer sighting locations, and *edge* shows edges between different vegetation covers. In this spatial join operation, *deer* is the source table and *edge* is the join table. *deer* must be on top of *edge* in the table of contents. Import *JoinByLocation* to Visual Basic Editor. Run the macro. The macro creates a new layer named *Spatial_Join*. The attribute table of *Spatial_Join* has the same number of records as the *deer* attribute table, but each record of *Spatial_Join* has attributes from *deer* and *edge* as well as a new distance field.

```
Private Sub JoinByLocation()
    ' Part 1: Define the source and join tables.
    Dim pMxDoc As IMxDocument
    Dim pMap As IMap
    Dim pSourceLayer As IFeatureLayer
    Dim pSourceFC As IFeatureClass
    Dim pJoinLayer As IFeatureLayer
    Dim pJoinFC As IFeatureClass
    Set pMxDoc = ThisDocument
    Set pMap = pMxDoc.FocusMap
    ' Define the source feature class.
    Set pSourceLayer = pMap.Layer(0)
    Set pSourceFC = pSourceLayer.FeatureClass
    ' Define the join feature class.
    Set pJoinLayer = pMap.Layer(1)
    Set pJoinFC = pJoinLayer.FeatureClass
```

Part 1 sets *pSourceFC* to be the feature class of the top layer in the active map, and *pJoinFC* to be the feature class of the second layer.

```
' Part 2: Define the output dataset.
Dim pOutWorkspaceName As IWorkspaceName
Dim pFCName As IFeatureClassName
Dim pDatasetName As IDatasetName
Set pFCName = New FeatureClassName
Set pDatasetName = pFCName
Set pOutWorkspaceName = New WorkspaceName
pOutWorkspaceName.WorkspaceFactoryProgID = "esriCore.ShapefileWorkspaceFactory.1"
pOutWorkspaceName.PathName = "c:\data\chap10"
pDatasetName.Name = "Spatial_Join"
Set pDatasetName.WorkspaceName = pOutWorkspaceName
```

Part 2 creates *pFCName* as an instance of the *FeatureClassName* class. The code then performs a QI for *IDatasetName* to define the workspace and name of *pFC-Name*.

```
' Part 3: Perform spatial join.
Dim pSpatialJoin As ISpatialJoin
Dim pOutputFeatClass As IFeatureClass
Dim maxMapDist As Double
' Create and define a spatial join object.
Set pSpatialJoin = New SpatialJoin
With pSpatialJoin
     .ShowProcess(True) = 0
     .LeftOuterJoin = False
     Set .SourceTable = pSourceFC
     Set .JoinTable = pJoinFC
End With
' Use infinity as the maximum map distance.
maxMapDist = -1
' Run the join nearest method.
Set pOutputFC = pSpatialJoin.JoinNearest(pFCName, maxMapDist)
```

Part 3 creates *pSpatialJoin* as an instance of the *SpatialJoin* class and defines its properties in a *With* block. The code specifies True for *ShowProcess* so that a message box will update the processing. The code specifies False for *LeftOuterJoin* so that a record will be added from the source table to the output only if the record has a match in the join table. Then the code uses the *JoinNearest* method on *ISpatialJoin* to create the output referenced by *pOutputFC*. The maximum map distance for the search radius for spatial join is −1, which means infinity.

```
' Part 4: Create the output layer and add it to the active map.
Dim pOutputFeatLayer As IFeatureLayer
Set pOutputFeatLayer = New FeatureLayer
Set pOutputFeatLayer.FeatureClass = pOutputFC
pOutputFeatLayer.Name = pOutputFC.AliasName
pMap.AddLayer pOutputFeatLayer
End Sub
```

Part 4 creates a feature layer from *pOutputFC*, and adds the layer to the active map.

10.6 MANIPULATING FEATURES

The *IBasicGeoprocessor* interface offers the methods of clip, dissolve, and merge for manipulating spatial features. This section covers dissolve and merge. Additionally, this section includes a macro that derives centroids for each polygon of a shapefile and saves the centroids into a new shapefile. The centroid is the geometric center of a polygon, a common measure that can be used as an input to vector data analysis.

10.6.1 *Dissolve*

Dissolve aggregates features of the input layer based on the values of an attribute. *Dissolve* has four parts. Part 1 defines the input table. Part 2 defines the workspace and name of the output. Part 3 performs dissolve. Part 4 creates a feature layer from the output and adds the layer to the active map.

> **Key Interfaces:** *ITable, IWorkspaceName, IFeatureClassName, IDatasetName, IBasicGeoprocessor*
> **Key Members:** *WorkspaceFactoryProgID, PathName, Name, WorkspaceName, Dissolve*
> **Usage:** Add *vulner.shp* to an active map. Open the attribute table of *vulner* and make sure that it contains the fields of Class and Shape_area. Class will be used as the dissolve field and shape_area, the summary field, in **Dissolve**. Import **Dissolve** to Visual Basic Editor. Run the macro. The macro adds a new layer named *Dissolve_result* to the active map. *Dissolve_result* is a shapefile that allows a polygon to have multiple components. Therefore, the attribute table of *Dissolve_result* has only five records, one for each class value. The field Shape_area in the attribute table summarizes the values of Shape_area in the input layer for each class value.

```
Private Sub Dissolve()
    ' Part 1: Define the input table.
    Dim pMxDoc As IMxDocument
    Dim pMap As IMap
    Dim pInputFeatLayer As IFeatureLayer
    Dim pInputTable As ITable
    Set pMxDoc = ThisDocument
    Set pMap = pMxDoc.FocusMap
    Set pInputFeatLayer = pMap.Layer(0)
    Set pInputTable = pInputFeatLayer
```

Part 1 sets *pInputFeatLayer* to be the top layer in the active map. The code then uses the *ITable* interface to set *pInputTable* to be the same as *pInputFeatLayer*.

```
    ' Part 2: Define the output dataset.
    Dim pNewWSName As IWorkspaceName
    Dim pFCName As IFeatureClassName
    Dim pDatasetName As IDatasetName
    Set pFCName = New FeatureClassName
    Set pDatasetName = pFCName
    Set pNewWSName = New WorkspaceName
```

```
pNewWSName.WorkspaceFactoryProgID = "esriCore.ShapefileWorkspaceFactory.1"
pNewWSName.PathName = "c:\data\chap10"
pDatasetName.Name = "Dissolve_result"
Set pDatasetName.WorkspaceName = pNewWSName
```

Part 2 creates *pFCName* as an instance of the *FeatureClassName* class and accesses *IDatasetName* to define the workspace and name for *pFCName*.

```
' Part 3: Perform dissolve.
Dim pBGP As IBasicGeoprocessor
Dim pOutputFC As IFeatureClass
Set pBGP = New BasicGeoprocessor
' Run dissolve.
Set pOutputFC = pBGP.Dissolve(pInputTable, False, "Class", "Dissolve.Shape, Sum.Shape_area", pFCName)
```

Part 3 creates *pBGP* as an instance of the *BasicGeoprocessor* class and uses the *Dissolve* method on *IBasicGeoprocessor* to create the output referenced by *pOutputFC*. *Dissolve* requires three arguments in addition to two object qualifiers already defined. The code specifies False for *useSelected* so that all records in the input table will be dissolved. The code specifies Class for *dissolveField*, or the field to dissolve. The argument *summaryFields* has a different syntax from the others. The first entry in the comma-delimited string, Dissolve.Shape, is required if the output is a shapefile (not a summary table). The second entry, Sum.Shape_area, specifies that the Shape_area values be summed for each value of the dissolve field. Besides sum, other available statistics are count, minimum, maximum, average, variance, and standard deviation.

```
' Part 4: Create the output feature layer and add the layer to the active map.
Dim pOutputFeatLayer As IFeatureLayer
Set pOutputFeatLayer = New FeatureLayer
Set pOutputFeatLayer.FeatureClass = pOutputFC
pOutputFeatLayer.Name = pOutputFC.AliasName
pMap.AddLayer pOutputFeatLayer
End Sub
```

Part 4 creates a feature layer from *pOutputFC*, and adds the layer to the active map.

Box 10.3 Dissolve_GP

Dissolve_GP uses the Dissolve tool in the Analysis toolbox to dissolve *vulner.shp* by using Class as the dissolve field. At the same time, the sum of Shape_area is calculated for each Class value in the output. Run the macro in ArcCatalog and examine the output (*Dissolve_result2.shp*) in the Catalog tree.

```
Private Sub Dissolve_GP()
    ' Create the Geoprocessing object.
    Dim GP As Object
    Set GP = CreateObject("esriGeoprocessing.GpDispatch.1")
    ' Dissolve <in_features> <out_feature_class> {dissolve_field;dissolve_field...}
    ' {field {Statistics_Type}; field{Statistics_Type}...}
```

```
' Execute the dissolve tool.
GP.Dissolve_management "c:\data\chap10\vulner.shp", "c:\data\chap10\Dissolve_result2.shp", "Class", _
"Shape_area Sum"
End Sub
```

10.6.2 *Merge*

Merge combines two line shapefiles into a single shapefile. *Merge* has four parts. Part 1 defines the two tables to be merged. Part 2 defines the output's workspace and name. Part 3 sets up a basic geoprocessor object to run Merge. Part 4 creates a feature layer from the output and adds the layer to the active map.

> **Key Interfaces:** *ITable, IWorkspaceName, IFeatureClassName, IDatasetName, IBasicGeoprocessor, IArray, IFeatureClass*
>
> **Key Members:** *WorkspaceFactoryProgID, PathName, Name, WorkspaceName, Array, Add, Merge, FeatureClass*
>
> **Usage:** Add *mtroads_idtm.shp* and *idroads.shp* to an active map. *mtroads_idtm* is a major road shapefile of Montana, and *idroads* is a major road shapefile of Idaho. Both shapefiles are projected onto the Idaho Transverse Mercator (IDTM) coordinate system. Because only one of the shapefiles can define the attributes for the output, *Merge* will use the attributes of *mtroads_idtm* for the output. Make sure that *mtroads_idtm* is on top of *idroads* in the table of contents. Import *Merge* to Visual Basic Editor. Run the macro. The macro creates *Merge_result.shp* and adds the shapefile to the active map.

```
Private Sub Merge()
    ' Part 1: Define the two input tables.
    Dim pMxDoc As IMxDocument
    Dim pMap As IMap
    Dim pFirstLayer As IFeatureLayer
    Dim pFirstTable As ITable
    Dim pSecondLayer As IFeatureLayer
    Dim pSecondTable As ITable
    Set pMxDoc = ThisDocument
    Set pMap = pMxDoc.FocusMap
    ' Define the first table.
    Set pFirstLayer = pMap.Layer(0)
    Set pFirstTable = pFirstLayer
    ' Define the second table.
    Set pSecondLayer = pMap.Layer(1)
    Set pSecondTable = pSecondLayer
```

Part 1 sets *pFirstLayer* to be the top layer in the active map and sets *pFirstTable* to be the same as *pFirstLayer*. Next, the code sets *pSecondLayer* to be the second layer and sets *pSecondTable* to be the same as *pSecondLayer*.

```
    ' Part 2: Define the output dataset.
    Dim pNewWSName As IWorkspaceName
    Dim pFCName As IFeatureClassName
```

```
Dim pDatasetName As IDatasetName
Set pFCName = New FeatureClassName
Set pDatasetName = pFCName
Set pNewWSName = New WorkspaceName
pNewWSName.WorkspaceFactoryProgID = "esriCore.ShapefileWorkspaceFactory.1"
pNewWSName.PathName = "c:\data\chap10"
Set pDatasetName.WorkspaceName = pNewWSName
pDatasetName.Name = "Merge_result"
```

Part 2 creates *pFCName* as an instance of the *FeatureClassName* class. The code then uses the *IDatasetName* interface to define the workspace and name for *pFC-Name*.

```
' Part 3: Perform merge.
Dim pBGP As IBasicGeoprocessor
Dim inputArray As IArray
Dim pOutputFC As IFeatureClass
' Add the two input tables to an array.
' Notice that esriSystem.Array replaces esriCore.Array in ArcGIS 9.2.
Set inputArray = New esriSystem.Array
inputArray.Add pFirstTable
inputArray.Add pSecondTable
Set pBGP = New BasicGeoprocessor
Set pOutputFC = pBGP.Merge(inputArray, pFirstTable, pFCName)
```

Part 3 first creates *inputArray* as an instance of the *esriSystem.Array* class and uses the *Add* method on *IArray* to add the two input tables to the array variable. Next, the code creates *pBGP* as an instance of the *BasicGeoprocessor* class and uses the *Merge* method to create the merged output referenced by *pOutputFC*. The object qualifier *pFirstTable* determines that the fields in the output correspond to those in the first input table.

```
' Part 4: Create the output layer and add it to the active map.
Dim pOutputFeatLayer As IFeatureLayer
Set pOutputFeatLayer = New FeatureLayer
Set pOutputFeatLayer.FeatureClass = pOutputFC
pOutputFeatLayer.Name = pOutputFC.AliasName
pMxDoc.FocusMap.AddLayer pOutputFeatLayer
End Sub
```

Part 4 creates a feature layer from *pOutputFC* and adds the layer to the active map.

10.6.3   *Centroid*

Centroid derives a centroid for each polygon of a shapefile and creates a new point shapefile that contains the centroids. *Centroid* has three parts. Part 1 defines the input dataset and calls the **CreateNewShapefile** function to create an empty output feature class. Part 2 derives the centroid of each polygon and stores the centroids in the feature class passed from **CreateNewShapefile**. Part 3 creates a feature layer from the output and adds the layer to the active map.

CreateNewShapefile requires the spatial reference of the input dataset as an argument and returns a feature class object to *Centroid*. *CreateNewShapefile* has three parts. Part 1 defines the output's workspace, Part 2 edits and defines a shape field and adds the field to a field collection, and Part 3 creates a new feature class and returns the feature class to *Centroid*. Two constants used in *CreateNewShapefile*, one for the output's workspace and the other for the output's name, are declared at the start of the module.

Key Interfaces: *ISpatialReference, IFeatureClass, IFeatureCursor, IArea, IFeature, IPoint, IWorkspaceFactory, IFeatureWorkspace, IFieldsEdit, IFieldEdit, IGeometryDef, IGeometryDefEdit*

Key Members: *SpatialReference, Search, NextFeature, Shape, QueryCentroid, CreateFeature, Store, OpenFromFile, Name, Type, GeometryType, GeometryDef, AddField, CreateFeatureClass*

Usage: Add *idcounty.shp* to an active map. *idcounty* shows 44 counties in Idaho. Import *Centroid* to Visual Basic Editor. Run the module. The module adds a new shapefile named *Centroid* to the active map. The attribute table of *Centroid* has three fields: FID, Shape, and ID. ArcGIS automatically adds ID because the software requires at least one field in addition to the object ID (FID) and the geometry field (Shape). ID has zeros for all 44 records but, if necessary, the values can be calculated to be FID + 1.

```
Const strFolder As String = "c:\data\chap10"
Const strName As String = "Centroid"

Private Sub Centroid()
    ' Part 1: Define the input dataset, and call a function to create the output.
    Dim pMxDoc As IMxDocument
    Dim pMap As IMap
    Dim pInputFeatLayer As IFeatureLayer
    Dim pGeoDataset As IGeoDataset
    Dim pSpatialReference As ISpatialReference
    Dim pOutputFeatClass As IFeatureClass
    Set pMxDoc = ThisDocument
    Set pMap = pMxDoc.FocusMap
    ' Define the input dataset.
    Set pInputFeatLayer = pMap.Layer(0)
    Set pGeoDataset = pInputFeatLayer
    Set pSpatialReference = pGeoDataset.SpatialReference
    ' Call the CreateNewShapefile function to create the output feature class.
    Set pOutputFeatClass = CreateNewShapefile(pSpatialReference)
```

Part 1 sets *pInputFeatLayer* to be the top layer in the active map and uses the *IGeoDataset* interface to derive its spatial reference, which is referenced by *pSpatialReference*. The code then calls the *CreateNewShapefile* function, which uses *pSpatialReference* as an argument and returns a feature class object referenced by *pOutputFeatClass*.

```
    ' Part 2: Derive centroids for each polygon and store them.
    Dim pFCursor As IFeatureCursor
```

```
Dim pInputFeature As IFeature
Dim pCentroidTemp As IPoint
Dim pArea As IArea
Dim pOutputFeature As IFeature
' Set up a cursor for all features.
Set pCentroidTemp = New Point
Set pFCursor = pInputFeatLayer.Search(Nothing, True)
Set pInputFeature = pFCursor.NextFeature
' Step through each polygon feature.
Do Until pInputFeature Is Nothing
    Set pArea = pInputFeature.Shape
    ' Get the centroid.
    pArea.QueryCentroid pCentroidTemp
    Set pOutputFeature = pOutputFeatClass.CreateFeature
    ' Store the centroid.
    Set pOutputFeature.Shape = pCentroidTemp
    pOutputFeature.Store
    Set pInputFeature = pFCursor.NextFeature
Loop
```

Part 2 derives centroids for each polygon and stores them as features in *pOutputFeatClass*. The code first creates *pCentroidTemp* as an instance of the *Point* class and creates a feature cursor referenced by *pFCursor*. Next the code uses a loop to step through each feature in the cursor, derive its centroid, and store the centroid as a new feature in *pOutputFeatClass*. The method for deriving the centroid is *QueryCentroid* on *IArea*, which is implemented by a polygon object. The method to store the centroid is *Store* on *IFeature*, which is implemented by a feature object.

```
' Part 3: Create the output feature layer, and add the layer to the active map.
Dim pFeatureLayer As IFeatureLayer
Set pFeatureLayer = New FeatureLayer
Set pFeatureLayer.FeatureClass = pOutputFeatClass
pFeatureLayer.Name = "Centroid"
pMap.AddLayer pFeatureLayer
End Sub
```

Part 3 creates a new feature layer from *pOutputFeatClass*, and adds the layer to the active map.

```
Public Function CreateNewShapefile(pSpatialReference As ISpatialReference) As IFeatureClass
    ' Part 1: Define the output's workspace.
    Dim pFWS As IFeatureWorkspace
    Dim pWorkspaceFactory As IWorkspaceFactory
    Set pWorkspaceFactory = New ShapefileWorkspaceFactory
    Set pFWS = pWorkspaceFactory.OpenFromFile(strFolder, 0)
```

Part 1 uses the *OpenFromFile* method and the constant *strFolder* to open a feature workspace for the new shapefile.

```
' Part 2: Edit and define a shape field.
Dim pFields As IFieldsEdit
Dim pField As IFieldEdit
Dim pGeomDef As IGeometryDef
Dim pGeomDefEdit As IGeometryDefEdit
' Make the shape field.
Set pField = New Field
pField.Name = "Shape"
pField.Type = esriFieldTypeGeometry
' Define the geometry of the shape field.
Set pGeomDef = New GeometryDef
Set pGeomDefEdit = pGeomDef
With pGeomDefEdit
    .GeometryType = esriGeometryPoint
    Set .SpatialReference = pSpatialReference
End With
Set pField.GeometryDef = pGeomDef
' Add the shape field to the field collection.
Set pFields = New Fields
pFields.AddField pField
```

Part 2 defines the shape field and adds it to a field collection. The code first creates *pField* as an instance of the *Field* class and defines its name and type properties. Next, the code creates *pGeomDef* as an instance of the *GeometryDef* class and uses the *GeometryDefEdit* interface to define the geometry type and spatial reference of the new geometry definition. After assigning *pGeomDef* to be the geometry definition of *pField*, the code creates *pFields* as an instance of the *Fields* class and adds *pField* to the new field collection.

```
' Part 3: Create the new feature class and return it.
Dim pFeatClass As IFeatureClass
Set pFeatClass = pFWS.CreateFeatureClass(strName, pFields, Nothing, Nothing, esriFTSimple, "Shape", "")
' Return the feature class.
Set CreateNewShapefile = pFeatClass
End Function
```

Part 3 uses the *CreateFeatureClass* method on *IFeatureWorkspace* to create a feature class object referenced by *pFeatClass*. **CreateNewShapefile** then returns *pFeatClass* to **Centroid**.

Raster Data Operations

The simple data structure of raster data is computationally efficient and well suited for a large variety of raster data operations. One typically starts a raster data operation by setting up an analysis environment that includes the area for analysis and the output cell size. Both parameters are important because the inputs to an operation may have different area extents and different cell sizes.

Raster data query selects cells that meet a query expression. The query expression may involve one factor from a single raster or multiple factors from multiple rasters. The output from a query is a new raster that separates cells that meet the query expression from those that do not.

Raster data analysis is traditionally grouped into local, neighborhood, zonal, and distance measure operations. A local operation computes the cell values of a new raster by using a function that relates the input to the output. The operation is performed on a cell-by-cell basis. A neighborhood operation uses a focal cell and a neighborhood. The operation computes the focal cell value from the cell values within the neighborhood. A zonal operation works with zones or groups of cells of same values. Given a single input raster, a zonal operation calculates the geometry of zones in the raster. Given a zonal raster and a value raster, a zonal operation summarizes the cell values in the value raster for each zone in the zonal raster. A distance measure operation involves an entire raster. The operation calculates for each cell the distance away from the closest source cell. If the distance is measured in cell units, the operation produces a series of wave-like distance zones over the raster. If the distance is measured in cost units, the operation produces for each cell the least accumulative cost to a source cell.

This chapter covers raster data operations. Section 11.1 reviews raster data analysis using ArcGIS. Section 11.2 discusses objects that are related to raster data operations. Section 11.3 includes macros and a Geoprocessing (GP) macro for saving, extracting, and querying raster data. Section 11.4 offers macros and two GP macros for cell-by-cell operations. Section 11.5 has a macro and a GP macro for a neighborhood operation. Section 11.6 has a macro and a GP macro for a zonal operation. Section 11.7 offers macros and a GP macro for physical distance and cost distance measure operations. All macros start with the listing of key interfaces and key members (properties and methods) and the usage. The Spatial Analyst extension must be checked in the Tools/Extensions menu before running the macros.

11.1 ANALYZING RASTER DATA IN ARCGIS

The Spatial Analyst extension to ArcGIS is designed for raster data operations. Commands of the extension are available through the extension menu and ArcTool-box. Discussions in this section refer to the extension menu. Spatial Analyst has an Options command that lets the user specify an analysis mask, an area for analysis, and an output cell size. An analysis mask limits analysis to cells that do not carry the cell value of no data. The area for analysis may correspond to a specific raster, an extent defined by a set of minimum and maximum x-, y-coordinates, a combination of rasters, or a mask. The output cell size can be at any scale deemed suitable by the user although, conceptually, it should be equal to the largest cell size among the input rasters for analysis.

Raster Calculator in Spatial Analyst incorporates arithmetic operators, logical operators, Boolean connectors, and mathematical functions in a dialog box. Raster Calculator is therefore useful for raster data queries as well as for cell-by-cell (local) operations. A raster data query can apply to a single raster or multiple rasters. The query result is saved into a raster in which cells that meet the condition are coded one and other cells are coded zero.

Besides Raster Calculator, the Cell Statistics and Reclassify commands can also perform local operations. The Cell Statistics command computes summary statistics such as maximum, minimum, and mean from multiple rasters. The Reclassify command reclassifies the values of the input cells on a cell-by-cell basis.

The Neighborhood Statistics command performs neighborhood operations. The command uses a dialog to gather the input data, field, statistic type, and neighborhood for computation. A neighborhood may be a rectangle, circle, annulus, or wedge. The statistic type includes maximum, minimum, range, sum, mean, standard deviation, variety, majority, and minority.

The Zonal Statistics command performs zonal operations on a zonal raster and a value raster. The command uses a dialog to gather the zone dataset, zone field, and value raster. Given a single raster with zones, one can use Raster Calculator with such functions as area, perimeter, centroid, and thickness to compute the zonal geometry.

The Distance command has the following selections: Straight Line Distance, Allocation, Cost Weighted, and Shortest Path. The first two use physical distance measures, and the last two use cost distance measures. Straight Line Distance creates an output raster containing continuous distance measures away from the source cells in a raster. Allocation creates a raster in which each cell is assigned the value of its closest source cell. Cost Weighted calculates for each cell the least accumulative cost, over a cost raster, to its closest source cell. Additionally, the Cost Weighted command can also create a direction raster that shows the direction of the least cost path from each cell to a source, and an allocation raster that shows the assignment of each cell to a source cell. Shortest Path uses the results from the Cost Weighted command to generate the least cost path from any cell or zone to the source cell.

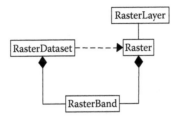

Figure 11.1 The relationship between *RasterDataset*, *RasterBand*, *Raster*, and *RasterLayer*.

11.2 ARCOBJECTS FOR RASTER DATA ANALYSIS

Raster data analysis involves objects that can be grouped into the categories of raster and operator.

11.2.1 Raster Objects

The basic raster objects are *RasterDataset*, *RasterBand*, *Raster*, and *RasterLayer* (Figure 11.1). A raster dataset object represents an existing dataset stored on disk in a particular raster format (for example, ESRI grid, TIFF). A raster band object represents an individual band of a raster dataset. The number of bands in a raster may vary; an ESRI grid typically contains a single band, whereas a satellite image has multiple bands. A raster object is a virtual representation of a raster dataset, useful for raster data operations. Created from an existing raster or a raster dataset, a raster layer object is a visual display of raster data.

One other raster object that needs to be mentioned is *RasterDescriptor*. When used for data conversion (Chapter 6), a raster descriptor object represents a raster that uses a field other than the default value field. When used for raster data query, a raster descriptor object is associated with a query filter and can be created from a raster or a raster's selection set (Figure 11.2).

11.2.2 Operator Objects

Operator objects provide methods for raster data analysis. A good reference for operator objects is the Spatial Analyst Functional Reference in the ArcGIS Desktop Help. The reference covers operator (Spatial Analyst) objects by type of analysis (for example, Local) and offers the ArcObjects syntax and example for each object.

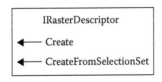

Figure 11.2 *IRasterDescriptor* has methods to create a raster descriptor from a raster or a selection set.

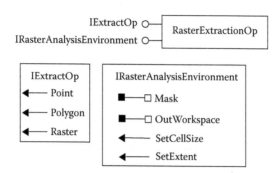

Figure 11.3 A *RasterExtractionOp* object supports *IRasterAnalysisEnvironment* and *IExtractOp*. *IRasterAnalysisEnvironment* has members for defining the analysis environment, and *IExtractOp* has methods to extract raster data.

This section focuses on operator objects that are used in this chapter's sample macros. A *RasterExtractionOp* object supports *IExtractOp*, which has methods for data extraction based on points, a polygon, or a raster (Figure 11.3). A *RasterReclassOp* object implements *IReclassOp*, which has methods for reclassifying raster data (Figure 11.4). A *RasterMathOp* object supports *ILogicalOp* and *IMathOp*. *ILogicalOp* has methods for logical (Boolean) operations, and *IMathOp* has methods for mathematical operations (Figure 11.5). Notice that every operator object also supports *IRasterAnalysisEnvironment*, which has members for defining the analysis environment.

Local, neighborhood, zonal, and distance measure operations are each covered by an operator object. A *RasterLocalOp* object supports *ILocalOp* (Figure 11.6), a *RasterNeighborhoodOp* object *INeighborhoodOp* (Figure 11.7), a *RasterZonalOp* object *IZonalOp* (Figure 11.8), and a *RasterDistanceOp* object *IDistanceOp* (Figure 11.9).

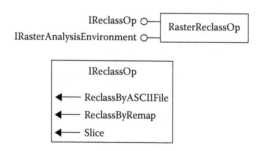

Figure 11.4 A *RasterReclassOp* object supports *IRasterAnalysisEnvironment* and *IReclassOp*. *IReclassOp* has methods for reclassifying and slicing raster data.

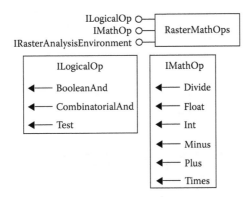

Figure 11.5 A *RasterMathOp* object supports *IRasterAnalysisEnvironment*, *ILogicalOp*, and *IMathOp*. *ILogicalOp* has methods for logical operations, and *IMathOp* has methods for mathematical operations.

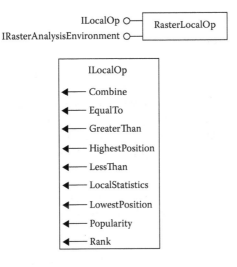

Figure 11.6 A *RasterLocalOp* object supports *IRasterAnalysisEnvironment* and *ILocalOp*. *ILocalOp* has methods for cell-by-cell operations.

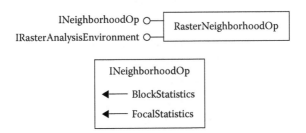

Figure 11.7 A *RasterNeighborhoodOp* object supports *IRasterAnalysisEnvironment* and *INeighborhoodOp*. *INeighborhoodOp* has methods for focal and block operations.

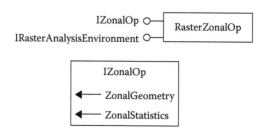

Figure 11.8 A *RasterZonalOp* object supports *IRasterAnalysisEnvironment* and *IZonalOp*. *IZonalOp* has methods for zonal operations using a single raster or a zonal raster and a value raster.

11.3 MANAGING RASTER DATA

This section covers the tasks of saving raster data, extracting raster data, and querying raster data.

11.3.1 *MakePermanent*

An output from a raster data operation is a temporary dataset in Spatial Analyst. The temporary dataset is lost unless a disk location and a file name are set up to make the dataset permanent. *MakePermanent* takes a raster layer, which represents a temporary dataset, and saves it to a permanent raster. The macro performs the same function as using the Make Permanent command from the layer's context menu in ArcMap. *MakePermanent* has two parts. Part 1 defines the temporary raster dataset, and Part 2 specifies a workspace and makes permanent the temporary raster dataset.

   Key Interfaces: *IRasterLayer, IRaster, IRasterBandCollection, IRasterDataset, IWorkspaceFactory, IWorkspace, ITemporaryDataset, IDataset*

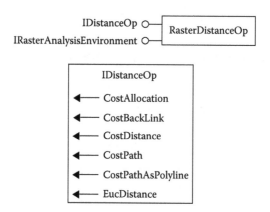

Figure 11.9 A *RasterDistanceOp* object supports *IRasterAnalysisEnvironment* and *IDistanceOp*. *IDistanceOp* has methods for physical distance or cost distance measure operations.

Key Members: *Raster, Item( ), RasterDataset, OpenFromFile, MakePermanentAS, Name*

Usage: Add *emidalat*, an elevation raster, to an active map. Click Spatial Analyst, point to Surface Analysis, and select Slope. A temporary grid *Slope of emidalat* is added to the active map. Right-click *Slope of emidalat* and select Properties. The Source tab of the Layer Properties dialog shows that the raster does have a temporary status and a name of SLOPEx. Import *MakePermanent* to Visual Basic Editor. Run the macro. The macro creates a permanent raster from SLOPEx. To verify the result, add SLOPEx and check its Layer Properties dialog.

```
Private Sub MakePermanent()
    ' Part 1: Define the temporary raster data.
    Dim pMxDoc As IMxDocument
    Dim pMap As IMap
    Dim pRasterLy As IRasterLayer
    Dim pRaster As IRaster
    Dim pRasBandC As IRasterBandCollection
    Dim pRasterDS As IRasterDataset
    Set pMxDoc = ThisDocument
    Set pMap = pMxDoc.FocusMap
    ' Get the raster from the raster layer.
    Set pRasterLy = pMap.Layer(0)
    Set pRaster = pRasterLy.Raster
    ' Set the raster dataset to be the first band of the raster
    Set pRasBandC = pRaster
    Set pRasterDS = pRasBandC.Item(0).RasterDataset
```

The code first sets *pRaster* to be the raster of the top layer in the active map. Next, the code performs a QueryInterface (QI) for the *IRasterBandCollection* interface and assigns the raster dataset associated with the first band of *pRaster* to *pRasterDB* (Figure 11.10).

```
    ' Part 2: Make permanent the raster dataset.
    Dim pWSF As IWorkspaceFactory
    Dim pWS As IRasterWorkspace
    Dim pTempDS As ITemporaryDataset
    Dim pDataset As IDataset
    Dim Name As String
    ' Define the workspace.
    Set pWSF = New RasterWorkspaceFactory
    Set pWS = pWSF.OpenFromFile("c:\data\chap11\", 0)
    ' Define the temporary dataset.
    Set pDataset = pRasterDS
    Name = pDataset.Name
    Set pTempDS = pRasterDS
    ' Make permanent the grid.
    Set pRasterDS = pTempDS.MakePermanentAs(Name, pWS, "GRID")
```

Part 2 makes *pRasterDS* a permanent raster. To do that, the code must first define a workspace and a name for the permanent raster. The workspace is specified using

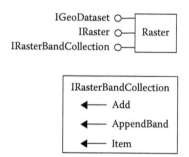

Figure 11.10   The diagram shows how to QI on *IRaster* for *IRasterBandCollection* so that a raster band can be added, appended, or extracted from a raster band collection object.

the *OpenFromFile* method on *IWorkspaceFactory*. The name is specified via the *Name* property on *IDataset*. Then the code accesses *ITemporaryDataset* and uses the *Make-PermanentAs* method to make *pRasterDS* a permanent ESRI grid (Figure 11.11).

11.3.2   *ExtractByMask*

ExtractByMask uses a mask raster to extract a data subset from an input raster. The macro performs the same function as using an analysis mask in Spatial Analyst to extract a new raster from an input raster. ***ExtractByMask*** has three parts. Part 1 defines the input raster, Part 2 prepares a mask raster and uses it to extract data, and Part 3 creates a raster layer from the extracted data and adds the layer to the active map.

> **Key Interfaces:** *IRasterLayer, IRaster, IGeoDataset, IExtractionOp*
> **Key Members:** *Raster*
> **Usage:** Add *splinegd* and *idoutlgd* to an active map. *splinegd* is an interpolated precipitation raster, whose extent is determined by the extent of *x*- and *y*-coordinates of the weather stations used in interpolation. *idoutlgd* is a raster showing the outline of Idaho. *splinegd* must be on top of *idoutlgd*. Import ***ExtractByMask*** to Visual Basic Editor. Run the macro. The macro uses *idoutlgd* as an analysis mask to extract a temporary raster from *splinegd*.

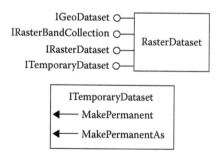

Figure 11.11   The diagram shows how to QI on *IRasterDataset* for *ITemporaryDataset* so that a temporary raster can be made permanent.

```
Private Sub ExtractByMask()
    ' Part 1: Define the input raster.
    Dim pMxDoc As IMxDocument
    Dim pInRasterLayer As IRasterLayer
    Dim pInRaster As IRaster
    Set pMxDoc = ThisDocument
    Set pInRasterLayer = pMxDoc.FocusMap.Layer(0)
    Set pInRaster = pInRasterLayer.Raster
```

Part 1 sets *pInRaster* to be the raster of the top layer in the active map.

```
    ' Part 2: Use a mask to extract the input raster.
    Dim pRasterLayer As IRasterLayer
    Dim pMaskDataset As IGeoDataset
    Dim pOutRaster As IRaster
    Dim pExtrOp As IExtractionOp
    ' Define the mask dataset.
    Set pRasterLayer = pMxDoc.FocusMap.Layer(1)
    Set pMaskDataset = pRasterLayer.Raster
    Set pExtrOp = New RasterExtractionOp
    ' Perform extraction.
    Set pOutRaster = pExtrOp.Raster(pInRaster, pMaskDataset)
```

Part 2 defines the mask dataset and performs an extraction operation. To define the mask raster, the code sets *pRasterLayer* to be the second layer of the active map and *pMaskDataset* to be the raster of *pRasterLayer*. Next, the code creates *pExtrOp* as an instance of the *RasterExtractOp* class and uses the *Raster* method on *IExtractOp* to create a raster referenced by *pOutRaster*.

```
    ' Part 3: Create the output layer and add it to the active map.
    Dim pRL As IRasterLayer
    Set pRL = New RasterLayer
    pRL.CreateFromRaster pOutRaster
    pMxDoc.FocusMap.AddLayer pRL
End Sub
```

Part 3 uses the *CreateFromRaster* method on *IRasterLayer* to create a raster layer from *pOutRaster*. The code then adds the raster layer to the active map.

Box 11.1 ExtractByMask_GP

ExtractByMask_GP uses the ExtractByMask tool in the Spatial Analyst toolbox to extract a raster (*extractgd*) from an input raster (*splinegd*) by using a mask (*idoutlgd*). Run the macro in ArcCatalog. The extracted raster should appear in the Catalog tree.

```
Private Sub ExtractByMask_GP()
    ' Create the Geoprocessing object and define its workspace.
    Dim GP As Object
```

```
Set GP = CreateObject("esriGeoprocessing.GpDispatch.1")
Dim filepath As String
filepath = "c:\data\chap11"
GP.Workspace = filepath
' ExtractByMask <in_raster> <in_mask_dat> <out_raster>
' Execute the extractbymask tool.
GP.ExtractByMask_sa "splinegd", "idoutlgd", "extractgd"
End Sub
```

11.3.3 *RasterQuery*

RasterQuery queries an input raster and produces an output raster that shows the query result. If the input raster has an attribute table, the output raster contains the value of one for cells that meet the expression and zero for cells that do not. If the input raster does not have an attribute table, the output raster retains the original cell values for those cells that meet the expression and no data for cells that do not. In the case of an integer raster, the macro performs a similar function as using Raster Calculator in Spatial Analyst to query a raster.

RasterQuery has three parts. Part 1 defines the input raster, Part 2 prepares a query filter and performs data query, and Part 3 creates a raster layer from the output and adds the layer to the active map.

> **Key Interfaces:** *IRaster, IQueryFilter, IRasterDescriptor, ILogicalOp, IGeoDataset*
> **Key Members:** *Raster, WhereClause, Create, Test, CreateFromRaster*
> **Usage:** Add *slopegrd*, a raster with four slope classes, to an active map. Import *RasterQuery* to Visual Basic Editor. Run the macro. The macro creates a temporary raster that shows cells in the slope class of 2, and adds it as a raster layer to the active map.

```
Private Sub RasterQuery()
    ' Part 1: Define the raster for query.
    Dim pMxDoc As IMxDocument
    Dim pMap As IMap
    Dim pLayer As IRasterLayer
    Dim pRaster As IRaster
    Set pMxDoc = ThisDocument
    Set pMap = pMxDoc.FocusMap
    Set pLayer = pMap.Layer(0)
    Set pRaster = pLayer.Raster
```

Part 1 sets *pRaster* to be the raster of the top layer in the active map.

```
    ' Part 2: Perform raster query.
    Dim pQFilter As IQueryFilter
    Dim pRasDes As IRasterDescriptor
    Dim pLogicalOp As ILogicalOp
    Dim pOutputRaster As IGeoDataset
    ' Prepare a query filter.
    Set pQFilter = New QueryFilter
```

```
pQFilter.WhereClause = "Value = 2"
' Prepare a raster descriptor.
Set pRasDes = New RasterDescriptor
pRasDes.Create pRaster, pQFilter, "value"
' Run a logical operation.
Set pLogicalOp = New RasterMathOps
Set pOutputRaster = pLogicalOp.Test(pRasDes)
```

Part 2 performs a raster data query and transforms the result into a raster with ones and zeros. The code first creates *pQFilter* as an instance of the *QueryFilter* class and defines its *WhereClause* condition. Next, the code sets *pRasDes* to be an instance of the *RasterDescriptor* class and uses the *Create* method to create the new raster descriptor from *pRaster*. Then the code creates *pLogicalOp* as an instance of the *RasterMathOp* class and uses the *Test* method on *ILogicalOp* to create a raster referenced by *pOutputRaster*.

```
' Part 3: Create the output raster layer, and add the layer to the active map.
Dim pOutputLayer As IRasterLayer
Set pOutputLayer = New RasterLayer
pOutputLayer.CreateFromRaster pOutputRaster
pOutputLayer.Name = "QueryOutput"
pMap.AddLayer pOutputLayer
End Sub
```

Part 3 creates a new raster layer from *pOutputRaster* and adds the layer to the active map. The temporary layer has the value of one for cells that have the slope value of two and zero for other cells.

11.3.4 *Query2Rasters*

Query2Rasters queries two rasters and produces an output raster with ones and zeros. The macro performs the same function as using Raster Calculator to query two rasters in Spatial Analyst. *Query2Rasters* has four parts. Part 1 queries the first input raster and saves the result into a dataset, and Part 2 queries the second raster and saves the result into another dataset. Part 3 uses the two datasets from the previous queries in a logical operation to create an output raster. Part 4 creates a raster layer from the output and adds the layer to the active map. Parts 1 and 2 contain basically the same code as *RasterQuery*.

Key Interfaces: *IRaster, IQueryFilter, IRasterDescriptor, ILogicalOp, IGeoDataset*
Key Members: *Raster, WhereClause, Create, Test, BooleanAnd, CreateFrom Raster, Name*
Usage: Add *slopegrd* and *aspectgrd* to an active map, with *slopegrd* on top in the table of contents. *slopegrd* contains four slope classes, and *aspectgrd* contains five aspect classes. Import *Query2Rasters* to Visual Basic Editor. Run the macro. The macro adds to the active map a temporary raster, which has the one and zero cell values. A cell value of one means that the cell has the slope class of 2 and the aspect class of 2.

```
Private Sub Query2Rasters()
    ' Part 1: Query the first raster.
    Dim pMxDoc As IMxDocument
    Dim pMap As IMap
    Dim pRLayer1 As IRasterLayer
    Dim pRaster1 As IRaster
    Dim pFilt1 As IQueryFilter
    Dim pDesc1 As IRasterDescriptor
    Dim pLogicalOp As ILogicalOp
    Dim pOutputRaster1 As IGeoDataset
    Set pMxDoc = ThisDocument
    Set pMap = pMxDoc.FocusMap
    ' Define the first raster.
    Set pRLayer1 = pMap.Layer(0)
    Set pRaster1 = pRLayer1.Raster
    ' Create the first query filter.
    Set pFilt1 = New QueryFilter
    pFilt1.WhereClause = "value = 2"
    ' Create the first raster descriptor.
    Set pDesc1 = New RasterDescriptor
    pDesc1.Create pRaster1, pFilt1, "value"
    ' Create the first output with 1's and 0's.
    Set pLogicalOp = New RasterMathOps
    Set pOutputRaster1 = pLogicalOp.Test(pDesc1)
```

Part 1 uses the same code as in ***RasterQuery*** to create *pOutputRaster1* from the first input raster and a query filter.

```
    ' Part 2: Query the second raster.
    Dim pRLayer2 As IRasterLayer
    Dim pRaster2 As IRaster
    Dim pFilt2 As IQueryFilter
    Dim pDesc2 As IRasterDescriptor
    Dim pOutputRaster2 As IGeoDataset
    ' Define the second raster.
    Set pRLayer2 = pMap.Layer(1)
    Set pRaster2 = pRLayer2.Raster
    ' Create the second query filter.
    Set pFilt2 = New QueryFilter
    pFilt2.WhereClause = "value = 2"
    ' Create the second raster descriptor.
    Set pDesc2 = New RasterDescriptor
    pDesc2.Create pRaster2, pFilt2, "value"
    ' Create the second output with 1's and 0's.
    Set pLogicalOp = New RasterMathOps
    Set pOutputRaster2 = pLogicalOp.Test(pDesc2)
```

Part 2 creates *pOutputRaster2* from the second input raster and a query filter.

```
    ' Part 3: Run a logical operation on the two query results.
    Dim pOutputRaster3 As IGeoDataset
```

```
Set pLogicalOp = New RasterMathOps
Set pOutputRaster3 = pLogicalOp.BooleanAnd(pOutputRaster1, pOutputRaster2)
```

Like Parts 1 and 2, Part 3 also performs a logical operation. But instead of using the *Test* method, which is limited to one input raster, the code uses the *BooleanAnd* method, which accepts two input rasters. The output referenced by *pOutputRaster3* has the cell value of one, where both *pOutputRaster1* and *pOutputRaster2* have the cell value of one, and zero elsewhere. The *BooleanAnd* method can also use *pDesc1* and *pDesc2*, instead of *pOutputRaster1* and *pOutputRaster2*, as the object qualifiers. But the output will have the cell values of one and no data, instead of one and zero.

```
' Part 4: Create the output raster layer, and add the layer to the active map.
Dim pRLayer As IRasterLayer
Set pRLayer = New RasterLayer
pRLayer.CreateFromRaster pOutputRaster3
pRLayer.Name = "QueryOutput2"
pMap.AddLayer pRLayer
End Sub
```

Part 4 creates a raster layer from *pOutputRaster3* and adds the layer to the active map.

11.4 PERFORMING LOCAL OPERATIONS

Local operations constitute the core of raster data analysis. A large variety of local operations are available in Spatial Analyst. Reclassify and combine are examples of local operations in this section. Reclassify operates on a single raster, whereas combine operates on two or more rasters.

11.4.1 *ReclassNumberField*

ReclassNumberField uses a number remap to reclassify an input raster. A remap has two columns. The first column lists a cell value or a range of cell values to be reclassified, and the second column lists the output value, including no data. A remap object can be either a number remap or a string remap.

ReclassNumberField performs the same function as using Reclassify in Spatial Analyst to create a classified integer raster. The macro has three parts. Part 1 defines the input raster, Part 2 performs the reclassification, and Part 3 creates a raster layer from the reclassify output and adds the layer to the active map.

> **Key Interfaces:** *IGeoDataset, IReclassOp, INumberRemap, IRaster*
> **Key Members:** *Raster, MapRange, ReclassByRemap, CreateFromRaster*
> **Usage:** Add *slope*, a continuous slope raster, to an active map. Import *ReclassNumberField* to Visual Basic Editor. Run the macro. The macro produces a temporary raster with the reclassification result and adds the raster to the active map.

```
Private Sub ReclassNumberField()
  ' Part 1: Define the raster for reclassify.
  Dim pMxDoc As IMxDocument
```

```
Dim pMap As IMap
Dim pRasterLy As IRasterLayer
Dim pGeoDs As IGeoDataset
Set pMxDoc = ThisDocument
Set pMap = pMxDoc.FocusMap
Set pRasterLy = pMap.Layer(0)
Set pGeoDs = pRasterLy.Raster
```

Part 1 sets *pGeoDs* to be the raster of the top layer in the active map.

```
' Part 2: Reclassify the input raster.
Dim pReclassOp As IReclassOp
Dim pNRemap As INumberRemap
Dim pOutRaster As IRaster
' Prepare a number remap.
Set pNRemap = New NumberRemap
pNRemap.MapRange 0, 10#, 1
pNRemap.MapRange 10.1, 20#, 2
pNRemap.MapRange 20.1, 30#, 3
pNRemap.MapRange 30.1, 90#, 4
' Run the reclass operation.
Set pReclassOp = New RasterReclassOp
Set pOutRaster = pReclassOp.ReclassByRemap(pGeoDs, pNRemap, False)
```

Part 2 reclassifies *pGeoDS* by using a remap that is built in code. The code first creates *pNRemap* as an instance of the *NumberRemap* class and uses the *MapRange* method on *INumberRemap* to set the output value based on a numeric range of the input values (Figure 11.12). A cell within the numeric range of 0 to 10.0 is assigned an output value of 1, 10.1 to 20.0 an output value of 2, and so on. Then, after having created *pReclassOp* as an instance of the *RasterReclassOp* class, the code uses the *ReclassByRemap* method to create a reclassified raster referenced by *pOutRaster*.

```
' Part 3: Create the output layer, and add it to the active map.
Dim pReclassLy As IRasterLayer
Set pReclassLy = New RasterLayer
```

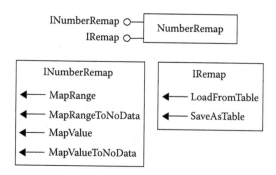

Figure 11.12   A *NumberRemap* object supports *INumberRemap* and *IRemap*. The interfaces have methods to define a remap object for reclassifying a raster.

```
    pReclassLy.CreateFromRaster pOutRaster
    pMap.AddLayer pReclassLy
End Sub
```

Part 3 creates a new raster layer from *pOutRaster* and adds the layer to the active map.

Box 11.2 ReclassNumberField_GP

ReclassNumberField_GP uses the Reclassify tool in the Spatial Analyst toolbox to reclassify *slope*, a slope raster, by using a remap, which is specified in the command line as an argument. Run the macro in ArcCatalog. The reclassed raster (*rec_slope*) should appear in the Catalog tree.

```
Private Sub ReclassNumberField_GP()
    ' Create the Geoprocessing object.
    Dim GP As Object
    Set GP = CreateObject("esriGeoprocessing.GpDispatch.1")
    ' Reclassify <in_raster> <reclass_field> <remap> <out_raster> {DATA | NODATA}
    ' Define a remap for the third parameter.
    Dim parameter3 As String
    parameter3 = "0.0 10.000 1;10.001 20.0 2;20.001 30.0 3;30.001 90.0 4"
    ' Execute the reclassify tool.
    GP.Reclassify_sa "c:\data\chap11\slope", "Value", parameter3, "c:\data\chap11\rec_slope"
End Sub
```

11.4.2 *Combine2Rasters*

Combine2Rasters combines two input rasters and produces an output raster with its cell values representing each unique combination of the input cell values. The macro performs the same function as using the combine function in Spatial Analyst's Raster Calculator. *Combine2Rasters* has three parts. Part 1 defines the two input rasters. Part 2 creates a new raster by adding to it two bands from the input rasters, and then performs a combine operation on the new raster. Part 3 creates a new raster layer from the output raster and adds the layer to the active map.

Key Interfaces: *IRaster, IRasterBandCollection, IRasterBand, ILocalOp*
Key Members: *Raster, Item( ), Add, Combine, CreateFromRaster*
Usage: Add *slopegrd* and *aspectgrd* to an active map. *slopegrd* shows four slope classes, and *aspectgrd* shows four principal directions and a fifth class for flat areas. Import *Combine2Rasters* to Visual Basic Editor. Run the macro. The macro adds to the active map a new temporary raster showing 19 unique combinations of slope and aspect values.

```
Private Sub Combine2Rasters()
    ' Part 1: Define the two rasters for combine.
    Dim pMxDoc As IMxDocument
    Dim pMap As IMap
    Dim pRLayer1 As IRasterLayer
```

```
Dim pRLayer2 As IRasterLayer
Dim pRaster1 As IRaster
Dim pRaster2 As IRaster
Dim pRasBC1 As IRasterBandCollection
Dim pBand1 As IRasterBand
Dim pRasBC2 As IRasterBandCollection
Dim pBand2 As IRasterBand
Set pMxDoc = ThisDocument
Set pMap = pMxDoc.FocusMap
' Derive the first band from the first raster.
Set pRLayer1 = pMap.Layer(0)
Set pRaster1 = pRLayer1.Raster
Set pRasBC1 = pRaster1
Set pBand1 = pRasBC1.Item(0)
' Derive the first band from the second raster.
Set pRLayer2 = pMap.Layer(1)
Set pRaster2 = pRLayer2.Raster
Set pRasBC2 = pRaster2
Set pBand2 = pRasBC2.Item(0)
```

Part 1 defines the raster bands for the combine operation. To define the first raster band, the code executes the following steps: sets *pRaster1* to be the raster of the top layer, switches to the *IRasterBandCollection* interface, and sets *pBand1* to be the first band of *pRaster1*. The same steps are taken to derive *pBand2*.

```
' Part 2: Create a new raster and perform combine.
Dim pRasBC As IRasterBandCollection
Dim pLocalOp As ILocalOp
Dim pOutRaster As IRaster
' Define a new raster and add to it two input bands.
Set pRasBC = New Raster
pRasBC.Add pBand1, 0
pRasBC.Add pBand2, 1
' Run the combine local operation.
Set pLocalOp = New RasterLocalOp
Set pOutRaster = pLocalOp.Combine(pRasBC)
```

Part 2 uses a multiband raster to perform the combine operation. The code first creates *pRasBC* as an instance of the *Raster* class. Next, the code uses the *Add* method on *IRasterBandCollection* to add *pBand1* and *pBand2* to *pRasBC*. The code then creates *pLocalOp* as an instance of the *RasterLocalOp* class and uses the *Combine* method on *ILocalOp* to create a combine raster referenced by *pOutRaster*.

```
' Part 3: Create the output layer, and add it to the active map.
Dim pRLayer As IRasterLayer
Set pRLayer = New RasterLayer
pRLayer.CreateFromRaster pOutRaster
pRLayer.Name = "slp_asp"
pMap.AddLayer pRLayer
End Sub
```

Part 3 creates *pRLayer* as a new raster layer from *pOutRaster* and adds the layer to the active map.

Box 11.3 Combine2Rasters_GP

Combine2Rasters_GP uses the Combine tool in the Spatial Analyst toolbox to combine *slopegrd* and *aspectgrd* in a local operation. Parameter 1 in the command line contains the input rasters. Run the macro in ArcCatalog and view the output raster (*combine2*) in the Catalog tree.

```
Private Sub Combine2Rasters_GP()
    ' Create the Geoprocessing object.
    Dim GP As Object
    Set GP = CreateObject("esriGeoprocessing.GpDispatch.1")
    ' Combine <in_raster;in_raster...> <out_raster>
    ' Define the input rasters for parameter 1.
    Dim parameter1 As String
    parameter1 = "c:\data\chap11\slopegrd;c:\data\chap11\aspectgrd"
    ' Execute the combine tool.
    GP.Combine_sa parameter1, "c:\data\chap11\combine2"
End Sub
```

11.4.3 Other Local Operations

Macros for other local operations involving multiple rasters have the same code structure as *Combine2Rasters*. A local mean operation (that is, deriving the mean on a cell-by-cell basis from two or more rasters) would change the last line of Part 2 in *Combine2Rasters* to read:

```
Set pOutRaster = pLocalOp.LocalStatistics(pRaster, esriGeoAnalysisStatsMean)
```

The line uses the *LocalStatistics* method on *ILocalOp* to compute the local mean from a raster with multiple raster bands and to save the results into an output raster.

Likewise, a local maximum operation would change the last line of Part 2 in *Combine2Rasters* to read:

```
Set pOutRaster = pLocalOp.HighestPosition(pRaster)
```

The line uses the *HighestPosition* method on *ILocalOp* to derive the highest value among the input rasters, which are stored in a multiband raster, and to save the results into an output raster.

11.5 PERFORMING NEIGHBORHOOD OPERATIONS

A Visual Basic for Applications (VBA) macro for a neighborhood operation must define the neighborhood to be used and the statistic to be derived. The statistics are computed by using either an overlapping neighborhood or a nonoverlapping neighborhood. An overlapping neighborhood occurs when the operation moves from cell to cell. A nonoverlapping neighborhood occurs when the operation moves from block to block.

11.5.1 *FocalMean*

FocalMean computes the mean using the cell values within a 3 × 3 neighborhood and assigns the mean to the focal cell. The macro performs the same function as using Neighborhood Statistics in Spatial Analyst. *FocalMean* has three parts. Part 1 defines the input raster. Part 2 prepares a neighborhood operator and a 3 × 3 neighborhood, and runs the focal mean operation. Part 3 creates a raster layer from the output raster and adds it to the active map.

> **Key Interfaces:** *IRaster, INeighborhoodOp, IRasterNeighborhood*
> **Key Members:** *Raster, FocalStatistics, CreateFromRaster*
> **Usage:** Add *emidalat* to an active map. Import *FocalMean* to Visual Basic Editor. Run the macro. The macro creates a temporary raster showing the neighborhood mean and adds the raster to the active map.

```
Private Sub FocalMean()
    ' Part 1: Define the input raster.
    Dim pMxDoc As IMxDocument
    Dim pMap As IMap
    Dim pRLayer As IRasterLayer
    Dim pRaster As IRaster
    Set pMxDoc = ThisDocument
    Set pMap = pMxDoc.FocusMap
    Set pRLayer = pMap.Layer(0)
    Set pRaster = pRLayer.Raster
```

Part 1 sets *pRaster* to be the raster of the top layer in the active map.

```
    ' Part 2: Perform focal mean.
    Dim pNbrOp As INeighborhoodOp
    Dim pNbr As IRasterNeighborhood
    Dim pOutputRaster As IRaster
    ' Define the neighborhood.
    Set pNbr = New RasterNeighborhood
    pNbr.SetRectangle 3, 3, esriUnitsCells
    ' Run the focal mean neighborhood operation.
    Set pNbrOp = New RasterNeighborhoodOp
    Set pOutputRaster = pNbrOp.FocalStatistics(pRaster, esriGeoAnalysisStatsMean, pNbr, False)
```

Part 2 creates *pNbr* as an instance of the *RasterNeighborhood* class and uses the *SetRectangle* method on *IRasterNeighborhood* to define a rectangular neighborhood with three cells for both its height and width. The code then creates *pNbrOp* as an instance of the *RasterNeighborhoodOp* class and uses the *FocalStatistics* method to create a focal mean raster referenced by *pOutputRaster*. (The *FocalStatistics* method uses an overlapping neighborhood, whereas the *BlockStatistics* method, also on *INeighborhoodOp*, uses a nonoverlapping neighborhood.)

```
' Part 3: Create the output layer and add it to the active map.
Dim pOutputLayer As IRasterLayer
Set pOutputLayer = New RasterLayer
pOutputLayer.CreateFromRaster pOutputRaster
pOutputLayer.Name = "FocalMean"
pMap.AddLayer pOutputLayer
End Sub
```

Part 3 creates a raster layer from *pOutputRaster* and adds the layer to the active map.

Box 11.4 FocalMean_GP

FocalMean_GP uses the FocalStatistics tool in the Spatial Analyst toolbox to derive a focal mean raster. Parameters 3 and 4 specify the neighborhood (3 × 3 rectangle) and the statistic (mean) respectively. Run the macro in ArcCatalog and view the output raster (*focalmean2*) in the Catalog tree.

```
Private Sub FocalMean_GP()
    ' Create the Geoprocessing object.
    Dim GP As Object
    Set GP = CreateObject("esriGeoprocessing.GpDispatch.1")
    ' FocalStatistics <in_raster> <out_raster> {neighborhood} {MEAN | MAJORITY | MAXIMUM | MEDIAN |
    ' MINIMUM | MINORITY | RANGE | STD | SUM | VARIETY} {DATA | NODATA}
    ' Execute the focalstatistics tool.
    GP.FocalStatistics_sa "c:\data\chap11\emidalat", "c:\data\chap11\focalmean2", RECTANGLE 3 3", "MEAN"
End Sub
```

11.6 PERFORMING ZONAL OPERATIONS

A VBA macro for a zonal operation involving two rasters must define a value raster and a zonal raster so that the cell values of the value raster can be summarized by zone. The zonal raster must be an integer raster. Like a neighborhood operation, a zonal operation offers various statistics.

11.6.1 *ZonalMean*

ZonalMean computes the mean precipitation for each watershed in Idaho. The macro performs the same function as using Zonal Statistics in Spatial Analyst. ***ZonalMean*** has three parts. Part 1 defines the rasters for the zonal operation and adds the zonal and value rasters to the active map, Part 2 defines and runs the zonal mean operation, and Part 3 creates the raster layer from the output and adds the layer to the active map.

> **Key Interfaces:** *IRaster, IGeoDataset, IZonalOp*
> **Key Members:** *Raster, CreateFromFilePath, ZonalStatistics, CreateFromRaster*
> **Usage:** Import ***ZonalMean*** to Visual Basic Editor in ArcMap. Run the macro. The macro adds *precipgd* and *hucgd*, the two input rasters, as well as *ZonalMean*, the output raster, to the active map. In this zonal operation, *precipgd* is the value raster,

which shows annual precipitation in hundredths of an inch, and *hucgd* is the zonal raster, which shows 13 six-digit watersheds in Idaho. *ZonalMean* contains the mean precipitation for each watershed.

```
Private Sub ZonalMean()
    ' Part 1: Define the zonal and value raster datasets.
    Dim pMxDocument As IMxDocument
    Dim pMap As IMap
    Dim pZoneRL As IRasterLayer
    Dim pZoneRaster As IRaster
    Dim pValueRL As IRasterLayer
    Dim pValueRaster As IRaster
    Set pMxDocument = Application.Document
    Set pMap = pMxDocument.FocusMap
    ' Define the zonal raster dataset.
    Set pZoneRL = New RasterLayer
    pZoneRL.CreateFromFilePath "c:\data\chap11\hucgd"
    Set pZoneRaster = pZoneRL.Raster
    ' Define the value raster dataset.
    Set pValueRL = New RasterLayer
    pValueRL.CreateFromFilePath "c:\data\chap11\precipgd"
    Set pValueRaster = pValueRL.Raster
    ' Add the zonal and value raster layers to the active map.
    pMap.AddLayer pZoneRL
    pMap.AddLayer pValueRL
```

Part 1 defines the zonal and value raster datasets to be used in the zonal operation. Instead of referring to raster layers in the active map, the code uses the *CreateFromFilePath* method on *IRasterLayer* to open a zonal layer referenced by *pZoneRL* and sets *pZoneRaster* to be its raster. The same procedure is followed to open *pValueRL* and to set *pValueRaster*. The code then adds *pZoneRL* and *pValueRL* to the active map.

```
    ' Part 2: Perform zonal mean.
    Dim pZonalOp As IZonalOp
    Dim pOutputRaster As IGeoDataset
    ' Run the zonal mean operation.
    Set pZonalOp = New RasterZonalOp
    Set pOutputRaster = pZonalOp.ZonalStatistics (pZoneDataset, pValueDataset, esriGeoAnalysisStatsMean, True)
```

Part 2 creates *pZonalOp* as an instance of the *RasterZonalOp* class and uses the *ZonalStatistics* method on *IZonalOp* to create a zonal mean raster referenced by *pOutputRaster*.

```
    ' Part 3: Create the output layer and add it to the active map.
    Dim pOutputLayer As IRasterLayer
    Set pOutputLayer = New RasterLayer
    pOutputLayer.CreateFromRaster pOutputRaster
    pOutputLayer.Name = "ZonalMean"
    pMap.AddLayer pOutputLayer
End Sub
```

Part 3 creates a raster layer from *pOutputRaster* and adds the layer to the active map.

Box 11.5 ZonalMean_GP

ZonalMean_GP uses the ZonalStatistics tool in the Spatial Analyst toolbox to derive a zonal mean raster from a zonal raster (*hucgd*) and a value raster (*precipgd*). Run the macro in ArcCatalog and view the output raster (*zonalmean2*) in the Catalog tree.

```
Private Sub ZonalMean_GP()
    ' Create the Geoprocessing object and define its workspace.
    Dim GP As Object
    Set GP = CreateObject("esriGeoprocessing.GpDispatch.1")
    Dim filepath As String
    filepath = "c:\data\chap11"
    GP.Workspace = filepath
    ' ZonalStatistics <in_zone_data> <zone_field> <in_value_raster> <out_raster> {MEAN | MAJORITY |
    ' MAXIMUM | MEDIAN | MINIMUM | MINORITY | RANGE | STD | SUM | VARIETY} {DATA | NODATA}
    ' Execute the zonalstatistics tool.
    GP.ZonalStatistics_sa "hucgd", "Value", "precipgd", "zonalmean2", "MEAN"
End Sub
```

11.7 PERFORMING DISTANCE MEASURE OPERATIONS

Distance measure operations can be based on physical distance or cost distance. Both types of operations calculate the distance from each cell of a raster to a source cell. The main difference is that cost distance measures are based on a cost raster. This section covers both types of distance measures.

11.7.1 *EucDist*

EucDist calculates continuous Euclidean distance measures away from a stream (source) raster. The macro performs the same function as using the Distance/Straight Line option in Spatial Analyst. *EucDist* has three parts. Part 1 defines the source raster, Part 2 defines and runs a distance measure operation, and Part 3 creates a raster layer from the output and adds the layer to the active map.

Key Interfaces: *IRaster, IGeoDataset, IDistanceOp*
Key Members: *Raster, EucDistance, CreateFromRaster*
Usage: Add *emidastrmgd* to an active map. Import *EucDist* to Visual Basic Editor. Run the macro. The macro creates and adds a new temporary raster named *EuclideanDistance* to the active map.

```
Private Sub EucDist()
    ' Part 1: Define the input raster dataset.
    Dim pMxDoc As IMxDocument
    Dim pMap As IMap
```

```
Dim pSourceRL As IRasterLayer
Dim pSourceRaster As IRaster
Set pMxDoc = ThisDocument
Set pMap = pMxDoc.FocusMap
Set pSourceRL = pMap.Layer(0)
Set pSourceRaster = pSourceRL.Raster
```

Part 1 sets the raster of the top layer to be the source raster, and references the raster by *pSourceRaster*.

```
' Part 2: Perform distance measures.
Dim pDistanceOp As IDistanceOp
Dim pOutputRaster As IGeoDataset
' Run the Euclidean distance operation.
Set pDistanceOp = New RasterDistanceOp
Set pOutputRaster = pDistanceOp.EucDistance(pSourceDataset)
```

Part 2 creates *pDistanceOp* as an instance of the *RasterDistanceOp* class and uses the *EucDistance* method on *IDistanceOp* to create a continuous distance measure raster referenced by *pOutputRaster*.

```
' Part 3: Create the output layer and add it to the active map.
Dim pOutputLayer As IRasterLayer
Set pOutputLayer = New RasterLayer
pOutputLayer.CreateFromRaster pOutputRaster
pOutputLayer.Name = "EuclideanDistance"
pMap.AddLayer pOutputLayer
End Sub
```

Part 3 creates a new raster layer from *pOutputRaster* and adds the layer to the active map.

11.7.2   Use of a Feature Layer as the Source in *EucDist*

EucDist uses a raster as the source. With some modification in Part 1, *EucDist* can also use a shapefile (*emidastrm.shp*) as the source. The shapefile must be converted into a raster before the distance measure operation starts. The following shows the change in Part 1 of *EucDist* to accommodate a shapefile source (EucDistToFeatLayer.txt on the companion CD is a complete macro to run the distance measure operation from *emidastrm.shp*):

```
' Part 1: Convert the input feature layer to a raster dataset.
Dim pMxDoc As IMxDocument
Dim pMap As IMap
Dim pSourceLayer As IFeatureLayer
Dim pSourceFC As IFeatureClass
Dim pConOp As IConversionOp
Dim pEnv As IRasterAnalysisEnvironment
Dim pWSF As IWorkspaceFactory
Dim pWS As IWorkspace
```

```
Dim pSourceDataset As IGeoDataset
Set pMxDoc = ThisDocument
Set pMap = pMxDoc.FocusMap
' Define the source feature class.
Set pSourceLayer = pMap.Layer(0)
Set pSourceFC = pSourceLayer.FeatureClass
' Define the workspace for the raster.
Set pWSF = New RasterWorkspaceFactory
Set pWS = pWSF.OpenFromFile("c:\data\chap11\", 0)
' Prepare a new conversion operator.
Set pConOp = New RasterConversionOp
Set pEnv = pConOp
pEnv.SetCellSize esriRasterEnvValue, 30
' Run the conversion operation.
Set pSourceDataset = pConOp.ToRasterDataset(pSourceFC, "GRID", pWS, "SourceGrid")
```

This revised Part 1 of **EucDist** first sets *pSourceFC* to be the feature class of the source layer. The conversion of *pSourceFC* to a raster involves three steps. First, the code defines the workspace for the raster. Second, the code creates *pConOp* as an instance of the *RasterConversionOp* class and uses the *IRasterAnalysisEnviron-ment* interface to define the output cell size. Third, the code uses the *ToRasterDataset* method on *IConversionOp* to create the source raster referenced by *pSourceDataset*.

Box 11.6 EucDist_GP

EucDist_GP uses the EucDistance tool in the Spatial Analyst toolbox and a shapefile (*emidastrm.shp*) to derive a distance measure raster (*eucdist2*) with a cell size of 30 meters. Run the macro in ArcCatalog and view the output raster in the Catalog tree.

```
Private Sub EucDist_GP()
    ' Create the Geoprocessing object and define its workspace.
    Dim GP As Object
    Set GP = CreateObject("esriGeoprocessing.GpDispatch.1")
    ' Define the workspace.
    Dim filepath As String
    filepath = "c:\data\chap11"
    GP.Workspace = filepath
    ' EucDistance_sa <in_source_data> <out_distance_raster> {maximum_distance}
    ' {cell_size} {out_direction_raster}
    ' Execute the eucdistance tool.
    GP.EucDistance_sa "emidastrm.shp", "eucdist2", "", "30"
End Sub
```

11.7.3 *Slice*

Slice calculates continuous Euclidean distance measures away from a stream (source) raster and reclassifies the distance measure raster into five equal intervals. **Slice** has three parts. Part 1 defines the source raster, Part 2 runs a distance measure operation and then a reclassify operation, and Part 3 creates new raster layers from the output rasters and adds them to the active map.

Key Interfaces: *IRaster, IGeoDataset, IDistanceOp, IReclassOp*
Key Members: *Raster, EucDistance, Slice, CreateFromRaster*
Usage: Add *emidastrmgd* to an active map. Import **Slice** to Visual Basic Editor. Run the macro. The macro creates and adds two new temporary rasters named *EuclideanDistance* and *EqualInterval* respectively to the active map.

```
Private Sub Slice()
    ' Part 1: Define the source raster.
    Dim pMxDoc As IMxDocument
    Dim pMap As IMap
    Dim pSourceRL As IRasterLayer
    Dim pSourceRaster As IRaster
    Set pMxDoc = ThisDocument
    Set pMap = pMxDoc.FocusMap
    Set pSourceRL = pMap.Layer(0)
    Set pSourceRaster = pSourceRL.Raster
```

Part 1 sets *pSourceRaster* to be the raster of the first layer in the active map.

```
    ' Part 2: Define and run the distance measure operation.
    Dim pDistanceOp As IDistanceOp
    Dim pOutputRaster As IGeoDataset
    Dim pReclassOp As IReclassOp
    Dim pSliceRaster As IGeoDataset
    ' Run a EucDistance operation.
    Set pDistanceOp = New RasterDistanceOp
    Set pOutputRaster = pDistanceOp.EucDistance(pSourceRaster)
    ' Run a slice operation.
    Set pReclassOp = New RasterReclassOp
    Set pSliceRaster = pReclassOp.Slice(pOutputRaster, esriGeoAnalysisSliceEqualInterval, 5)
```

Part 2 first creates *pDistanceOp* as an instance of the *RasterDistanceOp* class and uses the *EucDistance* method to create a continuous distance measure raster referenced by *pOutputRaster*. Then the code creates *pReclassOp* as an instance of the *RasterReclassOp* class and uses the *Slice* method on *IReclassOp* to create an equal-interval raster with five classes. The equal-interval or sliced raster is referenced by *pSliceRaster*.

```
    ' Part 3: Create the new raster layers and add them to the active map.
    Dim pOutputLayer As IRasterLayer
    Dim pSliceLayer As IRasterLayer
    ' Create and add the distance measure layer.
    Set pOutputLayer = New RasterLayer
    pOutputLayer.CreateFromRaster pOutputRaster
    pOutputLayer.Name = "EuclideanDistance"
    pMap.AddLayer pOutputLayer
    ' Create and add the slice layer.
    Set pSliceLayer = New RasterLayer
```

```
    pSliceLayer.CreateFromRaster pSliceRaster
    pSliceLayer.Name = "EqualInterval"
    pMap.AddLayer pSliceLayer
End Sub
```

Part 3 creates new raster layers from *pOutputRaster* and *pSliceRaster* respectively, and adds these two layers to the active map.

11.7.4 *CostDist*

CostDist uses a source raster and a cost raster to calculate the least accumulative cost distance. The macro performs the same function as using the Distance/Cost Weighted option in Spatial Analyst. *CostDist* has three parts. Part 1 defines the source and cost rasters, Part 2 runs a cost distance measure operation, and Part 3 creates a new raster layer from the output and adds it to the active map.

> **Key Interfaces:** *IRaster, IDistanceOp*
> **Key Members:** *Raster, CostDistance, CreateFromRaster*
> **Usage:** Add *emidastrmgd*, a source raster, and *emidacostgd*, a cost raster, to an active map. The source raster must be on top of the cost raster in the active map. Import *CostDist* to Visual Basic Editor. Run the macro. The macro creates and adds a new temporary raster named *CostDistance* to the active map.

```
Private Sub CostDist()
    ' Part 1: Define the source and cost rasters.
    Dim pMxDoc As IMxDocument
    Dim pMap As IMap
    Dim pSourceRL As IRasterLayer
    Dim pSourceRaster As IRaster
    Dim pCostRL As IRasterLayer
    Dim pCostRaster As IRaster
    Set pMxDoc = ThisDocument
    Set pMap = pMxDoc.FocusMap
    ' Define the source dataset.
    Set pSourceRL = pMap.Layer(0)
    Set pSourceRaster = pSourceRL.Raster
    ' Define the cost dataset.
    Set pCostRL = pMap.Layer(1)
    Set pCostRaster = pCostRL.Raster
```

Part 1 sets *pSourceRaster* to be the raster of the top layer in the active map, and *pCostRaster* to be the raster of the second layer.

```
    ' Part 2: Perform cost distance measures.
    Dim pDistanceOp As IDistanceOp
    Dim pOutputRaster As IRaster
    ' Run the cost distance operation.
    Set pDistanceOp = New RasterDistanceOp
    Set pOutputRaster = pDistanceOp.CostDistance(pSourceRaster, pCostRaster)
```

Part 2 creates *pDistanceOp* as an instance of the *RasterDistanceOp* class and uses the *CostDistance* method on *IDistanceOp* to create a least accumulative cost distance raster referenced by *pOutputRaster*.

```
' Part 3: Create the output layer and add it to the active map.
Dim pRLayer As IRasterLayer
Set pRLayer = New RasterLayer
pRLayer.CreateFromRaster pOutputRaster
pRLayer.Name = "CostDistance"
pMap.AddLayer pRLayer
End Sub
```

Part 3 creates a new raster layer from *pOutputRaster* and adds the layer to the active map.

11.7.5 *CostDistFull*

CostDistFull uses a source raster and a cost raster to calculate a back link raster and an allocation raster in addition to a least accumulative cost distance raster. The macro performs the same function as using the Distance/Cost Weighted command, with both Create direction and Create allocation checked, in Spatial Analyst. *CostDistFull* has four parts. Part 1 defines the source and cost rasters. Part 2 defines and runs a cost distance measure operation. From the output of Part 2, Part 3 derives the least accumulative cost, back link, and allocation rasters. Part 4 creates new raster layers of the least accumulative cost, back link, and allocation and adds these layers to the active map.

> **Key Interfaces:** *IRaster, IGeoDataset, IDistanceOp, IRasterBandCollection, IRaster-band*
>
> **Key Members:** *Raster, CostDistanceFull, Item( ), AppendBand, CreateFromRaster, Name*
>
> **Usage:** Add *emidastrmgd*, a source raster, and *emidacostgd*, a cost raster, to an active map. The source raster must be an integer raster for calculating the allocation output. In the active map, *emidastrmgd* must be on top of *emidacostgd*. Import *CostDistFull* to Visual Basic Editor in ArcMap. Run the macro. The macro creates and adds three temporary rasters: *CostDistance* for the least accumulative cost distance, *BackLink* for the back link, and *Allocation* for the allocation. *BackLink* contains cell values from zero through eight, which defines the next neighboring cell along the least accumulative cost path from a cell to reach its closest source cell. *Allocation* assigns each cell to its closest source cell.

```
Private Sub CostDistFull()
    ' Part 1: Define the source and cost rasters.
    Dim pMxDoc As IMxDocument
    Dim pMap As IMap
    Dim pSourceRL As IRasterLayer
    Dim pSourceRaster As IRaster
    Dim pCostRL As IRasterLayer
    Dim pCostRaster As IRaster
```

```
Set pMxDoc = ThisDocument
Set pMap = pMxDoc.FocusMap
' Define the source dataset.
Set pSourceRL = pMap.Layer(0)
Set pSourceRaster = pSourceRL.Raster
' Define the cost dataset.
Set pCostRL = pMap.Layer(1)
Set pCostRaster = pCostRL.Raster
```

Part 1 sets *pSourceRaster* to be the raster of the source layer and *pCostRaster* to be the raster of the cost layer.

```
' Part 2: Perform cost distance measures.
Dim pDistanceOp As IDistanceOp
Dim pOutputRaster As IGeoDataset
' Run the cost distance operation.
Set pDistanceOp = New RasterDistanceOp
Set pOutputRaster = pDistanceOp.CostDistanceFull (pSourceRaster, pCostRaster, True, True, True)
```

Part 2 creates *pDistanceOp* as an instance of the *RasterDistanceOp* class and uses the *CostDistanceFull* method on *IDistanceOp* to create the output referenced by *pOutputRaster*. The *CostDistanceFull* method creates the least accumulative cost distance, back link, and allocation rasters all at once.

```
' Part 3: Derive the least accumulative cost distance, backlink, and allocation rasters.
Dim pRasterBandCollection As IRasterBandCollection
Dim pCostDistRB As IRasterband
Dim pCostDistRBCollection As IRasterBandCollection
Dim pBackLinkRB As IRasterband
Dim pBackLinkRBCollection As IRasterBandCollection
Dim pAllocationRB As IRasterband
Dim pAllocationRBCollection As IRasterBandCollection
' Extract and create the least accumulative cost distance raster.
Set pRasterBandCollection = pOutputRaster
Set pCostDistRB = pRasterBandCollection.Item(0)
Set pCostDistRBCollection = New Raster
pCostDistRBCollection.AppendBand pCostDistRB
' Extract and create the backlink raster.
Set pBackLinkRB = pRasterBandCollection.Item(1)
Set pBackLinkRBCollection = New Raster
pBackLinkRBCollection.AppendBand pBackLinkRB
' Extract and create the allocation raster.
Set pAllocationRB = pRasterBandCollection.Item(2)
Set pAllocationRBCollection = New Raster
pAllocationRBCollection.AppendBand pAllocationRB
```

To extract the three rasters created in Part 2, Part 3 first performs a QI for the *IRasterBandCollection* interface and assigns the first band, which contains the least accumulative cost distance output to *pCostDistRB*. Next, the code creates *pCost-DistRBCollection* as an instance of the *Raster* class and uses the *AppendBand* method on *IRasterBandCollection* to append *pCostDistRB* to *pCostDistRBCollection*, thus completing the creation of the least accumulative cost distance raster. Part 3 repeats the same procedure to extract and create the back link raster referenced by *pBackLinkRBCollection* and the allocation raster referenced by *pAllocationRBCollection*.

```
' Part 4: Create the output layers and add them to the active map.
Dim pCostDistLayer As IRasterLayer
Dim pBackLinkLayer As IRasterLayer
Dim pAllocationLayer As IRasterLayer
' Add the least accumulative cost distance layer.
Set pCostDistLayer = New RasterLayer
pCostDistLayer.CreateFromRaster pCostDistRBCollection
pCostDistLayer.Name = "CostDistance"
pMap.AddLayer pCostDistLayer
' Add the back link layer.
Set pBackLinkLayer = New RasterLayer
pBackLinkLayer.CreateFromRaster pBackLinkRBCollection
pBackLinkLayer.Name = "BackLink"
pMap.AddLayer pBackLinkLayer
' Add the allocation layer.
Set pAllocationLayer = New RasterLayer
pAllocationLayer.CreateFromRaster pAllocationRBCollection
pAllocationLayer.Name = "Allocation"
pMap.AddLayer pAllocationLayer
End Sub
```

Part 4 creates new raster layers from *pCostDistRBCollection*, *pBackLinkRBCollection*, and *pAllocationRBCollection*, and adds these layers to the active map.

CHAPTER **12**

Terrain Mapping and Analysis

Terrain mapping refers to the use of techniques such as contours, hill shading, and perspective views to depict the land surface. Terrain analysis provides measures of the land surface such as slope and aspect. Terrain analysis also includes viewshed analysis and watershed analysis. A viewshed analysis predicts areas of the land surface that are visible from one or more observation points. A watershed analysis can derive watersheds and other topographic variables from an elevation raster. Terrain mapping and analysis are useful for a wide variety of applications.

A common data source for terrain mapping and analysis is the digital elevation model (DEM). A DEM consists of a regular array of elevation points compiled from stereo aerial photographs, satellite images, radar data, and other data sources. For terrain mapping and analysis, a DEM is first converted to an elevation raster. The simple data structure of an elevation raster makes it relatively easy to perform computations that are necessary for deriving slope, aspect, and other topographic parameters.

An alternative to the DEM is the triangulated irregular network (TIN). A TIN approximates the land surface with a series of nonoverlapping triangles. Elevation values and x-, y-coordinates are stored at nodes that make up the triangles. Many geographic information systems (GIS) users compile an initial TIN from a DEM or LIDAR (light detection and ranging) data and then use other data sources such as a stream network to modify and improve the TIN. In addition to flexible data sources, a TIN is also an excellent data model for terrain mapping and three-dimensional display. The triangular facets of a TIN create a sharper image of the terrain than a DEM does.

This chapter covers terrain mapping and analysis. Section 12.1 reviews terrain mapping and analysis using ArcGIS. Section 12.2 discusses objects that are related to terrain mapping and analysis. Section 12.3 includes macros for deriving contour, slope, aspect, and hillshade from an elevation raster. Three Geoprocessing (GP) macros are also introduced in Section 12.3 for deriving contour, slope, and aspect. The GP macro for deriving aspect is combined with a regular macro for displaying the aspect classes in color symbols. Section 12.4 has a macro for viewshed analysis.

Section 12.5 has a macro for watershed analysis. Section 12.6 offers macros and a GP macro for compiling and modifying a TIN and for deriving features of a TIN. All macros start with the listing of key interfaces and key members (properties and methods) and the usage. Make sure that the Spatial Analyst and 3D Analyst extensions are checked in the Tools/Extensions menu before running the macros.

12.1 PERFORMING TERRAIN MAPPING AND ANALYSIS IN ARCGIS

ArcGIS Desktop incorporates commands for terrain mapping and analysis in the Spatial Analyst and 3D Analyst extensions and ArcToolbox. Discussions in this section refer to the extensions. Both extensions have a surface analysis menu, which includes contour, slope, aspect, hillshade, and viewshed. The input to these analysis functions can be either an elevation raster or a TIN, and the output is in raster format except for the contour, which is in shapefile format.

Raster Calculator in Spatial Analyst is an important tool for terrain mapping and analysis because it can evaluate many surface analysis functions. For example, Raster Calculator can evaluate the slope function and create slope rasters directly, without going through the surface analysis menu. Raster Calculator can also evaluate watershed analysis functions that are not incorporated into Spatial Analyst's menu selections.

3D Analyst has menu selections for creating or modifying TINs. We can create an initial TIN from a DEM or feature data such as LIDAR data and contour lines, and then add point, line, and area features to modify the TIN. Additional point data may include surveyed elevation points and GPS (global positioning system) data. Line data may include breaklines such as streams, shorelines, ridges, and roads that represent changes of the land surface, and area data may include lakes and reservoirs. 3D Analyst also has a three-dimensional viewing application called ArcScene that lets the user prepare and manipulate perspective views, and three-dimensional draping and animation.

12.2 ARCOBJECTS FOR TERRAIN MAPPING AND ANALYSIS

Two primary components for terrain mapping and analysis are *RasterSurfaceOp* and *RasterHydrologyOp*. A *RasterSurfaceOp* object supports *IRasterAnalysisEnvironment* and *ISurfaceOp*. *IRasterAnalysisEnvironment* controls the analysis environment such as the output workspace and the output cell size. *ISurfaceOp* has methods for creating contour; calculating hillshade, slope, aspect, and curvature; and performing viewshed analysis (Figure 12.1).

A *RasterHydrologyOp* object supports *IRasterAnalysisEnvironment* and *IHydrologyOp*. *IHydrologyOp* has methods for filling sinks in a surface, creating flow direction, creating flow accumulation, assigning stream links, and delineating watersheds (Figure 12.2).

TIN is the primary component for three-dimensional applications. A TIN object implements *ITinEdit*, *ITinAdvanced*, and *ITinSurface* (Figure 12.3). *ITinEdit* has methods for constructing and editing TINs. *ITinAdvanced* has access to the underlying

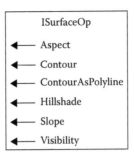

Figure 12.1 *ISurfaceOp* has methods for deriving contour, slope, aspect, hillshade, and view-shed from an elevation raster.

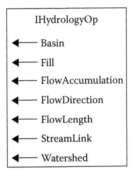

Figure 12.2 *IHydrologyOp* has methods for deriving hydrologic parameters for watershed analysis.

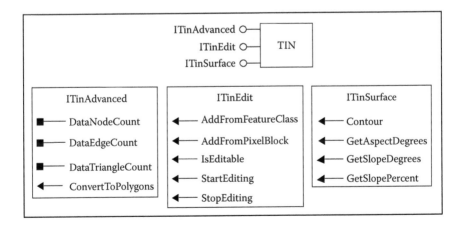

Figure 12.3 A *TIN* object supports *ITinAdvanced*, *ITinEdit*, and *ITinSurface*. These interfaces have members to create and edit a TIN, to derive topographic measures from a TIN, and to derive numbers of nodes, edges, and triangles from a TIN.

data structure of a TIN, including nodes, edges, and triangles. And *ITinSurface* provides surface analysis functions such as contouring and deriving slope and aspect from a TIN.

Raster objects covered in Chapter 11 may also be involved in terrain mapping and analysis. For example, to convert an elevation raster to a TIN requires the use of basic raster objects.

12.3 DERIVING CONTOUR, SLOPE, ASPECT, AND HILLSHADE

This section covers contour, slope, aspect, and hillshade. Besides showing how to derive these topographic measures from an elevation raster, this section also includes code fragments for selecting slope measurement units, classifying slope measures, and displaying aspect measures in eight principal directions.

12.3.1 *Contour*

Contour creates contour lines from an elevation raster by connecting points of equal elevation. The macro performs the same function as using the Surface Analysis/Contour command in Spatial Analyst to create a contour line shapefile. *Contour* has three parts. Part 1 defines the input elevation raster, Part 2 performs contouring and saves the output in a specified workspace, and Part 3 creates a feature layer from the output, prepares the contour line labels, and adds the layer to the active map.

> **Key Interfaces:** *ISurfaceOp, IRasterAnalysisEnvironment, IWorkspaceFactory, IGeo-Dataset, IGeoFeatureLayer*
> **Key Members:** *Raster, OpenFromFile, OutWorkspace, Contour, FeatureClass, DisplayField, DisplayAnnotation*
> **Usage:** Add *emidalat*, an elevation raster, to an active map. Import **Contour** to Visual Basic Editor. Run the macro. The macro adds *Contour* with labels to the active map.

```
Private Sub Contour()
    ' Part 1: Define the input raster.
    Dim pMxDoc As IMxDocument
    Dim pMap As IMap
    Dim pLayer As ILayer
    Dim pRasterLayer As IRasterLayer
    Dim pInputRaster As IRaster
    Set pMxDoc = ThisDocument
    Set pMap = pMxDoc.FocusMap
    Set pLayer = pMap.Layer(0)
    Set pRasterLayer = pLayer
    Set pInputRaster = pRasterLayer.Raster
```

Part 1 sets *pInputRaster* to be the raster of the top layer in the active map.

```
' Part 2: Perform contour.
Dim pSurfaceOp As ISurfaceOp
Dim pEnv As IRasterAnalysisEnvironment
Dim pWS As IWorkspace
Dim pWSF As IWorkspaceFactory
Dim pOutput As IGeoDataset
' Define a surface operation.
Set pSurfaceOp = New RasterSurfaceOp
Set pEnv = pSurfaceOp
Set pWSF = New RasterWorkspaceFactory
Set pWS = pWSF.OpenFromFile("c:\data\chap12", 0)
Set pEnv.OutWorkspace = pWS
' Run the contour surface operation.
Set pOutput = pSurfaceOp.Contour(pInputRaster, 50, 850)
POutput.Rename ("Contour")
```

Part 2 first creates *pSurfaceOp* as an instance of the *RasterSurfaceOp* class. Next, the code performs a QueryInterface (QI) for *IRasterAnalysisEnvironment* and uses the *OutWorkspace* property to assign *pWS* as the output workspace. The code then runs the *Contour* method on *ISurfaceOp* to create a geographic dataset referenced by *pOutput*. The *Contour* method uses the arguments of 50 (meters) for the contour interval and 850 (meters) for the base contour. The contour interval represents the vertical distance between contour lines, and the base contour is the contour line of the lowest elevation. The *Rename* method on *IDataset* changes the name of *pOutput* to *Contour*. (Without renaming, the contour dataset will have a default name of Shape*.shp.)

```
' Part 3: Create the output layer, and add it to the active map.
Dim pOutLayer As IFeatureLayer
Dim pGeoFeatureLayer As IGeoFeatureLayer
' Create the output feature layer.
Set pOutLayer = New FeatureLayer
Set pOutLayer.FeatureClass = pOutput
pOutLayer.Name = "Contour"
' Label the contour lines.
Set pGeoFeatureLayer = pOutLayer
pGeoFeatureLayer.DisplayField = "CONTOUR"
pGeoFeatureLayer.DisplayAnnotation = True
pMap.AddLayer pOutLayer
End Sub
```

Part 3 first creates a new feature layer from *pOutput* and names the layer Contour. The code then switches to the *IGeoFeatureLayer* interface to set up the contour labels, before adding the layer to the active map.

Box 12.1 Contour_GP

Contour_GP uses the Contour tool in the Spatial Analyst toolbox to derive a contour shapefile from an elevation raster. The contour interval (50) and the base contour (800) are specified as arguments. Run the macro in ArcCatalog and view the output (*contour2*) in the Catalog tree.

```
Private Sub Contour_GP()
    ' Create the Geoprocessing object.
    Dim GP As Object
    Set GP = CreateObject("esriGeoprocessing.GpDispatch.1")
    ' Contour <in_zone_data> <out_polyline_features> <contour_interval> {base_contour} {z_factor}
    ' Execute the contour tool.
    GP.Contour_sa "c:\data\chap12\emidalat", "c:\data\chap12\contour2", 50, 800
End Sub
```

12.3.2 *Slope*

Slope derives a temporary slope raster from an elevation raster. The macro performs the same function as using the Surface Analysis/Slope command in Spatial Analyst. *Slope* has three parts. Part 1 defines the input raster, Part 2 runs the slope operation and saves the output in a specified workspace, and Part 3 creates a raster layer from the output and adds the layer to the active map.

> **Key Interfaces:** *ISurfaceOp, IRasterAnalysisEnvironment, IWorkspace, IWorkspaceFactory*
> **Key Members:** *Raster, OpenFromFile, OutWorkspace, Slope, CreateFromRaster*
> **Usage:** Add *emidalat*, an elevation raster, to an active map. Import *Slope* to Visual Basic Editor. Run the macro. The macro adds a slope layer to the active map.

```
Private Sub Slope()
    ' Part 1: Define the input raster.
    Dim pMxDoc As IMxDocument
    Dim pMap As IMap
    Dim pLayer As ILayer
    Dim pRasterLayer As IRasterLayer
    Dim pInputRaster As IRaster
    Set pMxDoc = ThisDocument
    Set pMap = pMxDoc.FocusMap
    Set pLayer = pMap.Layer(0)
    Set pRasterLayer = pLayer
    Set pInputRaster = pRasterLayer.Raster
```

Part 1 sets *pInputRaster* to be the raster of the top layer in the active map.

```
    ' Part 2: Perform slope.
    Dim pSurfaceOp As ISurfaceOp
    Dim pEnv As IRasterAnalysisEnvironment
    Dim pWS As IWorkspace
    Dim pWSF As IWorkspaceFactory
    Dim pOutRaster As IRaster
    ' Prepare a raster surface operation.
    Set pSurfaceOp = New RasterSurfaceOp
```

```
Set pEnv = pSurfaceOp
Set pWSF = New RasterWorkspaceFactory
Set pWS = pWSF.OpenFromFile("c:\data\chap12", 0)
Set pEnv.OutWorkspace = pWS
' Run the slope surface operation.
Set pOutRaster = pSurfaceOp.Slope(pInputRaster, esriGeoAnalysisSlopePercentrise)
```

Part 2 first creates *pSurfaceOp* as an instance of the *RasterSurfaceOp* class. Next, the code accesses *IRasterAnalysisEnvironment* and sets the workspace for the output. Then the code uses the *Slope* method on *ISurfaceOp* to create a slope raster referenced by *pOutRaster*. The argument of esriGeoAnalysisSlopePercentrise specifies that the slope raster be measured in percentage of rise.

```
' Part 3: Create the output raster layer, and add it to the active map.
Dim pSlopelayer As IRasterLayer
Set pSlopeLayer = New RasterLayer
pSlopeLayer.CreateFromRaster pOutRaster
pSlopeLayer.Name = "Slope"
pMap.AddLayer pSlopeLayer
End Sub
```

Part 3 creates a new raster layer from *pOutRaster* and adds the layer to the active map.

Box 12.2 Slope_GP

Slope_GP uses the Slope tool in the Spatial Analyst toolbox to derive a percent slope raster from an elevation raster. Run the macro in ArcCatalog and view the output (*slope2*) in the Catalog tree.

```
Private Sub Slope_GP()
    ' Create the Geoprocessing object.
    Dim GP As Object
    Set GP = CreateObject("esriGeoprocessing.GpDispatch.1")
    ' Slope <in_zone_data> <out_raster> {DEGREE I PERCENT_RISE} {z_factor}
    ' Execute the slope tool.
    GP.Slope_sa "c:\data\chap12\emidalat", "c:\data\chap12\slope2", "PERCENT_RISE"
End Sub
```

12.3.3 Choice of Slope Measure

A slope raster can be measured in either degree slope or percent slope. Percent slope is 100 times the ratio of rise (vertical distance) over run (horizontal distance), whereas degree slope is the arc tangent of the ratio of rise over run. Because different projects may require different slope measures, a macro can use an input box and let the user choose one of these two options before running the slope operation. To offer these two options, one can replace Part 2 of *Slope* by the following code fragment (ChooseSlopeMeasure.txt on the companion CD incorporates the revision):

```
' Part 2: Perform slope.
    Dim pSurfaceOp As ISurfaceOp
    Dim pEnv As IRasterAnalysisEnvironment
    Dim pWS As IWorkspace
    Dim pWSF As IWorkspaceFactory
    Dim Message As String
    Dim Default As String
    Dim Choice As String
    Dim pOutRaster As IRaster
    ' Prepare a raster surface operation.
    Set pSurfaceOp = New RasterSurfaceOp
    Set pEnv = pSurfaceOp
    Set pWSF = New RasterWorkspaceFactory
    Set pWS = pWSF.OpenFromFile("c:\data\chap12", 0)
    Set pEnv.OutWorkspace = pWS
    ' Choose slope measures in degrees or in percentage of rise.
    Message = "Enter D for slope measures in degrees or P for slope measures in percentage of rise"
    Default = "D"
    Choice = InputBox(Message, , Default)
    ' Run the slope surface operation according to the choice.
    If Choice = "D" Or Choice = "d" Then
        Set pOutRaster = pSurfaceOp.Slope(pInputRaster, esriGeoAnalysisSlopeDegrees)
        ElseIf Choice = "P" Or Choice = "p" Then
            Set pOutRaster = pSurfaceOp.Slope(pInputRaster, esriGeoAnalysisSlopePercentrise)
        Else: Exit Sub
    End If
```

The code fragment runs the *Slope* method with the user's choice of the slope type (D for degrees and P for percentage of rise). D for degrees is the default.

12.3.4   *ReclassifySlope*

A slope raster is often classified into slope classes before it is used in a GIS project. *ReclassifySlope* is a sub that can be called at the end of *Slope* to group slope measures into five classes (ReclassifySlope.txt on the companion CD is a module that has both *Slope* and *ReclassifySlope*).

```
' Call the sub to display the classified slope layer in a defined color scheme.
Call ReclassifySlope(pSlopeLayer)
```

ReclassifySlope has three parts. Part 1 gets *pSlopeLayer* from *Slope* for the input raster, Part 2 performs reclassification, and Part 3 creates a raster layer from the output and adds the layer to the active map.

```
Private Sub ReclassifySlope(pSlopeLayer As IRasterLayer)
    ' Part 1: Define the raster for reclassify.
    Dim pMxDoc As IMxDocument
    Dim pMap As IMap
    Dim pRasterLy As IRasterLayer
```

```
Dim pGeoDs As IGeoDataset
Set pMxDoc = ThisDocument
Set pMap = pMxDoc.FocusMap
Set pRasterLy = pSlopeLayer
Set pGeoDs = pRasterLy.Raster
```

Part 1 gets *pSlopeLayer* from **Slope** and assigns the layer to *pRasterLy*. The code
then sets *pGeoDs* to be the raster of *pRasterLy*.

```
' Part 2: Reclassify the input raster.
Dim pReclassOp As IReclassOp
Dim pNRemap As INumberRemap
Dim pOutRaster As IRaster
' Prepare a number remap.
Set pNRemap = New NumberRemap
pNRemap.MapRange 0, 10#, 1
pNRemap.MapRange 10.1, 20#, 2
pNRemap.MapRange 20.1, 30#, 3
pNRemap.MapRange 30.1, 40#, 4
pNRemap.MapRange 40.1, 90#, 5
' Run the reclass operation.
Set pReclassOp = New RasterReclassOp
Set pOutRaster = pReclassOp.ReclassByRemap(pGeoDs, pNRemap, False)
```

Part 2 first creates *pNRemap* as an instance of the *NumberRemap* class and uses
the *MapRange* method on *INumberRemap* to set the output value based on a numeric
range of the input values. A cell within the numeric range of 0 to 10.0 is assigned
an output value of 1, 10.1 to 20.0 an output value of 2, and so on. Next, the code
creates *pReclassOp* as an instance of the *RasterReclassOp* class and uses the
ReclassByRemap method on *IReclassOp* to create a reclassified raster referenced by
pOutRaster. The reclassified raster has five slope classes.

```
' Part 3: Create the output layer, and add it to the active map.
Dim pReclassLy As IRasterLayer
Set pReclassLy = New RasterLayer
pReclassLy.CreateFromRaster pOutRaster
pReclassLy.Name = "Classified Slope"
pMap.AddLayer pReclassLy
End Sub
```

Part 3 creates a new raster layer from *pOutRaster* and adds the classified slope
layer to the active map.

12.3.5 Aspect

To derive a temporary aspect raster from an elevation raster, one can use the same
macro as **Slope** but change the last line statement in Part 2 to:

```
' Run the aspect surface operation.
Set pOutRaster = pSurfaceOp.Aspect(pInputRaster)
```

The *Aspect* method on *ISurfaceOp* uses *pInputRaster* as the only object qualifier.

12.3.6 *Aspect_Symbol*

Aspect is a circular measure, which starts with 0° at the north, moves clockwise, and ends with 360° also at the north. An aspect raster is typically classified into four or eight principal directions and an additional class for flat areas. *Aspect_Symbol* uses a random color ramp to display an aspect raster with eight principal directions plus flat areas. A good way to use *Aspect_Symbol* is to call the sub at the end of a macro that has derived an aspect layer. The macro can then pass the aspect layer as an argument to *Aspect_Symbol*. (Aspect_Symbol.txt on the companion CD includes *Aspect* for deriving aspect measures and *Aspect_Symbol* for displaying aspect measures.)

Aspect_Symbol has three parts. Part 1 creates a raster renderer, Part 2 specifies the color, break, and label for each aspect class, and Part 3 assigns the renderer to the raster layer and refreshes the active view.

```
Private Sub Aspect_Symbol(pAspectLayer As IRasterLayer)
    ' Part 1: Create a raster renderer.
    Dim pMxDoc As IMxDocument
    Dim pClassRen As IRasterClassifyColorRampRenderer
    Dim pRasRen As IRasterRenderer
    Dim pRaster As IRaster
    Set pRaster = pAspectLayer.Raster
    ' Prepare a raster classify renderer.
    Set pClassRen = New RasterClassifyColorRampRenderer
    pClassRen.ClassCount = 10
    ' Define the raster and update the renderer.
    Set pRasRen = pClassRen
    Set pRasRen.Raster = pRaster
    pRasRen.Update
```

Part 1 first sets *pRaster* to be the raster of the aspect layer passed to the sub as an argument. Next, the code creates *pClassRen* as an instance of the *RasterClassifyColorRampRenderer* class and specifies 10 for the number of classes. The code then accesses the *IRasterRenderer* interface, assigns *pRaster* to be the raster for *pRasRen*, and updates *pRasRen*.

```
    ' Part 2: Specify the color, break, and label for each aspect class.
    Dim pRamp As IRandomColorRamp
    Dim pColors As IEnumColors
    Dim pFSymbol As ISimpleFillSymbol
    ' Prepare a random color ramp.
    Set pRamp = New RandomColorRamp
    pRamp.Size = 10
```

```
pRamp.Seed = 100
pRamp.CreateRamp (True)
Set pColors = pRamp.Colors
' Define the symbol, break, and label for 10 aspect classes.
Set pFSymbol = New SimpleFillSymbol
pFSymbol.Color = pColors.Next
pClassRen.Symbol(0) = pFSymbol
pClassRen.Break(0) = -1
pClassRen.Label(0) = "Flat(-1)"
pFSymbol.Color = pColors.Next
pClassRen.Symbol(1) = pFSymbol
pClassRen.Break(1) = -0.01
pClassRen.Label(1) = "North(0-22.5)"
pFSymbol.Color = pColors.Next
pClassRen.Symbol(2) = pFSymbol
pClassRen.Break(2) = 22.5
pClassRen.Label(2) = "Northeast(22.5-67.5)"
pFSymbol.Color = pColors.Next
pClassRen.Symbol(3) = pFSymbol
pClassRen.Break(3) = 67.5
pClassRen.Label(3) = "East(67.5-112.5)"
pFSymbol.Color = pColors.Next
pClassRen.Symbol(4) = pFSymbol
pClassRen.Break(4) = 112.5
pClassRen.Label(4) = "Southeast(112.5-157.5)"
pFSymbol.Color = pColors.Next
pClassRen.Symbol(5) = pFSymbol
pClassRen.Break(5) = 157.5
pClassRen.Label(5) = "South(157.5-202.5)"
pFSymbol.Color = pColors.Next
pClassRen.Symbol(6) = pFSymbol
pClassRen.Break(6) = 202.5
pClassRen.Label(6) = "Southwest(202.5-247.5)"
pFSymbol.Color = pColors.Next
pClassRen.Symbol(7) = pFSymbol
pClassRen.Break(7) = 247.5
pClassRen.Label(7) = "West(247.5-292.5)"
pFSymbol.Color = pColors.Next
pClassRen.Symbol(8) = pFSymbol
pClassRen.Break(8) = 292.5
pClassRen.Label(8) = "Northwest(292.5-337.5)"
pFSymbol.Color = pColors.Next
pClassRen.Symbol(9) = pFSymbol
pClassRen.Break(9) = 337.5
pClassRen.Label(9) = "North(337.5-360)"
' Set Symbol 9 for north to be the same as Symbol 1 for north.
pClassRen.Symbol(9) = pClassRen.Symbol(1)
```

Part 2 creates *pRamp* as an instance of the *RandomColorRamp* class and specifies 10 for the number of colors to be generated, 100 for the seed of the generator, and

True to generate the color ramp. The rest of Part 2 assigns a random color, a class break, and a label for each of the ten classes. Because the north aspect includes classes (1) and (9) (0 to 22.5° and 337.5 to 360°), the symbol for class (9) is reset to be the same as the symbol for class (1).

```
' Part 3: Assign the renderer to the aspect layer, and refresh the active view.
pRasRen.Update
Set pAspectLayer.Renderer = pRasRen
Set pMxDoc = ThisDocument
pMxDoc.ActiveView.Refresh
pMxDoc.UpdateContents
End Sub
```

Part 3 assigns the updated *pRasRen* to *pAspectLayer*, refreshes the active view, and updates the contents of the map document.

Box 12.3 Aspect_GP

Aspect_GP uses the Aspect tool in the Spatial Analyst toolbox to derive an aspect raster (*aspect2*) from *emidalat*. Then the code defines the top layer of the active map as *pAspectLayer* and passes it to the sub *AspectSymbol*, which is the same as the sub in Section 12.3.6, for displaying the ten aspect classes in random colors. Add *emidalat* to the active map, before running *Aspect_GP* in ArcMap. The code adds *aspect2* with symbology to the map. This example shows how to combine a GP macro with a regular ArcObjects macro for the purpose of data display.

```
Private Sub Aspect_GP()
    ' Create the Geoprocessing object.
    Dim GP As Object
    Set GP = CreateObject("esriGeoprocessing.GpDispatch.1")
    ' Aspect <in_raster> <out_raster>
    ' Execute the aspect tool.
    GP.Aspect_sa "c:\data\chap12\emidalat", "c:\data\chap12\aspect2"
    ' Define the top layer of the active map as the newly created aspect map.
    Dim pMxDoc As IMxDocument
    Dim pMap As IMap
    Dim pLayer As ILayer
    Dim pAspectLayer As IRasterLayer
    Set pMxDoc = ThisDocument
    Set pMap = pMxDoc.FocusMap
    Set pLayer = pMap.Layer(0)
    Set pAspectLayer = pLayer
    ' Call the AspectSymbol sub and pass pAspectLayer to the sub.
    Call AspectSymbol(pAspectLayer)
End Sub

Private Sub AspectSymbol(pAspectLayer As IRasterLayer)
    ' Part 1: Create a color ramp for displaying the aspect raster.
    Dim pMxDoc As IMxDocument
```

```
Dim pClassRen As IRasterClassifyColorRampRenderer
Dim pRasRen As IRasterRenderer
Dim pRaster As IRaster
Set pRaster = pAspectLayer.Raster
' Prepare a raster classify renderer.
Set pClassRen = New RasterClassifyColorRampRenderer
pClassRen.ClassCount = 10
' Define the raster and update the renderer.
Set pRasRen = pClassRen
Set pRasRen.Raster = pRaster
pRasRen.Update

' Part 2: Specify the color, break, and label for each class.
Dim pRamp As IRandomColorRamp
Dim pColors As IEnumColors
Dim pFSymbol As ISimpleFillSymbol
' Prepare a random color ramp.
Set pRamp = New RandomColorRamp
pRamp.Size = 10
pRamp.Seed = 100
pRamp.CreateRamp (True)
Set pColors = pRamp.Colors
' Define the symbol, break, and label for 10 aspect classes.
Set pFSymbol = New SimpleFillSymbol
pFSymbol.Color = pColors.Next
pClassRen.Symbol(0) = pFSymbol
pClassRen.Break(0) = -1
pClassRen.Label(0) = "Flat(-1)"
pFSymbol.Color = pColors.Next
pClassRen.Symbol(1) = pFSymbol
pClassRen.Break(1) = -0.01
pClassRen.Label(1) = "North(0-22.5)"
pFSymbol.Color = pColors.Next
pClassRen.Symbol(2) = pFSymbol
pClassRen.Break(2) = 22.5
pClassRen.Label(2) = "Northeast(22.5-67.5)"
pFSymbol.Color = pColors.Next
pClassRen.Symbol(3) = pFSymbol
pClassRen.Break(3) = 67.5
pClassRen.Label(3) = "East(67.5-112.5)"
pFSymbol.Color = pColors.Next
pClassRen.Symbol(4) = pFSymbol
pClassRen.Break(4) = 112.5
pClassRen.Label(4) = "Southeast(112.5-157.5)"
pFSymbol.Color = pColors.Next
pClassRen.Symbol(5) = pFSymbol
pClassRen.Break(5) = 157.5
pClassRen.Label(5) = "South(157.5-202.5)"
pFSymbol.Color = pColors.Next
pClassRen.Symbol(6) = pFSymbol
```

```
pClassRen.Break(6) = 202.5
pClassRen.Label(6) = "Southwest(202.5-247.5)"
pFSymbol.Color = pColors.Next
pClassRen.Symbol(7) = pFSymbol
pClassRen.Break(7) = 247.5
pClassRen.Label(7) = "West(247.5-292.5)"
pFSymbol.Color = pColors.Next
pClassRen.Symbol(8) = pFSymbol
pClassRen.Break(8) = 292.5
pClassRen.Label(8) = "Northwest(292.5-337.5)"
pFSymbol.Color = pColors.Next
pClassRen.Symbol(9) = pFSymbol
pClassRen.Break(9) = 337.5
pClassRen.Label(9) = "North(337.5-360)"
' Set Symbol 9 for north to be the same as Symol 1 for north.
pClassRen.Symbol(9) = pClassRen.Symbol(1)

' Part 3: Assign the renderer to the aspect layer and refresh the active view.
pRasRen.Update
Set pAspectLayer.Renderer = pRasRen
Set pMxDoc = ThisDocument
pMxDoc.ActiveView.Refresh
pMxDoc.UpdateContents
End Sub
```

12.3.7 Hillshade

To derive a temporary hillshade raster from an elevation raster, one can use the same macro as **Slope** but change the last line statement in Part 2 to:

```
' Run the hillshade surface operation.
Set pOutRaster = pSurfaceOp.Hillshade(pInputRaster, 315, 30, True)
```

Besides the object qualifier *pInputRaster*, the *Hillshade* method on *ISurfaceOp* uses three arguments: 315 for the Sun's azimuth, 30 for the Sun's altitude, and True for the shaded relief type to include shadows. The Sun's azimuth is the direction of the incoming light, ranging from 0° (due north) to 360° in a clockwise direction. The Sun's altitude is the angle of the incoming light measured above the horizon between 0° and 90°.

12.4 PERFORMING VIEWSHED ANALYSIS

This section covers viewshed analysis. Inputs to a viewshed analysis include an elevation raster and a feature dataset containing one or more observation points. The analysis derives areas of the land surface that are visible from the observation point(s).

12.4.1 *Visibility*

Visibility creates a viewshed using an elevation raster and two observation points. The macro performs the same function as using the Surface Analysis/Viewshed command in Spatial Analyst. *Visibility* has three parts. Part 1 defines the elevation and lookout datasets, Part 2 runs the visibility analysis and saves the output in a specified workspace, and Part 3 creates a raster layer from the output and adds the layer to the active map.

> **Key Interfaces:** *IGeoDataset, ISurfaceOp, IRasterAnalysisEnvironment, IWorkspace, IWorkspaceFactory*
> **Key Members:** *FeatureClass, Raster, OpenFromFile, OutWorkspace, Visibility, CreateFromRaster*
> **Usage:** Add *plne*, an elevation raster, and *lookouts.shp*, a lookout point shapefile, to an active map. Make sure that *lookouts* is on top of *plne* in the table of contents. Import **Visibility** to Visual Basic Editor. Run the macro. The macro adds *Viewshed* to the active map.

```
Private Sub Visibility()
    ' Part 1: Define the elevation and lookout datasets.
    Dim pMxDoc As IMxDocument
    Dim pMap As IMap
    Dim pRasterLayer As IRasterLayer
    Dim pRaster As IRaster
    Dim pFeatureLayer As IFeatureLayer
    Dim pLookoutDataset As IGeoDataset
    Set pMxDoc = ThisDocument
    Set pMap = pMxDoc.FocusMap
    ' Define the elevation raster.
    Set pRasterLayer = pMap.Layer(1)
    Set pRaster = pRasterLayer.Raster
    ' Define the lookout dataset.
    Set pFeatureLayer = pMap.Layer(0)
    Set pLookoutDataset = pFeatureLayer.FeatureClass
```

Part 1 defines the elevation and lookout point datasets. The elevation dataset referenced by *pRaster* is the raster of the second layer in the active map. The lookout dataset referenced by *pLookoutDataset* is the feature class of the top layer.

```
    ' Part 2: Perform visibility analysis.
    Dim pSurfaceOp As ISurfaceOp
    Dim pEnv As IRasterAnalysisEnvironment
    Dim pWS As IWorkspace
    Dim pWSF As IWorkspaceFactory
    Dim pOutRaster As IGeoDataset
    ' Define a raster surface operation.
    Set pSurfaceOp = New RasterSurfaceOp
    Set pEnv = pSurfaceOp
    Set pWSF = New RasterWorkspaceFactory
```

```
Set pWS = pWSF.OpenFromFile("c:\data\chap12", 0)
Set pEnv.OutWorkspace = pWS
' Run the visibility surface operation.
Set pOutRaster = pSurfaceOp.Visibility(pRaster, pLookoutDataset, esriGeoAnalysisVisibilityFrequency)
```

Part 2 creates *pSurfaceOp* as an instance of the *RasterSurfaceOp* class. Next, the code uses the *IRasterAnalysisEnvironment* interface to set the workspace for the output. The *Visibility* method on *ISurfaceOp* uses *pRaster* and *pLookoutDataset* as the object qualifiers. In this case, the lookout dataset is a point shapefile. But it can also be a shapefile or coverage containing point or line features. The only other argument used by *Visibility* is the visibility type, which can be one of the following four choices:

- esriGeoAnalysisVisibilityFrequency to record the number of times each cell can be seen
- esriGeoAnalysisVisibilityObservers to record which lookout points can be seen
- esriGeoAnalysisVisibilityFrequencyUseCurvature to specify whether Earth curvature corrections will be used with frequency
- esriGeoAnalysisVisibilityObserversUseCurvature to specify whether Earth curvature corrections will be used with observers

Visibility uses the first visibility type. Therefore, the output shows the number of times each cell can be seen from the two lookout points.

```
' Part 3: Create the output layer and add it to the active map.
Dim pRLayer As IRasterLayer
Set pRLayer = New RasterLayer
pRLayer.CreateFromRaster pOutRaster
pRLayer.Name = "Viewshed"
pMap.AddLayer pRLayer
End Sub
```

Part 3 creates a new raster layer from *pOutRaster* and adds the layer to the active map.

12.5 PERFORMING WATERSHED ANALYSIS

To derive watersheds from an elevation raster requires creating the following intermediate rasters: a filled elevation raster, a flow direction raster, a flow accumulation raster, and a stream links raster. A filled elevation raster is void of depressions. A flow direction raster shows the direction water will flow out of each cell of a filled elevation raster. A flow accumulation raster tabulates for each cell the number of cells that will flow to it. In other words, a flow accumulation raster shows how many upstream cells will contribute drainage to each cell. Cells having high accumulation values generally correspond to stream channels. Therefore, a stream links raster can be derived from a flow accumulation raster by using some threshold accumulation value. Stream links and the flow direction raster are the inputs for deriving area-wide watersheds.

12.5.1 *Watershed*

Watershed delineates area-wide watersheds from an elevation raster. In the process, the macro also creates a filled elevation raster, a flow direction raster, a flow accumulation raster, a source raster, and a stream links raster. *Watershed* has eight parts. Part 1 defines the input elevation raster. Part 2 creates a hydrologic operation object and specifies the output workspace. Parts 3 through 8 perform the various hydrologic operations for the purpose of delineating watersheds. As each operation is completed, a raster layer is created and added to the active map. It will take a while to execute the macro, especially if the elevation dataset is large.

> **Key Interfaces:** *IHydrologyOp, IRasterAnalysisEnvironment, IWorkspace, IWorkspaceFactory, IQueryFilter, IRasterDescriptor, ILogicalOp, IRaster*
> **Key Members:** *Raster, OpenFromFile, OutWorkspace, Fill, CreateFromRaster, Flowdirection, FlowAccumulation, WhereClause, Create, Test, StreamLink, Watershed*
> **Usage:** Add *emidalat*, an elevation raster, to an active map. Import **Watershed** to Visual Basic Editor. Run the macro. The macro adds *Filled DEM, Flowdirection, Flowaccumulation, Source, Stream link*, and *Watershed* to the active map.

```
Private Sub Watershed()
    ' Part 1: Define the input raster.
    Dim pMxDoc As IMxDocument
    Dim pMap As IMap
    Dim pLayer As ILayer
    Dim pRasterLayer As IRasterLayer
    Dim pInputRaster As IRaster
    Set pMxDoc = ThisDocument
    Set pMap = pMxDoc.FocusMap
    Set pLayer = pMap.Layer(0)
    Set pRasterLayer = pLayer
    Set pInputRaster = pRasterLayer.Raster
```

Part 1 sets *pInputRaster* to be the raster of the top layer in the active map.

```
    ' Part 2: Create a new hydrology operation.
    Dim pHydrologyOp As IHydrologyOp
    Dim pEnv As IRasterAnalysisEnvironment
    Dim pWS As IWorkspace
    Dim pWSF As IWorkspaceFactory
    Set pHydrologyOp = New RasterHydrologyOp
    Set pEnv = pHydrologyOp
    Set pWSF = New RasterWorkspaceFactory
    Set pWS = pWSF.OpenFromFile("c:\data\chap12", 0)
    Set pEnv.OutWorkspace = pWS
```

Part 2 creates *pHydrologyOp* as an instance of the *RasterHydrologyOp* class and uses the *IRasterAnalysisEnvironment* interface to set the workspace for the output.

```
' Part 3: Fill sinks.
Dim pFillDS As IRaster
Set pFillDS = pHydrologyOp.Fill(pInputRaster)
' Add the filled DEM to the active map.
Set pRasterLayer = New RasterLayer
pRasterLayer.CreateFromRaster pFillDS
pRasterLayer.Name = "Filled DEM"
pMap.AddLayer pRasterLayer
```

Part 3 uses the *Fill* method on *IHydrologyOp* to create a filled elevation raster referenced by *pFillDS*. The code then creates a new raster layer from *pFillDS* and adds the layer to the active map.

```
' Part 4: Derive flow direction.
Dim pFlowdirectionDS As IRaster
Set pFlowdirectionDS = pHydrologyOp.Flowdirection(pFillDS, True, True)
' Add the flow direction layer to the active map.
Set pRasterLayer = New RasterLayer
pRasterLayer.CreateFromRaster pFlowdirectionDS
pRasterLayer.Name = "Flowdirection"
pMap.AddLayer pRasterLayer
```

Part 4 uses the *Flowdirection* method on *IHydrologyOp* to create a flow direction raster referenced by *pFlowdirectionDS*. Besides the object qualifier *pFillDS*, *Flowdirection* uses two other arguments: the first specifies whether or not an output raster will be created, and the second determines the flow direction at the edges of the elevation dataset. The code then adds a new flow direction raster layer to the active map.

```
' Part 5: Derive flow accumulation.
Dim pFlowAccumulationDS As IRaster
Set pFlowAccumulationDS = pHydrologyOp.FlowAccumulation(pFlowdirectionDS)
' Add the flow accumulation layer to the active map.
Set pRasterLayer = New RasterLayer
pRasterLayer.CreateFromRaster pFlowAccumulationDS
pRasterLayer.Name = "Flowaccumulation"
pMap.AddLayer pRasterLayer
```

Part 5 uses the *FlowAccumulation* method on *IHydrologyOp* to create a flow accumulation raster referenced by *pFlowAccumulationDS*. The code then adds a new flow accumulation raster layer to the active map.

```
' Part 6: Derive the source raster.
Dim pQFilter As IQueryFilter
Dim pRasDes As IRasterDescriptor
Dim pExtractOp As IExtractionOp
Dim pSourceDS As IRaster
' Use a minimum of 500 cells.
Set pQFilter = New QueryFilter
```

```
pQFilter.WhereClause = "Value > 500"
Set pRasDes = New RasterDescriptor
pRasDes.Create pFlowAccumulationDS, pQFilter, "Value"
' Run an extraction operation.
Set pExtractOp = New RasterExtractionOp
Set pSourceDS = pExtractOp.Attribute(pRasDes)
' Add the source layer to the active map.
Set pRasterLayer = New RasterLayer
pRasterLayer.CreateFromRaster pSourceDS
pRasterLayer.Name = "Source"
pMap.AddLayer pRasterLayer
```

Part 6 determines which cells in the flow accumulation raster are to be included in the source dataset. The code first creates *pQFilter* as an instance of the *Query-Filter* class and defines its *WhereClause* condition as "Value > 500." Next, the code uses the flow accumulation dataset, *pQFilter*, and the field name of value to create an instance of the *RasterDescriptor* class referenced by *pRasDes*. Then the code creates *pExtractOp* as an instance of the *RasterExtractionOp* class and uses the *Attribute* method to create a source dataset referenced by *pSourceDS*. The source dataset has cell values for those cells that have value > 500 and no data for those cells that have value <= 500. A new source raster layer is then added to the active map.

```
' Part 7: Derive stream links.
Dim pStreamLinkDS As IRaster
Set pStreamLinkDS = pHydrologyOp.StreamLink(pSourceDS, pFlowdirectionDS)
' Add the stream link layer to the active map.
Set pRasterLayer = New RasterLayer
pRasterLayer.CreateFromRaster pStreamLinkDS
pRasterLayer.Name = "Stream link"
pMap.AddLayer pRasterLayer
```

Part 7 uses the *StreamLink* method on *IHydrologyOp* to create a stream links raster referenced by *pStreamLinkDS*. The code then adds a new stream links raster layer to the active map.

```
' Part 8: Derive watersheds.
Dim pWatershed As IRaster
Set pWatershed = pHydrologyOp.Watershed(pFlowdirectionDS, pStreamLinkDS)
' Add the watershed layer to the active map.
Set pRasterLayer = New RasterLayer
pRasterLayer.CreateFromRaster pWatershed
pRasterLayer.Name = "Watershed"
pMap.AddLayer pRasterLayer
End Sub
```

Part 8 applies the *Watershed* method on *IHydrologyOp* to create a watershed raster referenced by *pWatershed*. The code then adds a new watershed raster layer to the active map.

12.6 CREATING AND EDITING TIN

This section includes three sample macros. The first macro converts an elevation raster to a TIN. The second macro modifies the initial TIN by using streams as breaklines. And the third macro reports the numbers of nodes and triangles that make up the modified TIN.

12.6.1 *RasterToTin*

RasterToTin converts an elevation raster to a TIN. The macro performs the same task as using the Convert/Raster to TIN command in 3D Analyst. *RasterToTin* has three parts. Part 1 defines the input raster band and initiates a new TIN, Part 2 creates a pixel block and reads into it raw pixels from the raster band, and Part 3 uses the pixel block to populate the TIN and adds the TIN layer to the active map.

> **Key Interfaces:** *IRasterBandCollection, IRasterBand, ITinEdit, IRawPixels, IRaster-Props, IPixelBlock, IEnvelope, ITinLayer*
>
> **Key Members:** *Item(), InitNew, SetCoords, CreatePixelBlock, Read, SafeArray(), Extent, AddFromPixelBlock, Dataset*
>
> **Usage:** Add *emidalat*, an elevation raster, to an active map. Import *RasterToTin* to Visual Basic Editor. Run the macro. The macro adds a new TIN to the active map.

```
Private Sub RasterToTin()
    ' Part 1: Define the input raster band and initiate a new TIN.
    Dim pMxDoc As IMxDocument
    Dim pMap As IMap
    Dim pInputRL As IRasterLayer
    Dim pInRaster As IRaster
    Dim pRasGDS As IGeoDataset
    Dim pRasterBandColl As IRasterBandCollection
    Dim pRasterBand As IRasterBand
    Dim pTin As ITinEdit
    Set pMxDoc = ThisDocument
    Set pMap = pMxDoc.FocusMap
    Set pInputRL = pMap.Layer(0)
    Set pInRaster = pInputRL.Raster
    ' Extract the first raster band.
    Set pRasterBandColl = pInRaster
    Set pRasterBand = pRasterBandColl.Item(0)
    ' Initiate a new TIN.
    Set pTin = New Tin
    Set pRasGDS = pInRaster
    pTin.InitNew pRasGDS.Extent
```

Part 1 sets *pInRaster* to be the raster of the first layer in the active map. Next, the code performs a QI for the *IRasterBandCollection* interface and uses the *Item* method to assign the first raster band of *pInRaster* to *pRasterBand*. The code then uses the *InitNew* method on *ITinEdit* to create *pTin* as an instance of the *TIN* class. The initialization of a TIN object requires an object qualifier that defines the data area, which corresponds to the extent of *pInRaster* in this case. The *InitNew* method also places *pTin* in edit mode.

```
' Part 2: Create a pixel block.
Dim pRawPixels As IRawPixels
Dim pProps As IRasterProps
Dim pBlockSize As IPnt
Dim pPixelBlock As IPixelBlock
Dim pBlockOrigin As IPnt
' Define a block size.
Set pProps = pRasterBand
Set pBlockSize = New Pnt
pBlockSize.SetCoords pProps.Width, pProps.Height
' Allocate a pixel block.
Set pRawPixels = pRasterBand
Set pPixelBlock = pRawPixels.CreatePixelBlock(pBlockSize)
' Define the block origin.
Set pBlockOrigin = New Pnt
pBlockOrigin.SetCoords 0, 0
' Read into the pixel block from the input raster band.
pRawPixels.Read pBlockOrigin, pPixelBlock
```

Part 2 reads *pRasterBand* from Part 1 into a block of pixels. The process involves defining a block size, allocating a pixel block, and reading the cell values into the pixel block. The process also involves *IRawPixels* and *IRasterProps* that a *Raster-Band* object supports (Figure 12.4). The code first accesses the *IRasterProps* interface and uses the width and height of *pRasterBand* to define a block size referenced by *pBlockSize*. Next, the code accesses *IRawPixels* and uses the *CreatePixelBlock* method to allocate a pixel block referenced by *pPixelBlock*. The allocation is based on the width and height of *pBlockSize*. After defining the block origin at (0, 0), the code uses the *Read* method on *IRawPixels* to read *pRasterBand* into *pPixelBlock*.

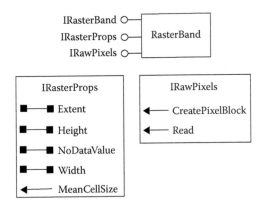

Figure 12.4 A *RasterBand* object supports *IRasterBand*, *IRasterProps*, and *IRawPixels*. These interfaces can be used to convert an elevation raster into a TIN.

```
' Part 3: Create the TIN and add the TIN layer to the active map.
Dim dataArray As Variant
Dim pEnv As IEnvelope
Dim zTol As Double
Dim pTinLayer As ITinLayer
' Use the pixel block and a z tolerance to create the TIN.
dataArray = pPixelBlock.SafeArray(0)
Set pEnv = pProps.Extent
zTol = 5
pTin.AddFromPixelBlock pEnv.XMin, pEnv.YMax, pProps.MeanCellSize.X, pProps.MeanCellSize.Y, _
pProps.NoDataValue, dataArray, zTol
pTin.SaveAs ("c:\data\chap12\emidatin")
' Create the TIN layer and add it to the active map.
Set pTinLayer = New TinLayer
With pTinLayer
    Set .Dataset = pTin
    .Visible = True
    .Name = "emidatin"
End With
pMxDoc.FocusMap.AddLayer pTinLayer
pMxDoc.UpdateContents
End Sub
```

Part 3 creates the TIN and adds the TIN layer to the active map. The code uses the *AddFromPixelBlock* method on *ITinEdit* to edit *pTin*, which has been set in edit mode in Part 1. The method has seven arguments. The first two are the coordinates of the TIN's origin. A raster's origin is at the upper left corner, whereas a TIN's origin is at the lower left corner. Therefore, the values entered are the *xmin* and *ymax* of the extent of *pRawPixels*. The third and fourth arguments are the pixel sizes in the *X* and *Y* dimensions, which are set to be the same as the mean cell sizes in *pRawPixels*. The fifth argument is the value of no data, which is set to be the same as the no data value in *pRawPixels*. The sixth argument is an array of pixel values to be read, which, in this case, is *dataArray* or the *SafeArray* property of *pPixelBlock*. The last argument is the *z*-tolerance used for converting a raster to a TIN. The *z*-tolerance determines the accuracy of a TIN in depicting the original elevation raster: The larger the *z*-tolerance is, the less accurate the TIN becomes. The code uses 5 meters as the *z*-tolerance. After *pTin* is created, it is saved as *emidatin* on disk. Finally, Part 3 creates a new TIN layer from *pTin* and adds the layer to the active map.

Box 12.4 RasterToTin_GP

RasterToTin_GP uses the RasterTin tool in the 3D Analyst toolbox to convert an elevation raster into a TIN. The *z*-tolerance is set to be 5 (meters). Run the macro in ArcCatalog and view the output TIN (*emidatin2*) in the Catalog tree.

```
Private Sub RasterToTin_GP()
    ' Create the Geoprocessing object.
    Dim GP As Object
```

```
  Set GP = CreateObject("esriGeoprocessing.GpDispatch.1")
  ' RasterTin <in_raster> <out_tin> {z_tolerance} {max_points} {z_factor}
  ' Execute the rastertin tool.
  GP.RasterTin_3d "c:\data\chap12\emidalat", "c:\data\chap12\emidatin2", 5
End Sub
```

12.6.2 *EditTin*

EditTin modifies an existing TIN by adding streams as hard breaklines. The macro
performs the same task as using the Create/Modify TIN/Add Features to TIN command in 3D Analyst. ***EditTin*** has two parts. Part 1 defines the TIN and the breaklines
dataset, and Part 2 adds streams as breaklines to the TIN and refreshes the map with
the modified TIN.

> **Key Interfaces:** *ITin, ITinEdit*
> **Key Members:** *Dataset, StartEditing, AddFromFeatureClass, StopEditing*
> **Usage:** Add *newtin*, a new TIN, and *emidastrm.shp*, a stream shapefile for modifying
> the TIN, to an active map. The stream shapefile must be on top of *newtin* in the
> table of contents. Import ***EditTin*** to Visual Basic Editor. Run the macro. The macro
> modifies *newtin* by adding *emidatstrm* as hardlines. The code then adds the mod-
> ified TIN as a new layer to the active map.

```
Private Sub EditTin()
  ' Part 1: Define the TIN and the layer for modifying the TIN.
  Dim pMxDoc As IMxDocument
  Dim pMap As IMap
  Dim pTinLayer As ITinLayer
  Dim pStreamFL As IFeatureLayer
  Dim pStreamFC As IFeatureClass
  Dim pTin As ITin
  Set pMxDoc = ThisDocument
  Set pMap = pMxDoc.FocusMap
  ' Define the TIN.
  Set pTinLayer = pMap.Layer(1)
  Set pTin = pTinLayer.Dataset
  ' Define the stream layer for modifying the TIN.
  Set pStreamFL = pMap.Layer(0)
  Set pStreamFC = pStreamFL.FeatureClass
```

Part 1 sets *pTin* to be the dataset of the TIN layer in the active map, and
pStreamFC to be the feature class of the stream feature layer.

```
  ' Part 2: Modify the TIN and add the modified TIN to the active map.
  Dim pTinEdit As ITinEdit
  Set pTinEdit = pTin
  ' Edit the TIN.
  pTinEdit.StartEditing
  pTinEdit.AddFromFeatureClass pStreamFC, Nothing, Nothing, Nothing, esriTinHardLine
  pTinEdit.StopEditing (True)
```

```
' Add the modified TIN to the active map.
Set pTinLayer = New TinLayer
With pTinLayer
    Set .Dataset = pTin
    .Visible = True
    .Name = "Modified Tin"
End With
pMxDoc.FocusMap.AddLayer pTinLayer
pMxDoc.UpdateContents
End Sub
```

Part 2 performs a QI for the *ITinEdit* interface and uses the *StartEditing* method to start the editing process, the *AddFromFeatureClass* method to add features from a feature class to the TIN, and the *StopEditing* method to stop editing. Besides the object qualifier *pStreamFC*, the *AddFromFeatureClass* method uses four other arguments. The first argument is a query filter object, which, if specified, uses selected features to modify the TIN. The second and third arguments are the fields for height and tag value, if they exist. The last is the surface type option. *EditTin* opts for hard breaklines as the surface type. Finally, the code adds a new TIN layer showing the modified TIN to the active map.

12.6.3   *TinNodes*

TinNodes prints the numbers of nodes and triangles of an existing TIN. The macro performs the same function as using the Source tab on the Layer Properties dialog. *TinNodes* has two parts. Part 1 defines the TIN dataset, and Part 2 derives and reports the numbers of nodes and triangles.

> **Key Interfaces:** *ITinAdvanced*
> **Key Members:** *Dataset, DataNodeCount, DataTriangleCount*
> **Usage:** Add *plnetin* to an active map. Import *TinNodes* to Visual Basic Editor. Run the macro. The macro reports the numbers of nodes and triangles in a message box.

```
Private Sub TinNodes()
    ' Part 1: Define the TIN.
    Dim pMxDoc As IMxDocument
    Dim pMap As IMap
    Dim pTinLayer As ITinLayer
    Dim pTin As ITin
    Set pMxDoc = ThisDocument
    Set pMap = pMxDoc.FocusMap
    Set pTinLayer = pMap.Layer(0)
    Set pTin = pTinLayer.Dataset
```

Part 1 defines *pTin* as the dataset of the TIN layer in the active map.

```
' Part 2: Derive and report numbers of nodes and triangles.
Dim pTinAdvanced As ITinAdvanced
Dim pNodeCount As Double
```

```
Dim pTriangleCount As Double
Set pTinAdvanced = pTin
pNodeCount = pTinAdvanced.DataNodeCount
pTriangleCount = pTinAdvanced.DataTriangleCount
MsgBox "The number of nodes is: " & pNodeCount & " The number of triangles is: " & pTriangleCount
End Sub
```

Part 2 accesses the *ITinAdvanced* interface, derives the number of nodes from the *DataNodeCount* property, and assigns the number to *pNodeCount*. The code also derives the number of triangles from the *DataTriangleCount* property and assigns the number to *pTriangleCount*. A message box then reports the numbers of nodes and triangles.

Spatial Interpolation

Spatial interpolation is the process of using points with known values to estimate values at other points. A geographic information system (GIS) typically applies spatial interpolation to a raster with estimates made for all cells. Spatial interpolation is therefore a means for converting point data to surface data so that the surface data can be used with other surfaces for analysis and modeling.

A variety of methods have been proposed for spatial interpolation. These methods can be categorized in several ways:

- Spatial interpolation methods can be grouped into global and local. A global interpolation method uses every known point available to estimate an unknown value, whereas a local interpolation method uses a sample of known points to estimate an unknown value.
- Spatial interpolation methods may be exact and inexact. Exact interpolation predicts a value at the point location that is the same as its known value, whereas inexact interpolation does not.
- Spatial interpolation methods may be deterministic or stochastic. A deterministic interpolation method provides no assessment of errors with predicted values, whereas a stochastic interpolation method does.

This chapter covers ArcObjects programming for spatial interpolation. Section 13.1 reviews spatial interpolation using ArcGIS. Section 13.2 discusses objects that are related to spatial interpolation. Section 13.3 includes macros and two Geoprocessing (GP) macros for using the methods of inverse distance weighted (IDW), spline, trend surface, and kriging for spatial interpolation. Both the IDW and spline methods are local, exact, and deterministic. Trend surface is a global, inexact, and deterministic method, and kriging is a local and stochastic method. Section 13.4 offers a macro for comparing different interpolation methods. All macros start with the listing of key interfaces and key members (properties and methods) and the usage. Make sure that the Spatial Analyst extension is checked in the Tools/Extensions menu before running the macros.

13.1 RUNNING SPATIAL INTERPOLATION IN ARCGIS

ArcGIS Desktop offers spatial interpolation through the Spatial Analyst and Geostatistical Analyst extensions and ArcToolbox. Discussions in this section refer to the Spatial Analyst extension, which has menu access to IDW, Spline (spline with tension and regularized spline), and Kriging (ordinary and universal).

Geostatistical Analyst's main menu has the selections of Explore Data, Geostatistical Wizard, and Create Subsets. The Explore Data command offers histogram, semivariogram, QQ plot, and other tools for exploratory data analysis. The Geostatistical Wizard offers a large variety of global and local interpolation methods including IDW, trend surface, local polynomial, radial basis function, kriging, and cokriging. The Create Subsets command handles model validation. Most ArcGIS Desktop users will probably use Geostatistical Analyst for spatial interpolation and take advantage of its data exploration and model validation capabilities.

13.2 ARCOBJECTS FOR SPATIAL INTERPOLATION

The *RasterInterpolationOp* coclass is the primary component for spatial interpolation. A *RasterInterpolationOp* object supports *IRasterAnalysisEnvironment* and *IInterpolationOp* (Figure 13.1). *IRasterAnalysisEnvironment* controls the analysis environment, such as analysis mask and output cell size. *IInterpolationOp* has the following methods for spatial interpolation: *IDW* for inverse distance weighted, *Krige* for kriging, *Spline* for spline, *Trend* for trend surface, and *Variogram* for kriging. *Krige* and *Variogram* differ in that the former uses a predefined type of semivariogram, whereas the latter uses a user-defined semivariogram.

A local interpolation operation using ArcObjects typically requires two object qualifiers to create the output (Figure 13.2). The first is a feature class descriptor, which provides the input data for interpolation. A feature class descriptor is a feature class based on a specific field. For example, the feature class descriptor to be used in this chapter's sample macros is based on a numeric field that records the annual

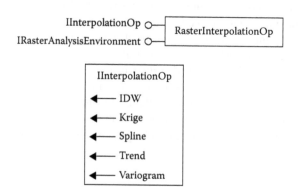

Figure 13.1 A *RasterInterpolationOp* object supports *IInterpolationOp* and *IRasterAnalysis-Environment. IInterpolationOp* has various methods for spatial interpolation.

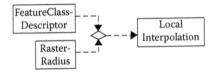

Figure 13.2 A *FeatureClassDescriptor* object and a *RasterRadius* object together create a *LocalInterpolation* object.

precipitation at weather stations. The second is a search radius, expressed as either a distance or a number, for selecting the sample points to be used in local interpolation.

Because spatial interpolation operations produce rasters, these rasters can be further analyzed by using objects covered in Chapter 11. For example, mathematical operations on rasters can compare and evaluate the results from two different interpolation methods.

13.3 PERFORMING SPATIAL INTERPOLATIONS

This section covers the spatial interpolation methods of IDW, spline, trend surface, and kriging. The first three methods have the same code structure. As the only stochastic method in the group, kriging produces an interpolated surface and an error measure surface.

13.3.1 *Idw*

Idw uses the IDW method to interpolate a precipitation surface from a point shapefile. The macro performs the same function as using the Interpolate to Raster/Inverse Distance Weighted command in Spatial Analyst. *Idw* has three parts. Part 1 defines the point shapefile and a mask dataset; Part 2 creates a raster interpolation object, sets the analysis environment, and runs the interpolation method; and Part 3 creates a raster layer from the output and adds the layer to the active map.

> **Key Interfaces:** *IFeatureClassDescriptor, IGeoDataset, IInterpolationOp, IRaster-AnalysisEnvironment, IRasterRadius, IRaster*
> **Key Members:** *Create, Raster, SetCellSize, Mask, SetVariable, IDW, CreateFromRaster*
> **Usage:** Add *idoutlgd* and *stations.shp* to an active map. *stations* must be on top of *idoutlgd* in the table of contents. *idoutlgd* is a raster showing the outline of Idaho, and *stations* is a point shapefile containing 105 weather stations in Idaho. The Ann_prec field in *stations* stores the annual precipitation values in inches. Import *Idw* to Visual Basic Editor. Run the macro. The macro creates a temporary precipitation surface raster named IDW.

```
Private Sub Idw()
    ' Part 1: Define the input and mask datasets.
    Dim pMxDoc As IMxDocument
```

```
Dim pMap As IMap
Dim pFeatureLayer As IFeatureLayer
Dim pFeatureClass As IFeatureClass
Dim sFieldName As String
Dim pFCDescr As IFeatureClassDescriptor
Dim pRasterLayer As IRasterLayer
Dim pMaskDataset As IGeoDataset
Set pMxDoc = ThisDocument
Set pMap = pMxDoc.FocusMap
Set pFeatureLayer = pMap.Layer(0)
Set pFeatureClass = pFeatureLayer.FeatureClass
' Use a value field to create a feature class descriptor.
sFieldName = "Ann_prec"
Set pFCDescr = New FeatureClassDescriptor
pFCDescr.Create pFeatureClass, Nothing, sFieldName
' Define a mask dataset.
Set pRasterLayer = pMap.Layer(1)
Set pMaskDataset = pRasterLayer.Raster
```

Part 1 defines the input and mask datasets. The code first sets *pFeatureClass* to be the feature class of the top layer in the active map. Next, the code creates *pFCDescr* as an instance of the *FeatureClassDescriptor* class and uses the *Create* method to create the feature class descriptor based on the Ann_prec field. The code also sets *pMaskDataset* to be the raster of the second layer in the active map.

```
' Part 2: Perform interpolation using IDW.
Dim pIntOp As IInterpolationOp
Dim pEnv As IRasterAnalysisEnvironment
Dim pRadius As IRasterRadius
Dim pOutRaster As IRaster
' Create a raster interpolation operation.
Set pIntOp = New RasterInterpolationOp
Set pEnv = pIntOp
pEnv.SetCellSize esriRasterEnvValue, pMaskDataset
Set pEnv.Mask = pMaskDataset
' Set the search radius for the input points.
Set pRadius = New RasterRadius
pRadius.SetVariable 12
' Run the IDW method.
Set pOutRaster = pIntOp.IDW(pFCDescr, 2, pRadius)
```

Part 2 creates *pIntOp* as an instance of the *RasterInterpolationOp* class, and performs a QueryInterface (QI) for the *IRasterAnalysisEnvironment* interface to set the analysis mask and cell size. Next, the code sets the search radius, referenced by *pRadius*, to include 12 sample points. The code then uses the *IDW* method on *IInterpolationOp* to create the interpolated raster referenced by *pOutRaster*. The *IDW* method uses *pFCDescr* and *pRadius* for the object qualifiers and specifies two for the weight (that is, exponent) of distance.

```
' Part 3: Create the output layer and add it to the active map.
Dim pRLayer As IRasterLayer
Set pRLayer = New RasterLayer
pRLayer.CreateFromRaster pOutRaster
pRLayer.Name = "IDW"
pMap.AddLayer pRLayer
End Sub
```

Part 3 creates a new raster layer from *pOutRaster* and adds the layer to the active map.

Box 13.1 Idw_GP

Idw_GP uses the IDW tool to create an interpolated surface from *stations.shp* and the ExtractByMask tool to clip the interpolated surface by using *idoutlgd* as the mask. Both tools reside in the Spatial Analyst toolbox. The code uses the input boxes to get the output cell size (e.g., 2000) and the power (e.g., 2) from the user. Run the macro in ArcMap. The macro adds the interpolated surface (*idw2*) and then the clipped surface (*idw2_extract*) to the map.

```
Private Sub Idw_GP()
    ' Create the Geoprocessing object.
    Dim GP As Object
    Set GP = CreateObject("esriGeoprocessing.GpDispatch.1")
    ' Define the workspace.
    Dim filepath As String
    filepath = "c:\data\chap13"
    GP.Workspace = filepath
    ' Get the output cell size from the user.
    Dim cellsize As Integer
    cellsize = InputBox("Enter the output cell size in meters")
    ' Get the IDW power from the user.
    Dim exponent As Integer
    exponent = InputBox("Enter the power of inverse distance weighted interpolation")
    ' IDW <in_point_features> <z_field> <out_raster> {cell_size} {power}
    ' Execute the idw tool.
    GP.IDW_sa "stations.shp", "ANN_PREC", "idw2", cellsize, exponent
    ' ExtractByMask <in_raster> <in_mask_data> <out_raster>
    ' Execute the ExtractByMask tool.
    GP.ExtractByMask_sa "idw2", "idoutlgd", "idw2_extract"
End Sub
```

13.3.2 Spline

To use spline as an interpolation method, the only change from *Idw* is the following line statement:

```
Set pOutRaster = pIntOp.Spline(pFCDescr, esriGeoAnalysisRegularizedSpline)
```

The *Spline* method on *IInterpolationOp* requires a feature class descriptor as an object qualifier and a selection on the type of spline. The two types of splines are esriGeoAnalysisRegularizedSpline for regularized spline and esriGeoAnalysisTensionSpline for spline with tension. The *Spline* method has two optional arguments for the weight to be used for interpolation and the number of sample points (with 12 as the default).

13.3.3 Trend Surface

To use trend surface as an interpolation method, the only change from *Idw* is the following line statement:

```
Set pOutRaster = pIntOp.Trend(pFCDescr, esriGeoAnalysisLinearTrend, 2)
```

The *Trend* method on *IInterpolationOp* requires a feature class descriptor as an object qualifier, a selection on the type of trend surface, and the order of the polynomial. For the type of trend surface, ArcObjects offers esriGeoAnalysisLinearTrend for least squares surface and esriGeoAnalysisLogisticTrend for logistic surface. The order of the polynomial can range from 1 through 12.

13.3.4 *Kriging*

Kriging uses the kriging method to interpolate a precipitation surface from a point shapefile. The macro performs the same function as using the Interpolate to Raster/Kriging command in Spatial Analyst. *Kriging* has four parts. Part 1 defines the input and mask datasets. Part 2 creates a raster interpolation object, sets the analysis environment, and runs kriging. Part 3 extracts the kriged and variance datasets from the interpolation output. Part 4 creates raster layers from the output datasets and adds the layers to the active map.

Unlike the previous interpolation methods, the kriging method apparently ignores the mask dataset for data analysis, but a mask dataset is still included in *Kriging* to specify the output cell size. A macro such as *ExtractByMask* in Chapter 11 can clip the output datasets from kriging to fit the study area.

> **Key Interfaces:** *IFeatureClassDescriptor, IGeoDataset, IInterpolationOp, IRasterAnalysisEnvironment, IRasterRadius, IRasterBandCollection, IRasterBand*
> **Key Members:** *Create, Raster, SetCellSize, SetVariable, Krige, Item( ), AppendBand, CreateFromRaster*
> **Usage:** Add *idoutlgd* and *stations.shp* to an active map. *stations* must be on top of *idoutlgd* in the table of contents. Import *Kriging* to Visual Basic Editor. Run the macro. The macro creates two temporary rasters: *Krige* represents the kriged surface, and *Variance* represents the variance surface.

```
Private Sub Kriging()
    ' Part 1: Define the input dataset.
    Dim pMxDoc As IMxDocument
    Dim pMap As IMap
    Dim pFeatureLayer As IFeatureLayer
    Dim pFeatureClass As IFeatureClass
```

```
Dim sFieldName As String
Dim pFCDescr As IFeatureClassDescriptor
Dim pRasterLayer As IRasterLayer
Dim pMaskDataset As IGeoDataset
Set pMxDoc = ThisDocument
Set pMap = pMxDoc.FocusMap
Set pFeatureLayer = pMap.Layer(0)
Set pFeatureClass = pFeatureLayer.FeatureClass
' Use a value field to create a feature class descriptor.
sFieldName = "Ann_prec"
Set pFCDescr = New FeatureClassDescriptor
pFCDescr.Create pFeatureClass, Nothing, sFieldName
' Define a mask dataset.
Set pRasterLayer = pMap.Layer(1)
Set pMaskDataset = pRasterLayer.Raster
```

Part 1 sets *pFeatureClass* to be the feature class of the top layer, and creates a feature class descriptor referenced by *pFCDescr*. Next, the code sets *pMaskDataset* to be the raster of the second layer.

```
' Part 2: Perform interpolation using kriging.
Dim pIntOp As IInterpolationOp
Dim pEnv As IRasterAnalysisEnvironment
Dim pRadius As IRasterRadius
Dim pOutRaster As IGeoDataset
' Create a raster interpolation operation.
Set pIntOp = New RasterInterpolationOp
Set pEnv = pIntOp
pEnv.SetCellSize esriRasterEnvValue, pMaskDataset
' Set the search radius for the input points.
Set pRadius = New RasterRadius
pRadius.SetVariable 12
' Run the Krige method.
Set pOutRaster = pIntOp.Krige(pFCDescr, esriGeoAnalysisCircularSemiVariogram, pRadius, True)
```

Part 2 creates *pIntOp* as an instance of the *RasterInterpolationOp* class, accesses the *IRasterAnalysisEnvironment* interface, and uses *pMaskDataset* as the cell size provider to set the analysis cell size. Next, the code sets the search radius to include 12 sample points. The code then uses the *Krige* method on *IInterpolationOp* to create the output raster referenced by *pOutRaster*. The *Krige* method requires a feature class descriptor, a selection on the type of semivariogram, a search radius, and the option for creating a variance dataset. ArcObjects offers the following types of semivariograms: circular, exponential, Gaussian, linear, spherical, and universal.

```
' Part 3: Extract the kriged surface and variance rasters.
Dim pRasterBandCollection As IRasterBandCollection
Dim pKrigeRB As IRasterBand
Dim pKrigeRBCollection As IRasterBandCollection
Dim pVarianceRB As IRasterBand
Dim pVarianceRBCollection As IRasterBandCollection
```

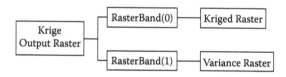

Figure 13.3 The output raster from *Krige* contains two raster bands, one for the kriged surface and the other for the variance surface.

```
Set pRasterBandCollection = pOutRaster
' Extract the kriged surface raster.
Set pKrigeRB = pRasterBandCollection.Item(0)
Set pKrigeRBCollection = New Raster
pKrigeRBCollection.AppendBand pKrigeRB
' Extract the variance raster.
Set pVarianceRB = pRasterBandCollection.Item(1)
Set pVarianceRBCollection = New Raster
pVarianceRBCollection.AppendBand pVarianceRB
```

Because Part 2 asks to create a variance dataset, the output raster referenced by *pOutRaster* actually contains two datasets. Part 3 is therefore designed to extract the two datasets (Figure 13.3). The code performs a QI for the *IRasterBandCollection* interface, extracts the first raster band of *pOutRaster*, and assigns the band to *pKrigeRB*. The code then creates *pKrigeRBCollection* as an instance of the *Raster* class and uses the *AppendBand* method on *IRasterBandCollection* to append *pKrigeRB* to *pKrigeRBCollection*. The raster band collection thus created has one raster band and represents the kriged surface. The code uses the same procedure to create *pVarianceRBCollection* for the variance surface.

```
' Part 4: Create the output layers and add them to the active map.
Dim pKrigeLayer As IRasterLayer
Dim pVarianceLayer As IRasterLayer
Set pKrigeLayer = New RasterLayer
pKrigeLayer.CreateFromRaster pKrigeRBCollection
pKrigeLayer.Name = "Krige"
pMap.AddLayer pKrigeLayer
Set pVarianceLayer = New RasterLayer
pVarianceLayer.CreateFromRaster pVarianceRBCollection
pVarianceLayer.Name = "Variance"
pMap.AddLayer pVarianceLayer
End Sub
```

Part 4 creates new raster layers from *pKrigeRBCollection* and *pVariance-RBCollection*, and adds the layers to the active map.

Box 13.2 Kriging_GP

Kriging_GP uses the Kriging tool in the Spatial Analyst toolbox to derive an interpolated surface (*krige2*) from *stations.shp*. The arguments specify that the semivariogram type be circular, the output cell size be 2000 (meters), the search radius be 12 known points, and the variance prediction raster (variance2) be created. After the kriged and variance rasters are created, the code uses the ExtractByMask tool and the mask of *idoutlgd* to clip the rasters. Run the macro in ArcMap. The macro adds four rasters to the map.

```
Private Sub Kriging_GP()
    ' Kriging <in_point_features> <z_field> <out_surface_raster> <semiVariogram_prop>
    ' {cell_size}{search_radius}{out_variance_prediction_raster}
    ' Create the Geoprocessing object and define its workspace.
    Dim GP As Object
    Set GP = CreateObject("esriGeoprocessing.GpDispatch.1")
    Dim filepath As String
    filepath = "c:\data\chap13"
    GP.Workspace = filepath
    ' Execute the kriging tool.
    GP.Kriging_sa "stations.shp", "ANN_PREC", "krige2", "CIRCULAR", 2000, "Variable 12", "variance2"
    ' ExtractByMask <in_raster> <in_mask_data> <out_raster>
    ' Execute the ExtractByMask tool.
    GP.ExtractByMask_sa "krige2", "idoutlgd", "krige2_clip"
    GP.ExtractByMask_sa "variance2", "idoutlgd", "var2_clip"
End Sub
```

13.4 COMPARING INTERPOLATION METHODS

A variety of factors can influence the result of spatial interpolation. They include the interpolation method, number of sample points, distribution of sample points, and quality of input data. This section shows how to create an output so that the results from two different interpolation methods can be compared visually. A more rigorous comparison of interpolation methods would involve cross validation and model validation techniques.

13.4.1 *Compare*

Compare compares the interpolated surfaces from the IDW method and the spline method, and highlights areas with large differences between the two (Figure 13.4). *Compare* has four parts. Part 1 defines the input and mask datasets. Part 2 performs the IDW and spline interpolation methods. Part 3 subtracts one interpolated surface from the other and reclassifies the surface representing the difference. Part 4 creates new raster layers from the outputs and adds them to the active map.

Key Interfaces: *IFeatureClassDescriptor, IGeoDataset, IInterpolationOp, IRaster-AnalysisEnvironment, IRasterRadius, IRaster, IMathOp, IRasterBandCollection, IRasterBand, IReclassOp, IRemap, INumberRemap*

Key Members: *Create, Raster, SetCellSize, Mask, SetVariable, IDW, Spline, Minus, Statistics, Minimum, Maximum, MapRange, ReclassByRemap, CreateFromRaster*

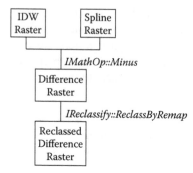

Figure 13.4   The flow chart shows how the **Compare** sub works.

> **Usage:** Add *idoutlgd* and *stations.shp* to an active map. *stations* must be on top of *idoutlgd* in the table of contents. Import **Compare** to Visual Basic Editor. Run the macro. The macro creates four temporary rasters named *IDW*, *Spline*, *Difference*, and *Reclassed Difference* and adds them to the active map.

```
Private Sub Compare()
    ' Part 1: Define the input and mask datasets.
    Dim pMxDoc As IMxDocument
    Dim pMap As IMap
    Dim pFeatureLayer As IFeatureLayer
    Dim pFeatureClass As IFeatureClass
    Dim sFieldName As String
    Dim pFCDescr As IFeatureClassDescriptor
    Dim pRasterLayer As IRasterLayer
    Dim pMaskDataset As IGeoDataset
    Set pMxDoc = ThisDocument
    Set pMap = pMxDoc.FocusMap
    Set pFeatureLayer = pMap.Layer(0)
    Set pFeatureClass = pFeatureLayer.FeatureClass
    ' Use a value field to create a feature class descriptor.
    sFieldName = "Ann_prec"
    Set pFCDescr = New FeatureClassDescriptor
    pFCDescr.Create pFeatureClass, Nothing, sFieldName
    ' Define a mask dataset.
    Set pRasterLayer = pMap.Layer(1)
    Set pMaskDataset = pRasterLayer.Raster
```

Part 1 defines a feature class descriptor and a mask dataset. The code creates the feature class descriptor from the feature class of the top layer based on the Ann_prec field. The feature class descriptor is referenced by *pFCDescr*. The code sets the mask dataset, referenced by *pMaskDataset*, to be the raster of the second layer.

```
    ' Part 2: Perform interpolations using IDW and Spline.
    Dim pIntOp As IInterpolationOp
```

```
Dim pEnv As IRasterAnalysisEnvironment
Dim pRadius As IRasterRadius
Dim pIdwRaster As IRaster
Dim pSplineRaster As IRaster
' Create a raster interpolation operation.
Set pIntOp = New RasterInterpolationOp
Set pEnv = pIntOp
pEnv.SetCellSize esriRasterEnvValue, pMaskDataset
Set pEnv.Mask = pMaskDataset
' Set the search radius for the input points.
Set pRadius = New RasterRadius
pRadius.SetVariable 12
' Run the IDW method.
Set pIdwRaster = pIntOp.IDW(pFCDescr, 2, pRadius)
' Run the Spline method.
Set pSplineRaster = pIntOp.Spline(pFCDescr, esriGeoAnalysisRegularizedSpline)
```

Before running the interpolation methods, Part 2 creates *pIntOp* as an instance of the *RasterInterpolationOp* class, sets the analysis cell size and mask, and defines the search radius. The code then runs the *IDW* method to create *pIdwRaster* and the *Spline* method to create *pSplineRaster*.

```
' Part 3: Compare the two interpolated rasters.
Dim pMathOp As IMathOp
Dim pDiffRaster As IRaster
Dim pRasBC As IRasterBandCollection
Dim pBand1 As IRasterBand
Dim pMinimum As Double
Dim pMaximum As Double
Dim pReclassOp As IReclassOp
Dim pRemap As IRemap
Dim pNRemap As INumberRemap
Dim pOutRaster As IRaster
' Subtract one raster from the other to create the difference raster.
Set pMathOp = New RasterMathOps
Set pDiffRaster = pMathOp.Minus(pIdwRaster, pSplineRaster)
' Derive the minimum and maximum values from the difference raster.
Set pRasBC = pDiffRaster
Set pBand1 = pRasBC.Item(0)
pMinimum = pBand1.Statistics.Minimum
pMaximum = pBand1.Statistics.Maximum
' Prepare a number remap.
Set pNRemap = New NumberRemap
pNRemap.MapRange pMinimum, -3.1, 1
pNRemap.MapRange -3, 3, 2
pNRemap.MapRange 3.1, pMaximum, 3
Set pRemap = pNRemap
' Reclassify the difference raster.
Set pReclassOp = New RasterReclassOp
Set pOutRaster = pReclassOp.ReclassByRemap(pDiffRaster, pRemap, False)
```

Part 3 derives a difference raster from *pIdwRaster* and *pSplineRaster*, and reclassifies the difference raster into three classes showing high negative difference, low difference, and high positive difference. To create the difference raster referenced by *pDiffRaster*, the code uses the *Minus* method on *IMathOp* and *pIdwRaster* and *pSplineRaster* as the object qualifiers. To reclassify *pDiffRaster*, the code first derives the minimum and maximum values in *pDiffRaster* by using the *Statistics* property on *IRasterBand*. The minimum and maximum values are then entered in a number remap referenced by *pNRemap* to set up the following three classes: minimum to −3.1 (high negative difference), −3 to 3 (low difference), and 3.1 to maximum (high positive difference). The result of applying the *ReclassByRemap* method on *IReclassOp* is a reclassified difference raster referenced by *pOutRaster*.

```
' Part 4: Create the output raster layers, and add them to the active map.
Dim pIdwLayer As IRasterLayer
Dim pSplineLayer As IRasterLayer
Dim pDiffLayer As IRasterLayer
Dim pOutLayer As IRasterLayer
' Add the IDW layer.
Set pIdwLayer = New RasterLayer
pIdwLayer.CreateFromRaster pIdwRaster
pIdwLayer.Name = "IDW"
pMap.AddLayer pIdwLayer
' Add the Spline layer.
Set pSplineLayer = New RasterLayer
pSplineLayer.CreateFromRaster pSplineRaster
pSplineLayer.Name = "Spline"
pMap.AddLayer pSplineLayer
' Add the Difference layer.
Set pDiffLayer = New RasterLayer
pDiffLayer.CreateFromRaster pDiffRaster
pDiffLayer.Name = "Difference"
pMap.AddLayer pDiffLayer
' Add the reclassed layer.
Set pOutLayer = New RasterLayer
pOutLayer.CreateFromRaster pOutRaster
pOutLayer.Name = "Reclassed Difference"
pMap.AddLayer pOutLayer
End Sub
```

Part 4 creates new raster layers from *pIdwRaster*, *pSplineRaster*, *pDiffRaster*, and *pOutRaster* respectively, and adds these layers to the active map.

Binary and Index Models

A model is a simplified representation of a phenomenon or a system. Many types of models exist in different disciplines. Two types of models that a geographic information system (GIS) can build are binary and index models, either vector-based or raster-based.

A binary model selects spatial features that meet a set of criteria. Those features that meet the selection criteria are coded one (true) and those that do not are coded zero (false). A common application of binary models is site analysis. A vector-based binary model requires that the overlay operations be performed to combine attributes (criteria) to be queried. A raster-based binary model, on the other hand, can be derived directly from querying multiple rasters, with each raster representing a criterion.

An index model calculates the index value for each unit area (i.e., polygon or cell) and produces a ranked map based on the index values. Index models are commonly used for suitability analysis and vulnerability analysis. Like a binary model, an index model also involves multicriteria evaluation. The weighted linear combination method is a popular method for developing an index model. The method calculates the index value by summing the weighted criterion values. The weight represents the relative importance of a criterion against other criteria, and the criterion values represent the standardized values, such as 0 to 1, 1 to 5, or 0 to 100, for each criterion.

This chapter covers binary models and index models. Section 14.1 reviews important commands in ArcGIS for building models. Section 14.2 discusses useful objects for building models, almost all of which have been covered in previous chapters. Section 14.3 includes sample modules and Geoprocessing (GP) macros for building binary and index models, both vector- and raster-based. Each module has a description of its usage. Make sure that the Spatial Analyst extension is checked in the Tools/Extensions menu before running the macros for raster-based models.

14.1 BUILDING MODELS IN ARCGIS

ArcGIS does not offer menu choices for building a binary or index model. The ModelBuilder extension can be used to build index models. Another option is to use various ArcGIS commands to construct binary and index models. Commands important to building a vector-based binary or index model are overlay (for example, union, intersect, and spatial join), attribute data manipulation (for example, adding fields and calculating field values), and attribute data query. Raster Calculator and Reclassify in Spatial Analyst are important commands for building raster-based binary or index models. If selection criteria involve buffer zones, then buffering using vector data or distance measuring using raster data becomes part of the model-building process.

14.2 ARCOBJECTS FOR GIS MODELS

ArcObjects does not have "model" objects. Many objects that have already been covered in Chapters 9, 10, and 11 can be used to build models.

14.3 BUILDING BINARY AND INDEX MODELS

This section covers sample VBA (Visual Basic for Applications) modules for building binary and index models, both vector- and raster-based. Because a model typically involves several separate tasks, each of the following modules consists of several subs and functions. And because many objects used have already been covered in previous chapters, they do not need detailed explanation.

14.3.1 *VectorBinaryModel*

Using an elevation zone shapefile and a stream shapefile as inputs, ***VectorBinaryModel*** produces a vector-based binary model that selects areas that are in elevation zone 2 and within 200 meters of streams.
   VectorBinaryModel has two subs and two functions (Figure 14.1):

   Start: a sub for managing the input and output layers and for attribute data query
   Intersect: a sub for the overlay/intersect operation
   Buffer: a function for creating a buffer zone
   SelectDataset: a function for selecting the input to the buffering and intersect operations

   Usage: Import ***VectorBinaryModel*** to Visual Basic Editor. Run the module. Select *stream* from the dialog box for the layer to be buffered. Then select *elevzone* for the overlay layer. The macro adds *stream*, *elevzone*, and the output shapefiles (*Buffer_Result* and *Intersect_Result*) to the active map, and highlights those areas that meet the selection criteria.

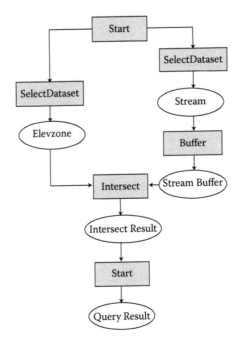

Figure 14.1 The flow chart shows the modular structure of *VectorBinaryModel*.

```
Private Sub Start()
    Dim pMxDoc As IMxDocument
    Dim pMap As IMap
    Dim Id As Long
    Dim Message As String
    Dim pBufferLayer As ILayer
    Dim pLayer1 As ILayer
    Dim pLayer2 As ILayer
    Dim pActiveView As IActiveView
    Dim pFeatureLayer As IFeatureLayer
    Dim pFeatureSelection As IFeatureSelection
    Dim pQueryFilter As IQueryFilter
    Set pMxDoc = ThisDocument
    Set pMap = pMxDoc.FocusMap
    ' Run buffering.
    MsgBox "Select the layer to be buffered"
    Set pBufferLayer = SelectDataset
    Set pLayer1 = Buffer(pBufferLayer, pMap)
    MsgBox "Select the overlay layer"
    Set pLayer2 = SelectDataset
    Call Intersect(pLayer1, pLayer2, pMap)
    ' Query the intersect result and highlight the final selection.
    Set pActiveView = pMap
    Set pFeatureLayer = pMap.Layer(0)
    Set pFeatureSelection = pFeatureLayer
```

```
Set pQueryFilter = New QueryFilter
pQueryFilter.WhereClause = "Zone = 2"
pActiveView.PartialRefresh esriViewGeoSelection, Nothing, Nothing
pFeatureSelection.SelectFeatures pQueryFilter, esriSelectionResultNew, False
pActiveView.PartialRefresh esriViewGeoSelection, Nothing, Nothing
End Sub
```

Start first uses a message box and *SelectDataset* to get *stream* as the input to *Buffer*. *Buffer* returns a buffer zone (*Buffer_Result*). *Start* then uses the buffer zone and *elevzone* as the inputs to *Intersect*. At the completion of the intersect operation, *Start* uses a query filter object to select areas that are in elevation zone 2. The code then highlights the selected areas by refreshing the active view.

```
Private Function Buffer(pBufferLayer As ILayer, pMap As IMap) As IFeatureLayer
    Dim pFeatureLayer As IFeatureLayer
    Dim pFCursor As IFeatureCursor
    Dim pSpatialReference As ISpatialReference
    Dim pBufWSName As IWorkspaceName
    Dim pBufDatasetName As IDatasetName
    Dim pBufFCName As IFeatureClassName
    Dim pFeatureCursorBuffer2 As IFeatureCursorBuffer2
    Dim pName As IName
    Dim pBufFC As IFeatureClass
    Dim pBufFL As IFeatureLayer
    Set pFeatureLayer = pBufferLayer
    ' Create a feature cursor.
    Set pFCursor = pFeatureLayer.Search(Nothing, False)
    Set pSpatialReference = pMap.SpatialReference
    ' Define the output.
    Set pBufWSName = New WorkspaceName
    pBufWSName.WorkspaceFactoryProgID = "esriCore.ShapeFileWorkspaceFactory.1"
    pBufWSName.PathName = "c:\data\chap14\ "
    Set pBufFCName = New FeatureClassName
    Set pBufDatasetName = pBufFCName
    Set pBufDatasetName.WorkspaceName = pBufWSName
    pBufDatasetName.Name = "Buffer_result"
    ' Perform buffering.
    ' Create a feature cursor buffer.
    Set pFeatureCursorBuffer2 = New FeatureCursorBuffer
    ' Define the feature cursor buffer.
    With pFeatureCursorBuffer2
        Set .FeatureCursor = pFCursor
        .Dissolve = True
        .ValueDistance = 200
        Set .BufferSpatialReference = pSpatialReference
        Set .DataFrameSpatialReference = pSpatialReference
        Set .SourceSpatialReference = pSpatialReference
        Set .TargetSpatialReference = pSpatialReference
    End With
    ' Run Buffer.
```

```
    pFeatureCursorBuffer2.Buffer pBufFCName
    ' Create the output layer from the output and add it to the active map.
    Set pName = pBufFCName
    Set pBufFC = pName.Open
    Set pBufFL = New FeatureLayer
    Set pBufFL.FeatureClass = pBufFC
    pBufFL.Name = "Buffer_Result"
    pMap.AddLayer pBufFL
    Set Buffer = pBufFL
End Function
```

Buffer gets the layer to be buffered and the active map as arguments from ***Start***. The code creates a feature cursor object by including all features in the input dataset, and defines the workspace and name for the output. Next, the code creates *pFeatureCursorBuffer2* as an instance of the *FeatureCursorBuffer* class, defines its properties, and uses the *Buffer* method to create the output referenced by *pBufFC-Name*. Then the code opens the name object, creates a new feature layer from the buffered feature class, and adds the layer to the active map. ***Buffer*** returns *pBufFL* as an *IFeatureLayer* to ***Start***.

```
Private Sub Intersect(pLayer1 As ILayer, pLayer2 As ILayer, pMap As IMap)
    Dim pInputLayer As IFeatureLayer
    Dim pOverlayLayer As IFeatureLayer
    Dim pInputTable As ITable
    Dim pOverlayTable As ITable
    Dim pNewWSName As IWorkspaceName
    Dim pFeatClassName As IFeatureClassName
    Dim pDatasetName As IDatasetName
    Dim pBGP As IBasicGeoprocessor
    Dim pOutputFeatClass As IFeatureClass
    Dim tol As Double
    Dim pOutputFeatLayer As IFeatureLayer
    ' Define the input and overlay tables.
    Set pInputLayer = pLayer1
    Set pOverlayLayer = pLayer2
    Set pInputTable = pInputLayer
    Set pOverlayTable = pOverlayLayer
    ' Define the output.
    Set pNewWSName = New WorkspaceName
    pNewWSName.WorkspaceFactoryProgID = "esriCore.ShapefileWorkspaceFactory"
    pNewWSName.PathName = "c:\data\chap14\ "
    Set pFeatClassName = New FeatureClassName
    Set pDatasetName = pFeatClassName
    pDatasetName.Name = "Intersect_result"
    Set pDatasetName.WorkspaceName = pNewWSName
    ' Perform intersect.
    Set pBGP = New BasicGeoprocessor
    tol = 0#
    Set pOutputFeatClass = pBGP.Intersect(pInputTable, False, pOverlayTable, False, tol, pFeatClassName)
    ' Create the output feature layer and add it to the active map.
```

```
    Set pOutputFeatLayer = New FeatureLayer
    Set pOutputFeatLayer.FeatureClass = pOutputFeatClass
    pOutputFeatLayer.Name = pOutputFeatClass.AliasName
    pMap.AddLayer pOutputFeatLayer
End Sub
```

Intersect gets the two layers to be intersected and the active map as arguments from *Start*. The code defines the input tables and the workspace and name of the intersect output. Next, the code creates *pBGP* as an instance of the *BasicGeoprocessor* class and uses the *Intersect* method to create the output referenced by *pOutputFeatClass*. *Intersect* then creates a feature layer from *pOutputFeatClass* and adds the layer to the active map.

```
Private Function SelectDataset() As IFeatureLayer
    ' Part 1: Prepare an Add Data dialog.
    Dim pGxDialog As IGxDialog
    Dim pGxFilter As IGxObjectFilter
    Dim pGxObjects As IEnumGxObject
    Dim pMxDoc As IMxDocument
    Dim pMap As IMap
    Dim pGxDataset As IGxDataset
    Dim pLayer As IFeatureLayer
    Set pGxDialog = New GxDialog
    Set pGxFilter = New GxFilterShapefiles
    ' Define the dialog's properties.
    With pGxDialog
        .AllowMultiSelect = False
        .ButtonCaption = "Add"
        Set .ObjectFilter = pGxFilter
        .StartingLocation = "c:\data\chap14\"
        .Title = "Add Data"
    End With
    ' Open the dialog.
    pGxDialog.DoModalOpen 0, pGxObjects
    Set pGxDataset = pGxObjects.Next
    ' Exit sub if no dataset has been added.
    If pGxDataset Is Nothing Then
        Exit Function
    End If
    ' Add the layers to the active map.
    Set pLayer = New FeatureLayer
    Set pLayer.FeatureClass = pGxDataset.Dataset
    pLayer.Name = pLayer.FeatureClass.AliasName
    Set pMxDoc = ThisDocument
    Set pMap = pMxDoc.FocusMap
    pMap.AddLayer pLayer
    pMxDoc.ActiveView.Refresh
    pMxDoc.UpdateContents
    ' Return pLayer to the Start sub.
    Set SelectDataset = pLayer
End Function
```

SelectDataset creates *pGxDialog* as an instance of the *GxDialog* class and defines its properties. The dialog shows only shapefiles and allows only one dataset to be selected. The code then creates *pLayer* as a new feature layer from the selected shapefile and adds the layer to the active map. *SelectDataset* returns *pLayer* to **Start**.

Box 14.1 VectorBinaryModel_GP

VectorBinaryModel_GP builds a vector-based binary model. The macro uses three tools in sequence: the Buffer tool in the Analysis toolbox, the Intersect tool in the Analysis toolbox, and the SelectLayerByAttribute tool in the Data Management toolbox. Run the macro in ArcMap. The macro adds *Buffer_Result_b* and *Intersect_Result_b*. Areas that meet the criteria are highlighted in *Intersect_Result_b*.

```
Private Sub VectorBinaryModel_GP()
    ' Run this macro in ArcMap.
    ' Create the Geoprocessing object and define its workspace.
    Dim GP As Object
    Set GP = CreateObject("esriGeoprocessing.GpDispatch.1")
    Dim filepath As String
    filepath = "c:\data\chap14\"
    GP.Workspace = filepath
    ' Execute the buffer tool.
    ' Buffer <in_features> <out_feature_class> <buffer_distance_or_field> {FULL | LEFT | RIGHT}
    ' {ROUND | FLAT} {NONE | ALL | LIST} {dissolve_field;dissolve_field...}
    GP.Buffer_analysis "stream.shp", "Buffer_Result_b.shp", 200, "FULL", "ROUND", "ALL"
    ' Execute the intersect tool.
    ' Intersect <features{Ranks};features{Ranks}...> <out_feature_class>
    ' {ALL | NO_FID | ONLY_FID} {cluster_tolerance} {INPUT | LINE | POINT}
    Dim parameter1 As String
    parameter1 = "elevzone.shp;Buffer_Result_b.shp"
    Dim parameter2 As String
    parameter2 = "Intersect_Result_b.shp"
    GP.Intersect_analysis parameter1, parameter2
    ' Execute selectlayerbyattribute
    ' SelectLayerByAttribute <in_layer_or_view> {NEW_SELECTION | ADD_TO_SELECTION |
    ' CLEAR_SELECTION} {where_clause}
    GP.SelectLayerByAttribute_management "Intersect_Result_b", NEW_SELECTION, "Zone = 2"
End Sub
```

14.3.2 *VectorIndexModel*

VectorIndexModel uses *soil*, *landuse*, and *depwater* as the inputs and produces a vector-based index model that shows the degree of susceptibility to groundwater contamination. The three feature classes are stored in a personal geodatabase. The criterion values have been computed and are stored in the feature classes: soilrate in *soil* for the soils criterion, lurate in *landuse* for the land use criterion, and dwrate in *depwater* for the depth to water criterion. **VectorIndexModel** calculates the index value using the following equation: $3 \times$ soilrate + lurate + dwrate. **VectorIndexModel**

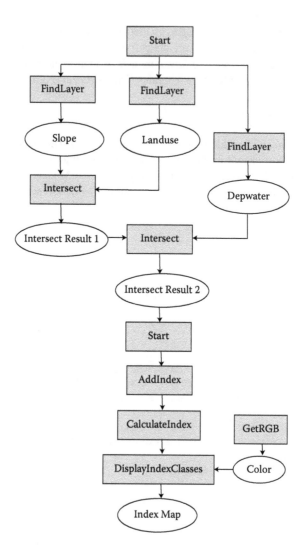

Figure 14.2   The flow chart shows the modular structure of *VectorIndexModel*.

then displays the index values in four classes. The index model does not apply to urban areas. Therefore, an additional task in building the model is to exclude urban areas from the analysis.

VectorIndexModel has five subs and two functions (Figure 14.2):

Start: a sub for managing the input and output layers and for calling the other subs
Intersect: a sub for performing the overlay operation
AddIndex: a sub for adding an index field to the overlay output
CalculateIndex: a sub for calculating the index values
DisplayIndexClasses: a sub for displaying the index values in four classes
FindLayer: a function for finding a specific layer
GetRGB: a function for defining a color symbol

Usage: Add *soil*, *landuse*, and *depwater* from *Index.mdb* to an active map. Import *VectorIndexModel* to Visual Basic Editor. Run the macro. The macro creates the feature classes of *Intersect_1* and *Intersect_2* and adds them to the active map. *Intersect_2* shows the index model in four classes.

```
Private Sub Start()
' Start uses the subs of FindLayer, Intersect, AddIndex, CalculateIndex, and DisplayIndexClasses.
    Dim pMxDoc As IMxDocument
    Dim pMap As IMap
    Dim Id As Long
    Dim Message As String
    Dim pLayer1 As ILayer
    Dim pLayer2 As ILayer
    Dim pOutName As String
    Dim pFLayer As IFeatureLayer
    Set pMxDoc = ThisDocument
    Set pMap = pMxDoc.FocusMap
    ' Run intersect using soil and landuse.
    Message = "soil"
    Id = FindLayer(Message)
    Set pLayer1 = pMxDoc.FocusMap.Layer(Id)
    Message = "landuse"
    Id = FindLayer(Message)
    Set pLayer2 = pMxDoc.FocusMap.Layer(Id)
    pOutName = "Intersect_1"
    Call Intersect(pLayer1, pLayer2, pOutName)
    ' Run intersect using the first overlay output and depwater.
    Message = "Intersect_1"
    Id = FindLayer(Message)
    Set pLayer1 = pMxDoc.FocusMap.Layer(Id)
    Message = "depwater"
    Id = FindLayer(Message)
    Set pLayer2 = pMxDoc.FocusMap.Layer(Id)
    pOutName = "Intersect_2"
    Call Intersect(pLayer1, pLayer2, pOutName)
    Set pFLayer = pMxDoc.FocusMap.Layer(0)
    ' Add a new field to the final output.
    Call AddIndex(pFLayer)
    ' Calculate the new field values.
    Call CalculateIndex(pFLayer)
    ' Display index values in classes.
    Call DisplayIndexClasses(pFLayer)
End Sub
```

Start runs the intersect operation twice to combine the three input layers of *soil*, *landuse*, and *depwater*. Names of the inputs and outputs for these operations are hard-coded and are used by the *FindLayer* function to locate them in the active map. *Start* then passes the output from the second intersect operation, which is referenced by *pFLayer*, as an argument to *AddIndex*, *CalculateIndex*, and *Display-IndexClasses*. These three subs perform the sequential tasks of adding a new (index) field, calculating the field values, and displaying the field values using a class breaks renderer.

```
Private Sub Intersect(pLayer1 As ILayer, pLayer2 As ILayer, pOutName As String)
    Dim pMxDoc As IMxDocument
    Dim pInputLayer As IFeatureLayer
    Dim pInputTable As ITable
    Dim pOverlayLayer As IFeatureLayer
    Dim pOverlayTable As ITable
    Dim pNewWSName As IWorkspaceName
    Dim pFeatClassName As IFeatureClassName
    Dim pDatasetName As IDatasetName
    Dim pBGP As IBasicGeoprocessor
    Dim tol As Double
    Dim pOutputFeatClass As IFeatureClass
    Dim pOutputFeatLayer As IFeatureLayer
    Set pMxDoc = ThisDocument
    ' Define the input and overlay tables.
    Set pInputLayer = pLayer1
    Set pInputTable = pInputLayer
    Set pOverlayLayer = pLayer2
    Set pOverlayTable = pOverlayLayer
    ' Define the output.
    Set pFeatClassName = New FeatureClassName
    Set pDatasetName = pFeatClassName
    Set pNewWSName = New WorkspaceName
    pNewWSName.WorkspaceFactoryProgID = "esriCore.AccessWorkspaceFactory"
    pNewWSName.PathName = "c:\data\chap14\Index.mdb"
    pDatasetName.Name = pOutName
    Set pDatasetName.WorkspaceName = pNewWSName
    ' Perform intersect.
    Set pBGP = New BasicGeoprocessor
    tol = 0#
    Set pOutputFeatClass = pBGP.Intersect(pInputTable, False, pOverlayTable, False, tol, pFeatClassName)
    ' Create the output feature layer and add it to the active map.
    Set pOutputFeatLayer = New FeatureLayer
    Set pOutputFeatLayer.FeatureClass = pOutputFeatClass
    pOutputFeatLayer.Name = pOutputFeatClass.AliasName
    pMxDoc.FocusMap.AddLayer pOutputFeatLayer
End Sub
```

Intersect gets two input layers and the output name as arguments from **Start**. **Intersect** first uses the input layers to set up the input and overlay tables. Next, the code defines the output's workspace and name. Then the code creates *pBGP* as an instance of the *BasicGeoprocessor* class and uses the *Intersect* method on *IBasic-Geoprocessor* to create the overlay output. Finally, the code creates a new feature layer from the overlay output and adds the layer to the active map.

```
Private Sub AddIndex(pFLayer As IFeatureLayer)
    Dim pFeatLayer As IFeatureLayer
```

```
        Dim pFClass As IFeatureClass
        Dim pField As IFieldEdit
        Set pFeatLayer = pFLayer
        Set pFClass = pFeatLayer.FeatureClass
        ' Create and define a new field.
        Set pField = New Field
        With pField
            .Name = "Total"
            .Type = esriFieldTypeDouble
            .Length = 8
        End With
        ' Add the new field.
        pFClass.AddField pField
End Sub
```

AddIndex gets the top layer in the active map (the layer from the second intersect operation) as an argument from **Start**. The code creates *pField* as an instance of the *Field* class and defines its properties, including the field name of Total. Then the code uses the *AddField* method on *IFeatureClass* to add *pField* to the feature class of the layer.

```
Private Sub CalculateIndex(pFLayer As IFeatureLayer)
        Dim pFeatLayer As IFeatureLayer
        Dim pFeatClass As IFeatureClass
        Dim pFields As IFields
        Dim ii As Integer
        Dim pQueryFilter As IQueryFilter
        Dim pCursor As ICursor
        Dim pCalc As ICalculator
        Set pFeatLayer = pFLayer
        Set pFeatClass = pFeatLayer.FeatureClass
        Set pFields = pFeatClass.Fields
        ii = pFields.FindField("Total")
        ' Prepare a cursor for features that have lurate < 99.
        Set pQueryFilter = New QueryFilter
        pQueryFilter.WhereClause = "lurate < 99"
        Set pCursor = pFeatClass.Update(pQueryFilter, True)
        ' Use the cursor to calculate the field values of Total.
        Set pCalc = New Calculator
        With pCalc
            Set .Cursor = pCursor
            .Expression = "([SOILRATE] * 3 + [LURATE] + [DWRATE]) / 250"
            .Field = "Total"
        End With
        pCalc.Calculate
End Sub
```

CalculateIndex gets the layer with the new field as an argument from **Start**. The code uses the *FindField* method on *IFields* to find the field Total. Then the code uses a query filter object to select those records that have lurate < 99 (nonurban land use) and saves them into a feature cursor. Next, the code creates *pCalc* as an instance of the *Calculator* class and defines its properties of cursor, expression, and field. Finally, **CalculateIndex** uses the *Calculate* method on *ICalculator* to populate the field values of Total.

```
Private Function FindLayer(Message As String) As Long
    Dim pMxDoc As IMxDocument
    Dim pMap As IMap
    Dim FindDoc As Variant
    Dim aLName As String
    Dim Name As String
    Dim i As Long
    Set pMxDoc = ThisDocument
    Set pMap = pMxDoc.FocusMap
    Name = Message
    For i = 0 To pMap.LayerCount - 1
        aLName = UCase(pMap.Layer(i).Name)
        If (aLName = (UCase(Name))) Then
            FindDoc = i
        End If
    Next
    FindLayer = FindDoc
End Function
```

FindLayer finds the index of the layer that matches the name from the message statement. The function makes sure that the appropriate layers are used in the intersect operations.

```
Private Sub DisplayIndexClasses(pFLayer As IFeatureLayer)
' DisplayIndexClasses uses the sub of GetRGB.
    Dim pMxDoc As IMxDocument
    Dim pLayer As ILayer
    Dim pGeoFeatureLayer As IGeoFeatureLayer
    Dim pClassBreaksRenderer As IClassBreaksRenderer
    Dim pFillSymbol As IFillSymbol
    Set pMxDoc = ThisDocument
    Set pLayer = pFLayer
    Set pGeoFeatureLayer = pLayer
    ' Define a class breaks renderer.
    Set pClassBreaksRenderer = New ClassBreaksRenderer
    pClassBreaksRenderer.Field = "Total"
    pClassBreaksRenderer.BreakCount = 4
    ' Hard code the symbol, break, and label for each class.
    Set pFillSymbol = New SimpleFillSymbol
    pFillSymbol.Color = GetRGBColor(255, 255, 255)
    pClassBreaksRenderer.Symbol(0) = pFillSymbol
    pClassBreaksRenderer.Break(0) = 0#
    pClassBreaksRenderer.Label(0) = "Urban Land Use"
    Set pFillSymbol = New SimpleFillSymbol
    pFillSymbol.Color = GetRGBColor(245, 175, 0)
```

```
pClassBreaksRenderer.Symbol(1) = pFillSymbol
pClassBreaksRenderer.Break(1) = 0.75
pClassBreaksRenderer.Label(1) = "0.60 - 0.75"
Set pFillSymbol = New SimpleFillSymbol
pFillSymbol.Color = GetRGBColor(245, 125, 0)
pClassBreaksRenderer.Symbol(2) = pFillSymbol
pClassBreaksRenderer.Break(2) = 0.85
pClassBreaksRenderer.Label(2) = "0.76 - 0.85"
Set pFillSymbol = New SimpleFillSymbol
pFillSymbol.Color = GetRGBColor(245, 0, 0)
pClassBreaksRenderer.Symbol(3) = pFillSymbol
pClassBreaksRenderer.Break(3) = 1#
pClassBreaksRenderer.Label(3) = "0.86 - 1.00"
' Assign the renderer to the layer and refresh the map.
Set pGeoFeatureLayer.Renderer = pClassBreaksRenderer
pMxDoc.ActiveView.PartialRefresh esriViewGeography, pLayer, Nothing
pMxDoc.UpdateContents
End Sub
```

DisplayIndexClasses gets the layer with the populated field Total as an argument from *Start*. The code first creates *pClassBreaksRenderer* as an instance of the *ClassBreaksRenderer* class and defines its properties of field and break count. *DisplayIndexClasses* then assigns the fill symbol, break, and label for each of the four classes. The color for the simple fill symbol is obtained by entering the red (R), green (G), and blue (B) values in the *GetRGBColor* function. Finally, the code assigns *pClassBreaksRenderer* to the layer, refreshes the map, and updates the contents of the map document.

```
Private Function GetRGBColor(R As Long, G As Long, B As Long)
    Dim pColor As IRgbColor
    Set pColor = New RgbColor
    pColor.Red = R
    pColor.Green = G
    pColor.Blue = B
    GetRGBColor = pColor
End Function
```

GetRGBColor gets the input values of R, G, and B from *DisplayIndexClasses* and returns a RGB color.

Box 14.2 VectorIndexModel_GP

VectorIndexModel_GP builds a vector-based index model. The macro first uses the Intersect tool in the Analysis toolbox to overlay three feature classes. (The Intersect tool for the ArcInfo version of ArcGIS can accept two or more input layers at one time.) Then it uses AddField and CalculateField, both tools in the Data Management toolbox, to add an index field and to populate the field values. Run the macro in ArcCatalog. The macro adds *Intersect2_b*. Open the attribute table of *Intersect2_b*, and examine the Total field.

```
Private Sub VectorIndexModel_GP()
    ' Create the Geoprocessing object and define its workspace.
    Dim GP As Object
    Set GP = CreateObject("esriGeoprocessing.GpDispatch.1")
    Dim filepath As String
    filepath = "c:\data\chap14\Index.mdb"
    GP.Workspace = filepath
    ' Execute the intersect tool.
    ' Intersect <features{Ranks};features{Ranks}...> <out_feature_class>
    ' {ALL | NO_FID | ONLY_FID} {cluster_tolerance} {INPUT | LINE | POINT}
    Dim parameter1 As String
    parameter1 = "depwater;landuse;soil"
    Dim parameter2 As String
    parameter2 = "Intersect2_b"
    GP.Intersect_analysis parameter1, parameter2
    ' Execute the addfield tool.
    ' AddField <in_table> <field_name> <LONG | TEXT | FLOAT | DOUBLE | SHORT |
    ' DATE | BLOB> {field_precision} {field_length} {field_alias} {NULLABLE | NON_NULLABLE}
    ' {NON_REQUIRED | REQUIRED} {field_domain}
    GP.AddField_management "Intersect2_b", "Total", "DOUBLE"
'CalculateField <in_table> <field> <expression> {VB | PYTHON} {code_block}
GP.CalculateField "Intersect2_b", "Total", "temp", "VB", "Dim temp As Double" & vbCrLf & _
"If [LURATE] = 99 Then" & vbCrLf & "temp = -1" & vbCrLf & "Else" & vbCrLf & _
"temp = ([SOILRATE] * 3 + [LURATE] + [DWRATE]) / 250" & vbCrLf & "End If"
End Sub
```

14.3.3 *RasterBinaryModel*

RasterBinaryModel performs the same task as *VectorBinaryModel* but in raster format. *stream_gd* and *elevzone_gd* are the raster equivalent of *stream.shp* and *elevzone.shp*, respectively. The binary model finds areas that are in elevation zone 2 and within 200 meters of streams. In building the binary model, *RasterBinary-Model* first creates a 200-meter distance measure raster from *stream_gd*, creates a descriptor of *elevzone_gd*, and then uses the distance raster and the descriptor in a raster data query. The output from the query shows selected areas as having the cell value of one.

RasterBinaryModel has three subs and one function (Figure 14.3):

Start: a sub for managing the input and output datasets and for calling the other subs
Distance: a sub for creating a 200-meter distance raster from streams
QueryGrids: a sub for raster data query of two processed rasters
FindLayer: a function for finding a specific layer

Usage: Add *stream_gd* and *elevzone_gd* to an active map. Import *RasterBina-ryModel* to Visual Basic Editor. Run the module. The module adds two temporary rasters to the active map. One is *Distance_To_Stream*, a continuous distance raster from streams. The other is *Model*, the raster-based binary model, which should look the same as the vector-based binary model created by *VectorBinaryModel*.

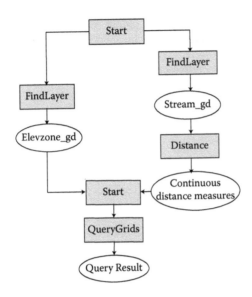

Figure 14.3 The flow chart shows the modular structure of *RasterBinaryModel*.

```
Private Sub Start()
' Start uses the subs of FindLayer, Distance, and QueryGrids.
    Dim pMxDoc As IMxDocument
    Dim pMap As IMap
    Dim Id As Long
    Dim Message As String
    Dim pRasterLy As IRasterLayer
    Dim pRasterLy1 As IRasterLayer
    Dim pRasterLy2 As IRasterLayer
    Set pMxDoc = ThisDocument
    Set pMap = pMxDoc.FocusMap
    ' Run Distance.
    Message = "stream_gd"
    Id = FindLayer(Message)
    Set pRasterLy = pMap.Layer(Id)
    Call Distance(pRasterLy)
    ' Run QueryGrids.
    Message = "Distance_to_Stream"
    Id = FindLayer(Message)
    Set pRasterLy1 = pMap.Layer(Id)
    Message = "elevzone_gd"
    Id = FindLayer(Message)
    Set pRasterLy2 = pMap.Layer(Id)
    Call QueryGrids(pRasterLy1, pRasterLy2)
End Sub
```

Start calls the ***Distance*** sub to run a 200-meter distance measure operation from *stream_gd*. Then the code calls the ***QueryGrids*** sub to run a query on *elevzone_gd* and *Distance_to_Stream*, which is the output from the ***Distance*** sub. The result of the query shows areas that are in elevation zone 2 and within 200 meters of streams.

```
Private Sub Distance(pRasterLy As IRasterLayer)
    Dim pMxDoc As IMxDocument
    Dim pMap As IMap
    Dim pSourceRL As IRasterLayer
    Dim pSourceRaster As IRaster
    Dim pDistanceOp As IDistanceOp
    Dim pOutputRaster As IRaster
    Dim pOutputLayer As IRasterLayer
    Set pMxDoc = ThisDocument
    Set pMap = pMxDoc.FocusMap
    ' Define the source raster.
    Set pSourceRL = pRasterLy
    Set pSourceRaster = pSourceRL.Raster
    ' Perform the Euclidean distance operation.
    Set pDistanceOp = New RasterDistanceOp
    Set pOutputRaster = pDistanceOp.EucDistance(pSourceRaster, 200)
    ' Create the raster layer and add it to the active map.
    Set pOutputLayer = New RasterLayer
    pOutputLayer.CreateFromRaster pOutputRaster
    pOutputLayer.Name = "Distance_To_Stream"
    pMap.AddLayer pOutputLayer
End Sub
```

Distance gets the source layer for distance measures as an argument from ***Start***. The code first sets *pSourceRaster* to be the raster of the stream layer. Next, the code creates *pDistanceOp* as an instance of the *RasterDistanceOp* class and uses the *EucDistance* method on *IDistanceOp*, with a maximum distance of 200 (meters), to create a distance measure raster referenced by *pOutputRaster*. Finally, the code creates a new layer from *pOutputRaster* and adds the layer to the active map.

```
Private Sub QueryGrids(pRasterLy1, pRasterLy2)
    Dim pMxDoc As IMxDocument
    Dim pMap As IMap
    Dim pRLayer1 As IRasterLayer
    Dim pRaster1 As IRaster
    Dim pRLayer2 As IRasterLayer
    Dim pRaster2 As IRaster
    Dim pFilt2 As IQueryFilter
    Dim pDesc2 As IRasterDescriptor
    Dim pLogicalOp As ILogicalOp
    Dim pOutputRaster As IRaster
    Dim pRLayer As IRasterLayer
    Set pMxDoc = ThisDocument
    Set pMap = pMxDoc.FocusMap
    ' Define the two rasters for query.
    Set pRLayer1 = pRasterLy1
```

```
    Set pRaster1 = pRLayer1.Raster
    Set pRLayer2 = pRasterLy2
    Set pRaster2 = pRLayer2.Raster
    ' Use the value field to create a raster descriptor from the elevzone raster.
    Set pFilt2 = New QueryFilter
    pFilt2.WhereClause = "value = 2"
    Set pDesc2 = New RasterDescriptor
    pDesc2.Create pRaster2, pFilt2, "value"
    ' Perform a logical operation.
    Set pLogicalOp = New RasterMathOps
    Set pOutputRaster = pLogicalOp.BooleanAnd(pRaster1, pDesc2)
    ' Create the output raster layer and add it to the active map.
    Set pRLayer = New RasterLayer
    pRLayer.CreateFromRaster pOutputRaster
    pRLayer.Name = "Model"
    pMap.AddLayer pRLayer
End Sub
```

QueryGrids gets two raster layers to be queried as arguments from *Start*: one is *Distance_to_Stream* and the other is *elevzone_gd*. To isolate areas within elevation zone 2 in *Elevzone_gd*, the code creates *pDesc2*, an instance of the *RasterDescriptor* class, by using a query filter object. Then the code uses the *BooleanAnd* method on *ILogicalOp* to create the logical query output referenced by *pOutputRaster*. Finally, the code creates a new layer from *pOutputRaster* and adds the layer to the active map.

```
Private Function FindLayer(Message As String) As Long
    Dim FindDoc As Variant
    Dim pMxDoc As IMxDocument
    Dim pMap As IMap
    Dim aLName As String
    Dim Name As String
    Dim i As Long
    Set pMxDoc = ThisDocument
    Set pMap = pMxDoc.FocusMap
    Name = Message
    For i = 0 To pMap.LayerCount - 1
        aLName = UCase(pMap.Layer(i).Name)
        If (aLName = (UCase(Name))) Then
            FindDoc = i
        End If
    Next
    FindLayer = FindDoc
End Function
```

FindLayer finds the index of the layer that matches the Message value. The function makes sure that the appropriate layers are entered as inputs to the subs and function.

Box 14.3 RasterBinaryModel_GP

RasterBinaryModel_GP builds a raster-based binary model. The macro first uses the EucDistance tool in the Spatial Analyst toolbox, with a maximum distance of 200 (meters), to create a distance measure raster. Then it uses the ExtractByAttributes tool to extract a raster within the elevation zone of 2. Finally, the macro uses the BooleanAnd tool to create the model raster by querying the distance measure raster and the extracted elevation raster. Run the macro in ArcMap. The macro adds *distance_b*, *elevzone_2b*, and *model_b* to the map. *model_b* is the final model.

```
Private Sub RasterBinaryModel_GP()
    ' Create the Geoprocessing object and define its workspace.
    Dim GP As Object
    Set GP = CreateObject("esriGeoprocessing.GpDispatch.1")
    Dim filepath As String
    filepath = "c:\data\chap14"
    GP.Workspace = filepath
    ' Execute the euclideandistance tool.
    ' EucDistance <in_source_data> <out_distance_raster> {maximum_distance}
    ' {cell_size} {out_direction_raster}
    GP.EucDistance_sa "stream_gd", "distance_b", 200
    ' Execute the extractbyattributes tool.
    ' ExtractByAttributes <in_raster> <where_clause> <out_raster>
    GP.ExtractByAttributes_sa "elevzone_gd", "value = 2", "elevzone_2b"
    ' Execute the booleanand tool.
    ' BooleanAnd <in_raster_or_constant1> <in_raster_or_constant2> <out_raster>
    GP.BooleanAnd_sa "elevzone_2b", "distance_b", "model_b"
End Sub
```

14.3.4   *RasterIndexModel*

RasterIndexModel performs the same task as ***VectorIndexModel*** in raster format. *soil_gd*, *landuse_gd*, and *depwater_gd* are the raster equivalents of *soil.shp*, *landuse.shp*, and *depwater.shp*. In building the index model, ***RasterIndexModel*** first reclassifies *landuse_gd* to exclude urban land use from the analysis. The module then performs a map algebra operation by using the three rasters as the inputs. The output from the map algebra operation is the index model. The cell values of the index model range from 0.58 to 1.0. Finally, ***RasterIndexModel*** uses a color ramp renderer to display the index model in three classes.

    RasterIndexModel has three subs and two functions (Figure 14.4):

    Start: a sub for performing the map algebra operation and for calling the other subs
    ReclassNumberField: a sub for reclassifying the landuse raster
    DisplayIndexClasses: a sub for displaying the index values in three classes
    FindLayer: a function for finding a specific layer
    GetRGB: a function for defining a color symbol

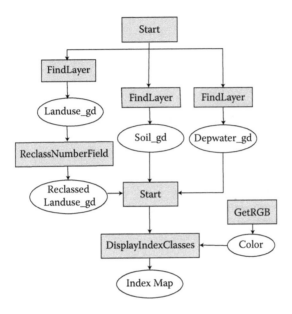

Figure 14.4   The flow chart shows the modular structure of ***RasterIndexModel***.

Usage: Add *soil_gd*, *landuse_gd*, and *depwater_gd* to an active map. Import ***RasterIndexModel*** to Visual Basic Editor. Run the module. The module adds two temporary rasters to the active map. One is *Reclass_Landuse*, a reclassed land use raster. The other is *Model*, the raster-based index model, shown in three classes. Urban land use is treated as no data and excluded from the legend. The raster-based index model should look very similar to the vector-based index model created by ***VectorIndexModel***.

```
Private Sub Start()
' Start uses the subs of FindLayer, ReclassNumberField, and DisplayIndexClasses.
    Dim pMxDoc As IMxDocument
    Dim pMap As IMap
    Dim Id As Long
    Dim Message As String
    Dim pRasterLy As IRasterLayer
    Dim pMapAlgebraOp As IMapAlgebraOp
    Dim pRasterAnalysisEnv As IRasterAnalysisEnvironment
    Dim pSoilLayer As IRasterLayer
    Dim pSoilRaster As IRaster
    Dim pLanduseLayer As IRasterLayer
    Dim pLanduseRaster As IRaster
    Dim pDepwaterLayer As IRasterLayer
    Dim pDepwaterRaster As IRaster
    Dim pModel As IRaster
    Dim pModelLayer As IRasterLayer
    Set pMxDoc = ThisDocument
```

```
    Set pMap = pMxDoc.FocusMap
    ' Reclass landuse_gd to exclude urban land use from analysis.
    Message = "landuse_gd"
    Id = FindLayer(Message)
    Set pRasterLy = pMap.Layer(Id)
    Call ReclassNumberField(pRasterLy)
    ' Create a map algebra operation.
    Set pMapAlgebraOp = New RasterMapAlgebraOp
    Set pRasterAnalysisEnv = pMapAlgebraOp
    ' Bind the symbol R1 to soil_gd.
    Message = "soil_gd"
    Id = FindLayer(Message)
    Set pSoilLayer = pMap.Layer(Id)
    Set pSoilRaster = pSoilLayer.Raster
    pMapAlgebraOp.BindRaster pSoilRaster, "R1"
    ' Bind the symbol R2 to reclassified landuse_gd.
    Message = "Reclass_Landuse"
    Id = FindLayer(Message)
    Set pLanduseLayer = pMap.Layer(Id)
    Set pLanduseRaster = pLanduseLayer.Raster
    pMapAlgebraOp.BindRaster pLanduseRaster, "R2"
    ' Bind the symbol R3 to depwater_gd.
    Message = "depwater_gd"
    Id = FindLayer(Message)
    Set pDepwaterLayer = pMap.Layer(Id)
    Set pDepwaterRaster = pDepwaterLayer.Raster
    pMapAlgebraOp.BindRaster pDepwaterRaster, "R3"
    ' Execute the map algebra operation.
    Set pModel = pMapAlgebraOp.Execute("([R1] * 3 + [R2] + [R3]) / 250")
    ' Create the output raster layer and add it to the active map.
    Set pModelLayer = New RasterLayer
    pModelLayer.CreateFromRaster pModel
    pModelLayer.Name = "Model"
    pMap.AddLayer pModelLayer
    Call DisplayIndexClasses(pModelLayer)
End Sub
```

Start first calls *ReclassNumberField* to reclassify urban areas in *landuse_gd* as no data, thus excluding urban areas from further analysis. Next, *Start* runs a raster map algebra operation. The operation requires that each input raster be given a "symbol." Therefore, the code calls the *FindLayer* sub to locate a layer in the active map and uses the *BindRaster* method on *IMapAlgebraOp* to bind the raster of the layer to a symbol. The R1 symbol binds *soil_gd*, the R2 symbol binds *reclass_landuse*, and the R3 symbol binds *depwater_gd*. These symbols of R1, R2, and R3 are then used in the expression that executes the map algebra operation. The result of the operation is an index model raster referenced by *pModel*. The code then creates a new layer named *Model* from *pModel* and adds the layer to the active map. *Start* concludes by calling *DisplayIndexClasses* to display the model layer in three classes.

```
Private Sub ReclassNumberField(pRasterLy As IRasterLayer)
    Dim pMxDoc As IMxDocument
    Dim pMap As IMap
    Dim pRaster2Ly As IRasterLayer
    Dim pGeoDs As IGeoDataset
    Dim pReclassOp As IReclassOp
    Dim pNRemap As INumberRemap
    Dim pOutRaster As IRaster
    Dim pReclassLy As IRasterLayer
    Set pMxDoc = ThisDocument
    Set pMap = pMxDoc.FocusMap
    ' Pass landuse_gd.
    Set pRaster2Ly = pRasterLy
    Set pGeoDs = pRaster2Ly.Raster
    ' Use a number remap to reclass landuse_gd.
    Set pReclassOp = New RasterReclassOp
    Set pNRemap = New NumberRemap
    With pNRemap
        .MapValue 20, 20
        .MapValue 40, 40
        .MapValue 45, 45
        .MapValue 50, 50
        .MapValueToNoData 99
    End With
    Set pOutRaster = pReclassOp.ReclassByRemap(pGeoDs, pNRemap, False)
    ' Create the output raster layer and add it to the active map.
    Set pReclassLy = New RasterLayer
    pReclassLy.CreateFromRaster pOutRaster
    pReclassLy.Name = "Reclass_Landuse"
    pMap.AddLayer pReclassLy
End Sub
```

ReclassNumberField gets *landuse_gd* as an argument from **Start**. Designed to exclude urban areas from analysis, the code creates *pNRemap* as an instance of the *NumberRemap* class and uses the *MapValueToNoData* method on *INumberRemap* to assign no data to urban areas (that is, with the map value of 99). Then the code uses the *ReclassByRemap* method on *IReclassOp* to create a reclassified raster referenced by *pOutRaster*. Finally the code creates a new layer from *pOutRaster* and adds the layer called *Reclass_Landuse* to the active map.

```
Private Function FindLayer(Message As String) As Long
    Dim FindDoc As Variant
    Dim pMxDoc As IMxDocument
    Dim pMap As IMap
    Dim aLName As String
    Dim Name As String
    Dim i As Long
    Set pMxDoc = ThisDocument
    Set pMap = pMxDoc.FocusMap
    Name = Message
```

```
    For i = 0 To pMap.LayerCount - 1
        aLName = UCase(pMap.Layer(i).Name)
        If (aLName = (UCase(Name))) Then
            FindDoc = i
        End If
    Next
    FindLayer = FindDoc
End Function
```

FindLayer finds the index of the layer that matches the Message value. The function makes sure that the appropriate layers are used in the reclassification and map algebra operations.

```
Private Sub DisplayIndexClasses(pModelLayer As IRasterLayer)
' DisplayIndexClasses uses the sub of GetRGB.
    Dim pMxDoc As IMxDocument
    Dim pMap As IMap
    Dim pRLayer As IRasterLayer
    Dim pRaster As IRaster
    Dim pClassRen As IRasterClassifyColorRampRenderer
    Dim pRasRen As IRasterRenderer
    Dim pFillSymbol As IFillSymbol
    Set pMxDoc = ThisDocument
    Set pMap = pMxDoc.FocusMap
    Set pRLayer = pModelLayer
    Set pRaster = pRLayer.Raster
    ' Define a raster classify color ramp renderer.
    Set pClassRen = New RasterClassifyColorRampRenderer
    Set pRasRen = pClassRen
    Set pRasRen.Raster = pRaster
    pClassRen.ClassCount = 3
    pRasRen.Update
    ' Hard code the symbol, break, and label for each class.
    Set pFillSymbol = New SimpleFillSymbol
    pFillSymbol.Color = GetRGBColor(245, 245, 0)
    pClassRen.Symbol(0) = pFillSymbol
    pClassRen.Break(0) = 0.58
    pClassRen.Label(0) = "0.58 - 0.75"
    pFillSymbol.Color = GetRGBColor(245, 175, 0)
    pClassRen.Symbol(1) = pFillSymbol
    pClassRen.Break(1) = 0.76
    pClassRen.Label(1) = "0.76 - 0.85"
    pFillSymbol.Color = GetRGBColor(245, 125, 0)
    pClassRen.Symbol(2) = pFillSymbol
    pClassRen.Break(2) = 0.86
    pClassRen.Label(2) = "0.86 - 1.00"
    ' Assign the renderer to the layer and refresh the map.
    pRasRen.Update
    Set pRLayer.Renderer = pRasRen
```

```
        pMxDoc.ActiveView.Refresh
        pMxDoc.UpdateContents
End Sub
```

DisplayIndexClasses gets the model layer as an argument from *Start*. The code creates *pClassRen* as an instance of the *RasterClassifyColorRampRenderer* class and uses *IRasterRenderer* and *IRasterClassifyColorRampRenderer* to define the renderer's raster and class count. The code then provides the fill symbol, break, and label for each class in *pClassRen*. The *GetRGBColor* function generates the color for the fill symbol. Finally, the code assigns *pRasRen* to be the renderer of *pRLayer*, refreshes the map, and updates the contents of the map document.

```
Private Function GetRGBColor(R As Long, G As Long, B As Long)
        Dim pColor As IRgbColor
        Set pColor = New RgbColor
        pColor.Red = R
        pColor.Green = G
        pColor.Blue = B
        GetRGBColor = pColor
End Function
```

GetRGBColor uses the input values of R, G, and B from *DisplayIndexClasses* and returns a RGB color.

Box 14.4 RasterIndexModel_GP

RasterIndexModel_GP builds a raster-based index model. The macro first uses the Reclassify tool in the Spatial Analyst toolbox to assign no data to urban areas in the land use raster. Then it uses the SingleOutputMapAlgebra tool to create the model raster. The expression for calculating the index value is included in parameter 1 of the command statement. Run the macro in ArcMap. The macro adds *reclass_lu_b* and *rastermodel_b* to the map.

```
Private Sub RasterIndexModel_GP()
        ' Create the Geoprocessing object and define its workspace.
        Dim GP As Object
        Set GP = CreateObject("esriGeoprocessing.GpDispatch.1")
        Dim filepath As String
        filepath = "c:\data\chap14"
        GP.Workspace = filepath
        ' Execute the reclassify tool.
        ' Reclassify <in_raster> <reclass_field> <remap> <out_raster> {DATA | NODATA}
        Dim parameter3 As String
        parameter3 = "20 20; 40 40; 45 45; 50 50; 99 NoData"
        GP.Reclassify_sa "landuse_gd", "Value", parameter3, "reclass_lu_b"
        ' Execute the singleoutputmapalgebra tool.
        ' SingleOutputMapAlgebra_sa <expression_string> <out_raster> {in_data}
        GP.SingleOutputMapAlgebra_sa "([soil_gd] * 3 + [depwater_gd] + [reclass_lu_b]) / 250", "rastermodel_b"
End Sub
```

Index

P